FEMINISM, SCIENCE, AND THE PHILOSOPHY OF SCIENCE

SYNTHESE LIBRARY

STUDIES IN EPISTEMOLOGY,

LOGIC, METHODOLOGY, AND PHILOSOPHY OF SCIENCE

VOLUME 256

FEMINISM, SCIENCE, AND THE PHILOSOPHY OF SCIENCE

Edited by

LYNN HANKINSON NELSON
Rowan College, Glassboro
and
JACK NELSON
Temple University, Philadelphia

KLUWER ACADEMIC PUBLISHERS
DORDRECHT / BOSTON / LONDON

Library of Congress Cataloging-in-Publication Data

A C.I.P. Catalogue record for this book is available from the Library of Congress.

ISBN: 0–7923–4162–7

Published by Kluwer Academic Publishers,
P.O. Box 17, 3300 AA Dordrecht, The Netherlands.

Kluwer Academic Publishers incorporates
the publishing programmes of
D. Reidel, Martinus Nijhoff, Dr W. Junk and MTP Press.

Sold and distributed in the U.S.A. and Canada
by Kluwer Academic Publishers,
101 Philip Drive, Norwell, MA 02061, U.S.A.

In all other countries, sold and distributed
by Kluwer Academic Publishers Group,
P.O. Box 322, 3300 AH Dordrecht, The Netherlands.

Printed on acid-free paper

Printed in Great Britain by Hartnolls Ltd., Bodmin, Cornwall

TABLE OF CONTENTS

PART IV: VIEWS FROM MULTICULTURAL AND GLOBAL FEMINISMS, AND FROM FEMINIST PHENOMENOLOGY

ACKNOWLEDGMENTS

As editors of this volume, we have incurred many debts. We are grateful to those who read drafts of the initial prospectus and suggested contributors, and to the anonymous reviewer who commented on earlier versions of these essays. Annie Kuipers of Kluwer Academic Publishers displayed patience and continuing encouragement. We thank Jim Fetzer for suggesting the volume and the contributors for agreeing to be part of this project and for receiving, in good humor, our comments on successive drafts of their essays. Finally, we thank our families, particularly our daughter Rebecca Watson, and our parents, Virginia and Donald Hankinson, and Valnette and Lowell Nelson, who maintained good humor when we worked through family reunions.

LYNN HANKINSON NELSON AND JACK NELSON

INTRODUCTION

The essays in this volume address issues at the intersections of two loosely-delineated and rapidly-evolving areas of inquiry, "mainstream" philosophy of science and feminist philosophy of science. We hope this collection will facilitate and encourage dialogue among feminists and their colleagues about the nature of science and the philosophy of science. Some recent analyses in feminist and mainstream philosophy of science suggest an unbridgeable chasm between these traditions. We demur. We take philosophers of science and scientists, feminists and non-feminists alike, to share an interest in the nature of objectivity, truth, evidence, cognitive agency, scientific method, and the relationship between science and values. We also take there to be substantive issues that divide feminists and their mainstream colleagues, not including interest in the notions just listed. And we believe that encouraging greater dialogue between mainstream and feminist philosophy of science will further our understandings of these notions.

We have invited philosophers and practicing scientists to explore parallels and tensions between feminist approaches to science and other approaches in the philosophy of science and science studies more broadly. Contributors have more than met our expectations. Their essays explore the notions at the heart of both mainstream and feminist philosophies of science just listed, the categories that should and should not figure in the explanatory principles employed in the philosophy of science, and the question of what, if any, implications feminist science scholarship carries for our understanding of these notions. The contributors explore parallels and tensions between, on the one hand, feminist approaches to the philosophy of science, and, on the other, the rationalist and empiricist traditions in mainstream philosophy of science, the core tenets of logical empiricism, Quinean holism, varieties of realism, naturalized philosophy of science, sociology of knowledge, varieties of social constructivism, postcolonial science studies, multicultural and global feminisms, and the phenomenological tradition in Continental philosophy. Several contributors also explore recent arguments that science studies, including feminist philosophy of science, fail to meet and/or explicitly reject standards of rational inquiry.

Mainstream philosophy of science serves as one source of the issues on which contributors focus. It is now common to describe the philosophy of science of recent decades as sharply divided into two loosely-delineated research programs or schools of thought: one devoted to the refinement and articulation of the program spawned by logical empiricism and its immediate heir (the post-positivist tradition represented by Rudolf Carnap, Carl Hempel, and Ernst Nagel) whose goal is to reconstruct the logic of science (the logic of explanation, justification, and so on);

L. H. Nelson and J. Nelson (eds), Feminism, Science, and the Philosophy of Science, ix–xix.

and the other constituted by reactions against the conceptualization of science as a body of theories and assumed primacy of the "logic of justification" in the philosophy of science. This latter school of thought is typically associated with the work of Paul Feyerabend and Thomas A. Kuhn, but it in fact includes a variety of approaches to understanding science in the philosophy of science, united by the emphasis on science as an activity rather than as a body of theories.

Indeed, it is common to describe post-positivist philosophy of science as characterized by three broad epistemological positions: contemporary versions of empiricism (constructive empiricism, for example), neo-Kantian social constructivism (a position typically attributed to Thomas A. Kuhn, for example), and scientific realism.[1] While more detailed than the scheme outlined in the preceding paragraph, this classificatory scheme is also problematic. For one thing, it does not identify naturalized philosophy of science, a research tradition in which there is substantial interest and work. As importantly, the distinctions the scheme presupposes are by no means universally applicable. Among naturalized philosophers of science, for example, one finds advocates of scientific realism and of empiricism (Ronald Giere and W. V. Quine, respectively, have offered such arguments). So, too, "social constructivism" has been used both by its advocates and its critics in enough different ways – some vague and others quite specific, some generous and others quite limited – as to make the usefulness of the category without further explication dubious. Similarly, although "scientific realism" is understood by some to constitute the primary (or only) alternative to "social constructivism", "realism" is itself used in a variety of ways that are not obviously compatible (as attested to by its role as a component of "scientific realism", "naturalistic realism", and "perspectival realism"). And the relationship of some of the views associated with social constructivism to constructive empiricism or to the naturalistic realism advocated by Quine is itself hardly obvious. Finally, there seems to be no single thread running through the work of those who use "realism", "social constructivism", and/or "empiricism" to describe their positions.

Notwithstanding these complexities and ambiguities, the "isms", distinctions, and oppositions we have mentioned are assumed and appealed to in a good deal of recent work in the philosophy of science, including many of the essays in this volume. We suggest that readers look to see what content an individual author attributes to the notions in question, rather than superimpose such content or implications.

The second source of the issues addressed by contributors to this volume is feminist science scholarship, an area of inquiry also marked by significant development in the last two decades. Feminists commonly describe this development in terms of an evolution, from an initial emphasis on the sociology of science (with the latter construed narrowly to include social arrangements within science communities, and particularly women's positions and relatively low numbers within them), to critiques of the methods, theories, and research projects of various sciences (e.g., of androcentric methods and theories), to more general questions about the social processes in which scientific knowledge is generated, and (eventually) to critiques and analyses of theories about science, including theories developed by philosophers of science, and the development of alternative theories about science. While such descriptions are useful starting points, they are potentially misleading. The "levels" of analysis just outlined have been evolving apace, each informing the other, and the ability to maintain the boundaries traditionally assumed between "sociological" issues and "epistemological" issues is one issue about which feminist scientists and philosophers of science, and their colleagues, often deeply disagree – as the essays in this volume attest.

Feminist critiques of mainstream philosophical positions and feminist alternatives to these positions are commonly classified using the categories "feminist empiricism", "feminist standpoint theory", and "feminist postmodernism", with each category understood to locate a feminist methodological approach at the intersections of a specific philosophical tradition and feminist theory. Cautions paralleling those offered earlier in the discussion of mainstream philosophy of science are appropriate here. Specifically, while the terms "realism", "empiricism", and "social constructivism" figure in feminist analyses, authors do not always understand the terms in the same ways. Nor do feminists always see these epistemological positions as mutually exclusive. Further, there is more than one version of feminist empiricism, of feminist standpoint theory, and of feminist postmodernism; and there are feminists who advocate versions of realism, others who advocate phenomenological approaches to science, and varieties of feminism. Finally, feminist naturalized philosophy of science is a current research program, but as in "mainstream" naturalized philosophy of science, considerable disagreement remains in feminist philosophy of science about what "naturalized" means in this context. Attending to the details of the analyses undertaken in individual essays is appropriate.

The essays address overlapping themes, and the way in which they are presented is but one of several possible ways of organizing them. We have selected sections based on our sense of larger and significant topics. In some sections, there is significant disagreement among the authors about a specific issue and/or about the implications of feminist science scholarship for the traditional categories and emphases of the philosophy of science; in others, there are areas of substantive agreement.

The essays included in Part One together provide a historical overview of how opposing views of the relationship between, on the one hand, good science and, on the other hand, social and political values, developed within logical empiricism (Giere) and contemporary feminist empiricism (Tuana).

In "The Feminism Question in Science", Ronald N. Giere explores the grounds for the resistance among many philosophers of science to admitting any influence of gender within good science. He maintains that an important source of this resistance is to be found in the presuppositions of post-war logical empiricism, presuppositions shaped not by argument but by the specific circumstances within which logical empiricism developed, and the carry over of such presuppositions to post-positivist philosophy of science. Giere includes among such presuppositions the alleged distinction between the contexts of discovery and justification and the insistence that cultural factors (like gender) could not play any role in establishing the legitimacy of scientific claims. Ironically, Giere argues, important developments within post-positivist philosophy of science itself, particularly the turn to the notion of "rational progress within a research tradition" that is associated with the work of Imre Lakatos and other historically-oriented philosophers of science, do establish the *theoretical* possibility that gender could influence good science. Hence, Giere maintains, the existence of such influence is an empirical question to be considered by examining particular cases. Giere concludes by offering "perspectival realism", a view that builds on the semantic view of theories, as compatible with and promising for feminist empiricism.

In the second essay of this section, "Revaluing Science: Starting from the Practices of Women", Nancy Tuana explores the relationship between views in feminist empiricism of the relationship between cultural values and science, and an evolving body of feminist science scholarship. Tuana analyzes specific critiques advanced by feminist scientists and uses these case studies to argue that feminist scientists draw both on the methodological norms of their disciplines and feminist values. Tuana maintains that one implication of her case studies is that cognitive agents, including scientists, are engaged members of epistemic communities whose values and affective interests can have positive epistemic value. Building on the epistemology she attributes to primary care providers in the medical profession, Tuana sketches a view of the relationship between "knowers" and "known" that she maintains will allow for a more complex notion of scientific objectivity than conceptions that privilege disinterestedness and value-neutrality allow.

There are intriguing parallels between Giere's and Tuana's treatments of the relationship between the cultural values philosophers of science espouse and the theories of science they advocate, in their arguments for a relationship between such values and a then current state of science and science scholarship, and in their insistence that the question of what role cultural factors have in science is an empirical question. Giere's analysis of the values informing logical empiricist arguments and Tuana's of the values informing feminist empiricism suggest more continuities between the two research programs than is usually assumed. But their analyses also suggest that a fundamental difference between these traditions is that the relationship between cultural values and philosophical positions was downplayed in the first and is emphasized in the second, and that this difference is important for understanding the formulation of a discovery/justification context distinction in logical empiricism and its rejection in feminist empiricism. Finally, differing views about the relationship between relativism and cultural values in science and in the philosophy of science emerge as relevant to the differences between logical and feminist empiricism just mentioned, as both Giere and Tuana address.

Relativism is also a central concern in the essays included in Part Two. It is perhaps not surprising that some of the strongest disagreements among contributors are to be found in this section, in which mainstream philosophers of science (Haack, J. Nelson, and Niiniluoto) and feminist philosophers of science (Longino,

L. H. Nelson, and Potter) consider the relationships between values and science, the nature of evidence, the loci of scientific knowledge, the viability and implications of Quine's underdetermination thesis, realism, and the implications or lack thereof of feminist science scholarship for each.

In "Cognitive and Non-Cognitive Values in Science: Rethinking the Dichotomy", Helen E. Longino questions whether values long construed as cognitive really *are* cognitive (or at least "purely cognitive"), where "cognitive" is taken to mean something like conducive to the discovery of truth. On the basis of a juxtaposition of traditional cognitive virtues (simplicity, external consistency, breadth of scope...) with virtues she finds implicit in the practice of feminist scientists (ontological heterogeneity, novelty, mutuality of interaction...), Longino argues that the former are at least not purely cognitive and, like those that inform the practice of feminist scientists, presently carry political valence. Assuming a view of evidential relations that emphasizes the role of social processes, Longino concludes that the epistemic weight attributed to a particular theoretical virtue in the choice between methods, theories, or research programs, is determined by local, negotiated, and ideally pluralistic, considerations, and that it is time to rethink the distinction between the "cognitive" and "political" salience of these virtues.

In "The Last Dogma of Empiricism?", Jack Nelson addresses the broader issue of whether the science/values distinction can be maintained given a reasonably holistic view of evidence. Nelson argues that there are strong Quinean grounds for rejecting the distinction and that it is far from clear that doing so would vitiate the objectivity of science. Nelson concludes that whether or not the distinction is abandoned, the failure to take value issues seriously works against the doing of good science.

We find interesting contrasts between Longino's and J. Nelson's approaches to and conclusions about the science/values dichotomy. While both question the viability of the traditional understandings of and arguments for the dichotomy, Nelson argues for holism "with distinctions" and does not assume it is impossible or unimportant to distinguish between value theory and empirical theory. Longino's analysis seems, for the reasons earlier outlined, to question the kinds of distinction Nelson's analysis suggests can and should be maintained.

In "Science as Social? – Yes and No", Susan Haack also addresses the relationship of values to science, as well as the nature of cognitive agency and the relationship between these issues. Haack acknowledges that the doing of science is in many ways a social or communal activity; indeed, she maintains, intersubjectivity and cooperative engagement of many researchers across generations are important contributors to science's notable success. But recognizing the social character of science in these terms, Haack argues, does not entail the thesis that non-constitutive values either inevitably do, or should be allowed to, influence science. There is an important distinction, Haack maintains, between the evidential warrant for a hypothesis or theory and its acceptance by individual scientists. Further, Haack argues, "science is social" is either a genuine insight, but not a feminist one, or not a genuine insight. On the basis of these arguments, Haack concludes that feminist science scholarship carries no new or significant implications for a theory of science.

In "Empiricism without Dogmas", Lynn Hankinson Nelson advocates the adoption within naturalized philosophy of science of methodological principles that incorporate a modest and inclusive holistic account of evidence, and that take science communities as the primary loci of philosophical explanations of scientific practice. Building on developments in naturalized and feminist philosophies of science, Nelson maintains that "explanations" of good science that do not recognize more kinds of social factors to be constitutive of science than are allowed for by traditional methodologies are empirically inadequate, and that there are cases of good science that cannot be adequately explained without including a substantive role for social beliefs and values. She advocates a social empiricism, the key epistemic notion of which is evidential warrant. Nelson concludes with an argument from the perspective of social empiricism to the effect that there are normative questions about the social processes characterizing science that should be pursued in naturalized philosophy of science: precisely those questions that are simultaneously questions about the bodies of evidence that support scientific theories and research programs.

In their competing arguments for epistemological individualism and social empiricism, Haack and L. H. Nelson appeal to views about the nature and scope of the evidence that supports theories in the sciences and in the philosophy of science. We suspect that broad differences in their accounts of evidential warrant (differences not limited to those concerning the role of social beliefs and values) are one source of their incompatible conclusions about the relationship between social processes and evidential warrant, the appropriate loci of philosophical explanations of science, and the implications of feminist science scholarship for a theory of science.

W. V. Quine's underdetermination thesis is a presupposition of Longino's analysis of the role of theoretical virtues in determining theory choice and is rejected by Haack who also rejects arguments that she takes to be based on that thesis to the effect that social and political values inevitably will (and should) influence such choice. The viability of the underdetermination thesis and its role in feminist science scholarship are the focus of the next essay in this section.

In "Underdetermination Undeterred", Elizabeth Potter explores the issues separating proponents and critics of the underdetermination thesis and the recent arguments against the thesis advanced by Larry Laudan and Jarrett Leplin. Maintaining that Laudan's and Leplin's arguments against the thesis are not successful, Potter also argues that one issue at stake in the debate over underdetermination is a broader disagreement between those subscribing to rationalist approaches to science, and those (some empiricists, for example, and some pragmatists) who reject such approaches as well as the thesis that all science that is influenced by political factors and/or nonconstitutive values is *ipso facto* bad science. In rejecting the narrow definitions of rationality advocated by rationalist philosophers of science and recognizing that social or political considerations sometimes constrain good scientific decisions, Potter maintains, feminist science scholars do not thereby reject scientific rationality. They are presumed to do so by rationalists, Potter argues, because the latter mistakenly assume that the rationality and cognitive authority possessed by science depend upon the rationalist project.

In "The Relativism Question in Feminist Epistemology", Ilkka Niiniluoto considers the questions of whether and how feminist epistemology can avoid relativism, and maintains that realism is necessary to doing so. After surveying varieties of relativism, Niiniluoto asserts that feminist epistemologies have not, to date, sufficiently engaged the issue of realism. As feminist epistemologies also seem to presuppose gender relativism, Niiniluoto argues, they have not yet, in their empiricist, standpoint, or postmodernist formulations, precluded relativism. The issue is important, Niiniluoto argues, because embracing relativism would undermine the feminist project of criticizing "'male bias' in science" and the emancipatory potential of feminism. Accordingly, he advocates that feminist philosophers of science adopt "critical fallibilist scientific realism", according to which "all factual beliefs in science are fallible, liable to error", *and* classifiable as "uncertain, probable, or truthlike". Niiniluoto also suggests that a feminist epistemology informed by a commitment to realism of this sort and tied to the applied social sciences represents an appropriate incorporation of feminist values in epistemology and science, and that political issues important to women would be served by feminist applied research.

There are other significant contrasts among the essays included in this section. We find a contrast in the approaches Haack and J. Nelson take to the science/values distinction, and in what these authors perceive as the consequences of its abandonment. Haack's analysis suggests that maintaining the dichotomy as an ideal is necessary to the future of good science *and* to the critiques feminist scientists have leveled against androcentric and sexist science. J. Nelson's analysis suggests that from the perspective of holism, abandoning the dichotomy might allow for the kind of attention to values that would contribute to *better* science. Several distinguishable philosophical traditions are drawn upon and appealed to in these essays. Haack's analysis reflects the core research questions and emphases of traditional epistemology, J. Nelson and L. H. Nelson locate their analyses in naturalized philosophy of science, and L. H. Nelson maintains this is a distinguishable discipline from traditional epistemology. Longino's "contextual empiricism" presumes a semantic view of theories in keeping with the view outlined by Giere as the basis for "perspectival realism"; the analyses undertaken by Haack, J. Nelson, L. H. Nelson, Niiniluoto, and Potter, on the other hand, presuppose a syntactic view of theories. Finally, as the foregoing suggests, several distinguishable versions of realism are advocated in these essays.

The disagreements we have noted notwithstanding, we find the rejection of relativism and the assumption that the philosophy of science is a normative enterprise common to all of the essays in Parts One and Two. These issues are also addressed in the essays included in Part Three, in which the discussion is broadened to include parallels and discontinuities between feminist approaches to science, and sociology of science and varieties of social constructivism, and recent charges by critics of science studies that feminist scientists and philosophers reject both the actuality and possibility of objectivity in the sense traditionally attributed to science.

In "Meeting the Universe Halfway", theoretical physicist Karen Barad builds on the epistemology she attributes to Niels Bohr to develop "agential realism": a view she describes as "a social constructivist view" that constitutes a new form of realism, makes room for a robust notion of objectivity, and is compatible with feminist insights into science. Taking the rejection of relativism to be a common theme in feminist science studies, the first three sections of Barad's essay are devoted to explicating Bohr's views and agential realism. Bohr's philosophy of physics, Barad argues, constitutes an alternative to the dualisms of subject-object, culture-nature, and word-world that are criticized by feminist science scholars, for it conceptualizes "objects" and "agencies of observation" as forming "a non-dualistic whole". This view, Barad argues, also serves as an alternative to versions of social constructivism that reduce knowledge to power plays or to language. After explicating agential realism, Barad relates it to arguments in feminist science studies that knowledge is situated and that objectivity requires critical reflexivity, rather than the separation of an observer from the objects observed.

An argument common to Barad's and Niiniluoto's essays is that realism is important to feminist science critiques and to feminist epistemology. We find intriguing contrasts in the views of realism these authors advocate and in what they take to be necessary if relativism is to be avoided. Barad maintains that the form of social constructivism she advocates, according to which the objects of science are inseparable from the theories and theorizers that posit them, is sufficient to avoid relativism. In contrast, Niiniluoto's analysis suggests that realism of a traditional kind, in which the objects of science are assumed to exist independently of the scientists who posit them and truth is defined as "correspondence between a statement and reality", are required. Another intriguing contrast concerns the appropriate scope of feminist theorizing within and about science. Niiniluoto's analysis suggests that this scope may be limited to the social sciences. Barad argues that "agential realism" is an appropriate epistemology for the natural sciences, e.g., physics, and commensurate with the practices of feminist scientists in a variety of sciences.

In "Feminism and the Social Construction of Scientific Knowledge", Joseph Rouse compares the conceptions of science presupposed by and being developed in feminist science studies with those informing the sociology of science. While Rouse sees both research programs as representing explicit challenges to the epistemological individualism that has dominated the philosophy of science, he maintains that the differences between feminist philosophy of science and sociology of science are at least as significant. Specifically, Rouse argues that the supposedly sharp differences between normative, mainstream philosophy of science and descriptive sociology of science actually presuppose a shared conception of knowledge according to which "'knowledge' demarcates a coherent, surveyable domain of inquiry". In contrast, according to Rouse, feminist science studies emphasize science and the philosophy of science as practices rather than bodies of theory, and construe knowing as concretely situated and more interactive than representational. In addition, Rouse argues, the rejection of methodological relativism, the willingness to retain and employ revised conceptions of evidence and objectivity, and the insistence on engaging in normative evaluations of scientific practice further distinguish feminist science studies from the sociology of science.

In "Science and Anti-Science: Objectivity and Its Real Enemies", Elisabeth A. Lloyd explores recent charges that feminist scientists and feminist analyses of science are "anti-scientific". Surveying specific charges made in several books and articles, Lloyd identifies what she takes to be central assumptions and concerns motivating these charges. Among the more important, she argues, are assumptions that counterpose scientific and scholarly objectivity with overtly political goals; and that presume a dichotomy between, on the one hand, social investigations and explanations of scientific processes and products, and, on the other hand, investigations and explanations presented in terms of standards of evidence, theories, testing, and the like – categories figuring in "internal" scientific evaluations of knowledge. Lloyd maintains that neither the first nor the second assumption stands up to scrutiny, and she uses specific analyses offered by feminist scientists to argue that critics of feminist science studies often distort the arguments made by feminist scientists and philosophers of science so that they appear to be "anti-science", downplay or ignore the substantive contributions feminist scientists have made to research in their disciplines, and/or attempt to exclude feminist scientists from those in a position to engage science by ignoring the scientific credentials of these scientists.

Substantive areas of agreement among the essays in this section have emerged. Perhaps most obviously, Barad, Rouse, and Lloyd reject the claims that feminist scientists and science scholars advocate relativism, reject standards of rational inquiry, and/or are aptly described as "anti-science". Relatedly, each author seems to reject the assumption that explanatory principles that incorporate social factors and processes are necessarily incompatible with principles that incorporate the traditional categories of evidence and objectivity. These issues link the essays of this section to essays in earlier sections, with the question of what kind of explanatory principles are appropriately employed in the philosophy of science and/or science studies more broadly emerging as one of the most pervasive and contested issues in the volume as a whole.

There is a difference worth noting among the essays in this section. While Rouse argues that a significant difference between feminist science studies and sociology of science is that the former rejects methodological relativism, and Barad distinguishes the kind of social constructivism she advocates from at least some versions advocated in the sociology of science also by reference to her rejection of relativism, Lloyd's analysis suggests that the charge of relativism made against sociology of science is itself misplaced.

The essays in Part Four further broaden the discussion by bringing traditions outside British and American philosophy of science to bear on feminist and mainstream philosophy of science.

In "Multicultural and Global Feminist Philosophies of Science: Resources and Challenges", Sandra Harding argues that significant philosophical issues concerning science emerge when themes in multicultural and global feminisms, and in postcolonial science studies, are brought to bear on Northern philosophy of science, including Northern feminist philosophy of science. Among the issues Harding identifies and considers are relationships between androcentrism and Eurocentrism in Northern philosophies of science, the expansion of Northern sciences and technologies to developing countries, and gender relations within global political economies. Viewed from the perspective of postcolonial science studies and multicultural and global feminisms, Harding maintains, Northern sciences can be seen to constitute "local", rather than universally applicable, knowledge systems. Attention to the distinctive philosophical issues raised by multicultural and global feminisms, she concludes, can expand the concerns of postcolonial and Northern feminist philosophies of science in ways that support the development of epistemologies and ontologies capable of detecting the androcentrism and Eurocentrism of dominant frameworks in the sciences and in the philosophy of science.

In "Woman – Nature, Product, Style?" Sara Heinämaa brings the phenomenological tradition in Continental philosophy to bear on conceptions of the sex/gender distinction that she takes to be functioning in feminist science critiques. Understanding phenomenology as critically attending to the basic concepts at work in the natural and human sciences, Heinämaa describes feminist phenomenology as attending to the meanings of the basic concepts and terms presumed in feminist philosophy of science. Focusing on the role of the sex/gender distinction in feminist science critique, Heinämaa argues that the phenomenology of the body provides the resources for a more fundamental challenge to the dichotomies of biological/cultural, corporeal/intentional, and object/subject presupposed in the "(bio)-scientific conception of the body" and criticized by feminist scientists and science scholars. Indeed, Heinämaa maintains, as it has functioned in feminist theorizing within and about the sciences, the sex/gender distinction has served to reinforce these dichotomies and contributed to inadequate conceptions of gender and of women.

A common theme in the essays by Harding and Heinämaa is that while feminist philosophies of science have challenged important features of the philosophical traditions that serve as part of their origins, they have retained other features that can and should be challenged. Another common theme is that gender has been inadequately conceptualized, not only in the sciences, but in British and American feminist theory; both authors offer substantive suggestions for how feminist philosophy of science might proceed so as to overcome these limitations.

We note additional contrasts and parallels among the essays in the volume. We suspect that the understandings of "social constructivism" at work in Barad's and Rouse's essays are different in significant ways, and that each is different from the understandings of that notion at work in the essays by Haack and Lloyd, which also stand in sharp contrast to one another. We find interesting parallels between the rejection of the epistemology/metaphysics distinction that Barad attributes to Bohr, the perspectival realism Giere advocates, and the naturalistic realism J. Nelson and L. H. Nelson attribute to Quine and advocate. And the emphasis on knowing as an interactive relationship between scientists and nature is common to the essays by Barad, Rouse, and Tuana.

We suspect there are substantive differences in the content of "epistemic" at work in Longino's and Haack's approaches to the "social nature of science", in Potter's exploration of the issues dividing rationalist and anti-rationalist philosophies of science, and in Rouse's view of the contrasts between feminist philosophies of science and both mainstream philosophy and sociology of science. And we find Giere's, Longino's, J. Nelson's, L. H. Nelson's, Potter's, and Tuana's understandings of empiricism to be significantly different from the view of empiricism at work in Harding's essay as well, in some cases, as from one another.

Concerns about gender essentialism are common to the essays by Haack, Harding, Heinämaa, Niiniluoto, and Tuana, and the relationships between feminist philosophy of science and relativism that concern Niiniluoto are addressed in virtually all of the other essays. In addition, we find interesting parallels in Heinämaa's explication of Merleau-Ponty's phenomenology of the body, Tuana's exploration of the epistemological significance of species-specific embodiment, and the view of knowing as embodied and situated that is an important feature of Barad's account of agential realism.

We have noted that questions concerning the explanatory principles that should figure in the philosophy of science are among the more pervasive and contested issues in the volume. We conclude by noting broad continuities. Many contributors advance arguments that presuppose both the empirical success of science and science's central role in our broader social and world communities. Additionally, the arguments we have summarized concerning the explanatory principles appropriate to the philosophy of science presuppose a shared view of the philosophy of science as a normative enterprise that can contribute substantive insights into the nature of science. We view these continuities as indicating that further dialogue among feminists and their colleagues about the nature of science and the philosophy of science is both possible and desirable.

Lynn Hankinson Nelson
Jack Nelson

NOTES

[1] This is the schema laid out, for example, in the introduction to Richard Boyd, Philip Gasper, and J. D. Trout (eds): *The Philosophy of Science* (Cambridge, MA and London: MIT Press, 1991).

PART I

HISTORICAL OVERVIEW: LOGICAL EMPIRICISM AND FEMINIST EMPIRICISM

RONALD N. GIERE

THE FEMINISM QUESTION IN THE PHILOSOPHY OF SCIENCE

INTRODUCTION

My title is a reflection of Sandra Harding's *The Science Question in Feminism* (1986). Her science question in feminism is this: Feminist claims of masculine bias in science are often themselves based on scientific studies, particularly the findings of various social sciences. But if the claims or methods of science are in general as suspect as many feminists claim, then appeals to scientific findings to support charges of bias are undercut. In short, is it possible simultaneously to appeal to the authority of science while issuing general challenges to that same authority?

My feminism question in the philosophy of science is this: To what extent is it possible to incorporate feminist claims about science within the philosophy of science? Are feminist claims about science compatible with a philosophy of science that rejects relativism? Are they compatible with a philosophy of science that embraces realism? In short, how seriously should philosophers of science, in general, take the claims of feminists that the philosophy of science should incorporate feminist claims about science? The answer to my question, of course, depends both on what feminist claims one considers and on one's conception of the philosophy of science.

From the standpoint of the philosophy of science, the most significant claim of feminist scholars is that the very *content* of accepted theory in many areas of science reveals the gender bias of the mostly male scientists who created it. Moreover, the theories in question came to be accepted through the application of accepted methodological practices. So the sciences and scientists involved cannot be written off as obviously biased or otherwise marginal. Thus, gender bias in the content of accepted science is both possible and, in some cases, actual.

CASE STUDIES

An appropriate starting point for an examination of feminist critiques of science is with the many case studies of actual scientific research purporting to demonstrate masculine bias in the results of what had been regarded as clear cases of acceptable scientific practice. Investigating such cases, however, is much more difficult than one might think. Before explaining why, I will provide a rough taxonomy of cases and mention a few examples.

L. H. Nelson and J. Nelson (eds), Feminism, Science, and the Philosophy of Science, 3–15.

The most convincing cases are those in which the subject matter of the science consists either of real human beings or higher mammals, and the theories in question focus on aspects of life in which gender is obviously a variable. This includes parts of many sciences such as anthropology, sociology, ethology, and primate evolution. Standard examples of these sorts of cases include theories of human evolution based on a model of "man the hunter." According to these theories, the evolution from higher primates to humans was driven by selective forces operating in small groups of male hunters. The use of tools, the development of language, and particularly human forms of social organization, have all been claimed to have evolved in the context of hunting by males. This theory has been the standard theory in many fields for several generations. This approach was not seriously challenged until women entered these fields in more than token numbers and began developing an alternative model of "woman the gatherer." These women have argued that gathering and elementary agriculture likewise require complex skills, social organization, communication, and the development of basic tools. And, they argue, the evidence for this theory is at least as good as that for the standard "man the hunter" paradigm. The lesson drawn is that the "man the hunter" account was the accepted theory for so long at least in part because it was developed and sustained by scientific communities dominated by men with masculine values and experiences. Developing a plausible rival required women with female values and experiences.[1] The investigations of Longino and Doell (Longino, 1990, Chs. 6 and 7) into theories of the biological origin of sex differences in humans provides another outstanding example of this type of case.

A second category consists of cases in which the subjects are humans or primates, but the theories are not directly about obviously gendered aspects of their lives. Here a good example comes from the field of psychological and moral development. The standard theories for most of the twentieth century were those developed by Freud, Erikson, and Kohlberg. These theories purported to be theories of "human development" but were in fact based primarily on studies of boys and men. When studies of girls and women were made, observed differences were treated as "deviations" from the established norm, or even as evidence of failure by girls to reach the higher stages of development. A contrary view emerged in the 1970s through the work of female psychologists such as Carol Gilligan as reported in her now classic book *In a Different Voice* (1982). Gilligan studied moral development in both men and women, but concentrated on women. Her conclusion was that women are neither deviant nor lagging in their moral development, just different. The lesson is the same as in the "man the hunter" model.

A third category of cases involves living, but non-mammalian subjects, and theories in which sex is not a salient variable. A good example here is Barbara McClintock's work on genetic transposition as interpreted by Evelyn Fox Keller in her 1983 book, *A Feeling for the Organism*. Keller argues that McClintock approached her subject with values and interests that were connected with the fact that she was not a man in a profession dominated by men. McClintock had an appreciation for complexity, diversity, and individuality, and an interest in functional organization and development, which was at variance with the desire for simple mechanical structures that motivated most of her male colleagues. That, according to Keller, explains both why McClintock was able to make the discoveries she did, and why her mostly male colleagues failed for so long to understand or appreciate what she had done.

The fourth and most difficult category for the feminist critique involves non-living subjects, and theories that obviously do not explicitly incorporate gender as a relevant variable. This includes sciences from molecular biology to high energy physics. Here Keller (1985, 1992, 1995) and a few others have argued that the influence of gender can be seen in the metaphors that, they claim, both motivate and give meaning to the theories that are generally accepted. DNA, for example, is thought of as a kind of genetic control center issuing orders along a hierarchical chain of command – a clearly male, military, or corporate, metaphor.

For any of these cases to be effective as a critique of science, one must maintain *both* that they exhibit a clear masculine bias *and* that they nevertheless constitute examples of acceptable scientific practice. To dismiss the cases, therefore, one can argue either that the case for masculine bias is not sufficiently substantiated, or that bias does exist, but the cases are not acceptable science. The power of the anti-feminist position lies in the fact that one can use the argument for gender bias as itself grounds for concluding that the case is one of bad science, thus undercutting the feminist critique. And this strategy is likely to be most successful in the examples where the prima facie case for masculine bias seems strongest. Suspicion of the scientific credibility of such "soft" sciences as anthropology and cognitive development long antedated feminist critiques of theories in these fields.

I believe that a credible case for the feminist position has been made in at least some of these examples, but this claim can only be substantiated by a detailed examination of the cases themselves. So, rather than engage the debate at this level, I will shift my attention to the question whether it is *theoretically possible* that the feminist conclusion is correct. Could there be gender bias in what by all other criteria must count as good science? There is a rhetorical as well as a theoretical reason for raising this question. Many philosophers and philosophers of science simply do not regard it as theoretically possible that the feminist critique could be correct. For these philosophers, looking carefully at the cases is merely an academic exercise. To be convinced, therefore, that it is worth even considering the implications of the feminist critique for the philosophy of science, one must first be convinced that it is at least theoretically possible that the critique is correct. That is what I hope to do here – make a convincing case that it is theoretically possible.

SOME SOURCES OF THE ANTI-FEMINIST POSITION

I will consider several sources of the assumption that the feminist position is theoretically impossible. If it can be shown that the anti-feminist position rests on inadequate foundations, that would undercut the assumption that the feminist position is theoretically impossible.

One source is the enlightenment ideal of science. The cornerstone of the enlightenment ideal is the view that the ability to acquire genuine knowledge of the world is independent of personal virtue or social position. Popes and Kings, Bishops and Knights, have no special access to genuine knowledge. What matters is the correct employment of natural reason, and that is, in principle, within the grasp of any normal person. The irrelevance of gender was presumed, although too of ten because women were deemed not capable of exercising the powers of natural reason. In the present-day philosophical canon, most of the thinkers between Descartes and Kant held an enlightenment picture of science, even if, like Descartes, they were precursors rather than participants in the enlightenment as such. To a large extent, much of contemporary philosophy simply presupposes this enlightenment ideal. And that at least partly explains why so many contemporary philosophers and philosophers of science find it simply impossible that gender might matter for what counts as legitimate scientific knowledge.

Feminists, not surprisingly, tend to take a dim view of the enlightenment. I would urge a middle ground, insisting that the enlightenment was a genuine advance over what came before, but recognizing that its presumption of the gender neutrality of human reason was *merely* a presumption, and not based on any firm grounds, particularly not the sorts of empirical investigations now common in the cognitive and social sciences. But I do not want to dwell on the enlightenment. There are sources much closer to our own time for the view that the feminist critique could not possibly be correct.

The current configuration of views within philosophy of science in the United States derives mainly from European sources transmitted by refugees displaced by World War II. For the most part, these influential refugees were German speaking members of a loosely knit group advocating a scientific philosophy, a "Wissenschaftliche Weltauffassung." These thinkers were repelled by the various neo-Kantian idealisms then dominant within German philosophy, and in German intellectual life generally. And they were simultaneously inspired by the new physics associated above all others with the work of Einstein.

In a nutshell, the position of the scientific philosophers was that to understand the nature of fundamental categories like space and time, one should look to Einstein's relativity theory, not to the *a priori* theorizing of neo-Kantian philosophers. Similarly, to understand the nature of causality, one should look at the new quantum theory. Their program was a radical program, a program to *replace* much of philosophy as it was generally practiced in Germany with a new scientific philosophy. It is thus not surprising that none of these philosophers occupied positions of great influence, whether intellectual or institutional, within the German speaking philosophical world.

The most prominent at the time was Moritz Schlick, Professor of Philosophy at the University of Vienna. But he was not really part of the Viennese philosophical establishment. The chair he held had earlier belonged to Ernst Mach, a philosopher–scientist of radical empiricist persuasions. Schlick himself was murdered by a former student, under somewhat shadowy circumstances, in 1936.

Before his death, however, Schlick had provided both philosophical inspiration and institutional support for the Vienna Circle. It was he who, in 1926, brought the young Rudolf Carnap to Vienna as an instructor in philosophy. And it was Schlick who maintained contact with Wittgenstein, who had his own program for a philosophy to end all philosophies. But it was Carnap who became the intellectual leader of the Vienna Circle, a heterogeneous group of mathematicians, natural scientists, social scientists, and scientifically trained philosophers like himself.

The 1920s and early 1930s were disquieting times in Germany and Austria. Political life, often played out in the streets, was fractured left and right. The threat of anarchy ended abruptly on January 30, 1933 when Hitler came to power in Germany. The scientific philosophers were overwhelmingly internationalist in outlook; liberal, socialist, or even communist in political orientation; and many were Jews. For such people, life in Germany, and even in Austria, became increasingly difficult.

Most prominent among the scientific philosophers outside of Vienna was Hans Reichenbach in Berlin. While a student of physics and mathematics in the teens, Reichenbach was active in socialist student movements. That ended when he began teaching science and mathematics in various Technische Hochschule. He also began publishing logical–philosophical analyses of Einstein's theory of relativity. In 1927, Einstein, together with Planck and von Laue, arranged for Reichenbach to be offered an untenured position in the physics department at the University of Berlin, the pinnacle of German, and, at that time, world, physics. The philosophers in Berlin voted not to admit Reichenbach as a member of their department, but Einstein, at least initially, welcomed his help in carrying on his own intellectual battles with the neo-Kantians over the nature of space, time and causality. Reichenbach relished the role.

With the imposition of the Nazi racial laws in the spring of 1933, Reichenbach, along with hundreds of other German professors, was dismissed from his post. Einstein, having resigned from abroad, found safe haven at the newly created Institute for Advance Study in Princeton. Reichenbach was among fifty or so former German professors who accepted five-year contracts at the University of Istanbul. This was part of Kemal Atatürk's effort to bring Turkey into the modern world. Even before his call to Berlin, Reichenbach had been exploring the possibility of emigrating to the United States. Now he resumed these efforts in earnest. As part of his plan to find a position in the United States, he put aside his technical work both on relativity and on the theory of probability, and began writing, in English, a general work on scientific epistemology. That work, *Experience and Prediction*, was completed in 1937 and published by the University of Chicago Press in 1938 – the year Reichenbach began his tenure at UCLA.

In the very first section of that book, titled "The Three Tasks of Epistemology," Reichenbach introduces his distinction between "the context of discovery" and "the context of justification," remarking that "epistemology is only occupied in constructing the context of justification" (p. 7). The introduction of the distinction is not the conclusion of any argument. It is a precondition for the analysis to follow.

In fact, this distinction, though of course not in these words, had existed in German philosophy for half a century. But this seems to be the first time it appeared in Reichenbach's writings. It reappears only once in *Experience and Prediction*, near the end of the final chapter on probability and induction, where he writes (p. 382):

> What we wish to point out with our theory of induction is the logical relation of the new theory to the known facts. We do not insist that the discovery of the new theory is performed by a reflection of a kind similar to our expositions; we do not maintain anything about the question of how it is performed – what we maintain is nothing but a relation of a theory to facts, *independent of the man who found the theory*. There must be some definite relation of this kind, or there would be nothing to be discovered by the man of science. Why was Einstein's theory of gravitation a great discovery, even before it was confirmed by astronomical observations? Because Einstein saw – as his predecessors had not seen – that the known facts indicate such a theory. ... (emphasis added).

Here I wish to indulge in a bit of historical speculation. The speculation is this: When Reichenbach writes of "a relation of theory to facts, independent of the man who found the theory," he is thinking primarily of Einstein, whose views were vilified in the Nazi press not because of any lack of a proper logical relation between Einstein's theories and the facts, but simply because of a personal fact about the man with whom those theories originated – he was a Jew. Reichenbach's own personal situation differed from Einstein's mainly in that his accomplishments and, consequently, his reputation, were less exalted.[2]

One can now see a clear connection between contemporary feminist critiques of science and Reichenbach's use of the distinction between discovery and justification. Reichenbach, I believe, made it a precondition for doing scientific epistemology that the very notion of "Jewish science" be philosophically inadmissible. The Nazi racial laws were not only a crime against humanity, they were a crime against philosophical principle. The feminist notion of "masculine science," or any sort of gendered science, is not a principle any different. It makes the epistemological status of a scientific theory dependent on facts about the scientists themselves, as historical persons, quite apart from internal, logical, relations between fact and theory.

Even if I am mistaken about the personal motivation behind Reichenbach's use of a then well-known distinction in his first general epistemological work, there is no doubt that his understanding of the distinction rules out the relevance of gender to any philosophically correct understanding of legitimate scientific knowledge. Moreover, this understanding of the task of scientific epistemology was shared by most of the European scientific philosophers. And it was these philosophers who came to dominate philosophical thought about science in the United States in the post-war period.

One might object that this is all just so much history of the philosophy of science. Where are the arguments? I hope it is clear that this response begs the question at issue. The validity of the discovery–justification distinction was not established by argument. It was, as is clear in Reichenbach's book, part of the initial statement of the task of a scientific epistemology. It is part of that conception of scientific epistemology that gender or other cultural factors cannot possibly play any role in establishing the legitimacy of scientific claims. My "argument" has been that it is to a large extent due to the legacy of those whose conception of the philosophy of science was formed in the war against Nazi power and ideology that the idea of gendered science still seems to many as being simply impossible.

The point of my historical remarks can be put more sharply. The insistence on the irrelevance of origins which has characterized logical empiricism in America is refuted by the history of that movement itself. The prominence of many doctrines, like the discovery–justification distinction, was not the result of argument, but an assumption forming the conceptual context within which arguments were formulated. The only way to understand why those doctrines were held is to inquire into the historical origins of their role in that movement. Indeed, it is a revealing irony that later criticisms of the discovery–justification distinction focused exclusively on its validity or usefulness, not on its origins.

THE POSSIBILITY OF GENDER BIAS IN POST-POSITIVIST PHILOSOPHY OF SCIENCE

The contemporary feminist movement in America has its own roots in the civil rights movement and the anti-war movement of the 1960s. That was a different war, a different generation, and a different set of political circumstances. The major influence on the philosophy of science of that decade was Thomas Kuhn's *The Structure of Scientific Revolutions*. Kuhn clearly did not set out to become a hero of the 1960s cultural revolution. Nor could one who wrote so unselfconsciously about "the man of science" have been promoting a feminist agenda. Yet his work has, I think correctly (e.g., by Keller, 1985), been seen as providing support for the possibility of gendered science.

In Kuhn's book, the distinction between "the context of discovery" and "the context of justification," in just those words, appears again in the very first chapter. Here, however, Kuhn himself remarks that the distinction seems not to have been the result of any investigation into the nature of science. Rather, he claims, it was part of a framework within which the study of science had been carried out. He makes clear that his own inquiry does not presuppose any such distinction. And, indeed, Kuhn's own theory of science, with its emphasis on the role of individual judgment exercised by scientists in communities, yields nothing that would rule out the possible influence of gender on the eventual beliefs of a typical scientific community.

In the philosophical profession at large, it is widely believed that Kuhn was part of a historical turn in the philosophy of science which superseded logical empiricism. That belief is mistaken on at least two counts. First, the historical tradition within the philosophy of science did not supersede logical empiricism. It was, rather, a rival philosophical tradition which emerged around 1960 and was in part stimulated by Kuhn's work. Logical empiricism continued to evolve both in terms of the study of particular scientific theories and in terms of general methodological inquiries. Both sorts of developments are exemplified, for example, in the works of Bas van Fraassen (1980, 1989, 1991). Second, Kuhn himself was only marginally a part of the historical tradition within the philosophy of science. Most of the philosophers of science associated with that tradition, including Paul Feyerabend, N. R. Hanson, Imre Lakatos, Larry Laudan, Ernan McMullin, Dudley Shapere, and Stephen Toulmin, shared Kuhn's rejection of logical empiricism. And they agreed with his focus on scientific development as the central notion for the study of science. But, for the most part, they also shared a rejection of Kuhn's own theory of science.

With the obvious exception of Feyerabend, these historically oriented philosophers of science sought not to reject the Logical Empiricist idea of an objective connection between data and theory, but to *replace* the idea of a *logical connection* between data and theory with that of *rational progress* within a research tradition. This shift is clearest in the case of Lakatos. For Lakatos, a research program is progressive to the extent that it generates successful novel predictions yielding new confirmed empirical content. There appears to be no room in this definition for any influence from cultural variables such as gender. I will now argue that the apparent impossibility of gender bias in post-positivist philosophical theories of rational progress is only apparent. It is possible even on Lakatos' hard-line account.

One of the many lessons Kuhn claimed to have learned from his study of the history of science was that scientists rarely abandon a research tradition unless they first can at least imagine a promising alternative. Both Lakatos and Laudan explicitly adopted this idea, arguing that the evaluation of a research tradition is not based on a two place relationship between data and a theory, but on a three place relationship between data and at least two rival research traditions.

There cannot be many examples in the history of science where the existing rival research programs exhaust all the logical possibilities. So it is typically possible that the theories making up the existing rival research programs are in fact all false. Nevertheless, as Kuhn argued, and almost everyone else agreed, it is rare to have a scientific field in which there is no clearly favored research program. There is typically an establishment position. It follows that, at any particular time, which research program is most progressive by any proposed criteria depends on which of the logically possible research programs are among the actually existing rival programs. Against other logically possible rivals, the current favorite might not have fared so well. Moreover, Lakatos and Laudan, but most others as well, retain a distinction between discovery and justification to the extent that their accounts of rational progress place few if any constraints on how a possible research program comes to be an active contender. There is little to rule out this process being driven by gender bias or any other cultural value.

So, for any leading research program, it is possible that its position as the current leading contender is in part a result of gender or some other cultural bias. If these biases had been different, other programs might have been considered, and a different program might have turned out to be comparatively more progressive at the time in question. In short, the fact that a given program is judged normatively most progressive by stated criteria might possibly be due, at least in part, to the operation of gender biases in the overall process of scientific inquiry. And that is enough to establish the possibility that the feminist critique is correct in at least some cases.

A POPPERIAN RESPONSE

My earlier survey of leading scientific philosophers omitted any mention of Karl Popper. That was deliberate, because, as I see it, Popper had little influence on what became logical empiricism, particularly in America, until after publication of the 1959 English edition of his 1935 monograph, *Logic der Forschung*, under the even more misleading title, *The Logic of Scientific Discovery*. Despite his own claims that it was he who killed positivism (1974), the accidental fact that the English edition of Popper's book appeared shortly before Kuhn's put him in a position to become a primary defender of the positivist faith against the Kuhnian heresy.

The titles of Popper's book are misleading because, on his account of science, there is no such thing as a "logic" of research or of scientific discovery. The main role for logic in science is the use of *modus tollens* in the refutation of a universal generalization by a statement describing a negative instance. This form of inference requires no reference to alternative hypotheses. So, apart from questions about how one establishes the truth of the required singular "observation statement," this form of inference would seem to be immune to gender or any other cultural influences. Popper's work thus shows that it is possible to construct a theory of science which maintains a strong enough distinction between the contexts of discovery and justification to eliminate the possibility of gender bias. But it also shows how very difficult it is to construct a *good* theory of science that fulfills this requirement. No one better exhibited the shortcomings, not to say the utter implausibility, of Popper's theory of science than his successor, Imre Lakatos – and Lakatos borrowed heavily from Kuhn. It should be noted that the approaches to scientific justification taken by both Carnap and Reichenbach would, if successful, also eliminate any possibility for gender or other cultural biases. For both, theory evaluation is not comparative, at least not in any obvious way. I will not elaborate this point further because these approaches have few defenders today.

The successor to Carnap's conception of inductive logic is a subjective probability logic, as championed, for example, by Carnap's associate, Richard Jeffrey (1965). Theories of subjective probability, however, place only minimal constraints on how an individual assigns initial probabilities to any theory. This leaves lots of room for individual scientists to assign high initial probabilities to theories reflecting their own particular gender biases. The best the probabilistic approach can offer is proof of the diminishing influence of the initial probability assignment in the face of increasing observational evidence. But there is no way of knowing, in this framework, how much the probability assigned a particular theory at any given time might be the product of some form of bias, including gender bias. That leaves feminist critiques as much room as they need.

In sum, there is little in current philosophical theories of science that supports the widespread opinion that gender bias is impossible within the legitimate practice of science. That opinion seems mainly the product of a traditional adherence to an enlightenment ideal of science strongly reinforced by the historical origins of twentieth century scientific philosophy in Europe and its rebirth as logical empiricism in America. As disquieting as it may seem to many, we shall have to learn to live with the empirical possibility of "Jewish science." That is, for any particular scientific theory, it must be an *empirical* question whether its acceptance as the best available account of nature might be due at least in part to its having been created and developed by Jewish scientists rather than scientists embodying some other religious tradition. In another cultural context in which science as we know it is generally practiced, some other theory might now be the accepted theory. Whether or not this is true for any *particular* theory can only be determined empirically by looking in detail at the history of how that theory achieved its present status. The irrelevance of religious origins cannot be guaranteed *a priori*. The same holds for gender.

PERSPECTIVAL REALISM

In countenancing the relevance of cultural forces in the acceptance of scientific theories, have we not moved too far in the direction of *relativism*? In particular, is this position compatible with a reasonable scientific realism? I think it is, but the issue is complex. If we suppose that the world is organized in a way that might be mirrored in a humanly constructable linguistic system, then there is indeed a problem. For then realism seems to require that we could have reason to believe that our theories are literally *true* of the world. The objects in the world are grouped as our theories say they are and behave as our theories say they should behave. If, however, what we *take to be* true of the world is influenced by cultural factors, there is no reason to think that this influence would promote the development of actually true theories and considerable reason to suspect that it would do just the opposite. That sounds like relativism, not realism.

Radical though it may seem, I think the resolution of this problem is to reject the usefulness of the notion of *truth* in understanding scientific realism. I do not mean that we cannot use an everyday notion of truth, as when asserting that it is indeed true that the earth is round. Here truth may be understood as no more than a device for linguistic ascent. Rather, it is the *analysis* of truth developed in the foundations of logic and mathematics, and used in formal semantics, that we should reject in our attempts to understand modern science. But if we reject the standard analyses of truth and reference, what resources have we left with which even to formulate claims of realism for science? The answer is that the notion of linguistic truth is but one form of the more general notion of *representation*. What realism requires is only that our theories well *represent* the world, not that they be true in some technical sense. So we need a notion of representation for science that does not rest on the usual analyses of truth for linguistic entities. What might that be?

A first step is to reject the analysis of scientific theories as sets of statements in favor of a model-based account which makes non-linguistic models the main vehicles for representing the world, and places language in a supporting role.[3] We may, of course, use language to characterize our models, and what we say of the models is true. But this is merely the truth of definition, and requires little analysis. The important representational relationship is something like *fit* between a model and the world. Unlike truth, fit is a more qualitative relationship, as clothes may be said to fit a person more or less well. Of course we can say it is true that the clothes fit, but this is again merely the everyday use of the notion of truth.

Here I can offer no general analysis of the notion of fit, only a further analogy – maps. There are many different kinds of maps: road maps, topological maps, subway maps, plat maps, etc. And it can hardly be denied that maps do genuinely represent at least some aspects of the world. How else can we explain their usefulness in finding one's way in otherwise unfamiliar territory? Moreover, the idea of mapping the world has long been present in science. There were star charts before there were world atlases, and scientists around the world are now busy "mapping" the human genome. Maps have many of the representational virtues we need for understanding how scientists represent the world. There is no such thing as a universal map. Neither does it make sense to question whether a map is true or false. The representational virtues of maps are different. A map may, for example, be more or less accurate, more or less detailed, of smaller or larger scale. Maps require a large background of human convention for their production and use. Without such they are no more than lines on paper. Nevertheless, maps do manage to correspond in various ways with the real world.

Since no map can include *every* feature of the terrain to be mapped, what determines which features are to be mapped, and to what degree of accuracy? Obviously these specifications cannot be read off the terrain itself. They must be imposed by the mapmakers. Presumably which set of specifications gets imposed is a function of the *interests* of the intended users of the maps.

Among cartographers, those whose job it is to make maps, it is assumed that constructing a map requires a prior selection of features to be mapped. Another aspect of mapmaking emphasized by cartographers is *scale*, particularly for linear dimensions. How many units of length in the actual terrain are represented by one unit on the map? These two aspects of mapmaking, feature selection and scale, are related. The greater the scale the more features that can be represented. The required trade-offs again typically would reflect the interests of the intended users.

It is not stretching an analogy too far to say that the selection of scale and of features to be mapped determines the *perspective* from which a particular map represents the intended terrain. Photographs taken from different locations provide more literal examples of different perspectives on a terrain or a building. In any case, given a perspective in this sense, it is an empirical question whether a particular map successfully represents the intended terrain. If it does, we can reasonably claim a form of *realism* for the relationship between the map and the terrain mapped. I will call this form of realism *perspectival realism*.[4]

Standard analyses of reference and truth suggest a metaphysics in which the domain of interest consists of discrete objects grouped into sets defined by necessary and sufficient conditions. Likewise, there is a metaphysics suggested by perspectival realism. Rather than thinking of the world as packaged into sets of objects sharing definite properties, perspectival realism presents it as highly complex and exhibiting many qualities that at least appear to vary continuously. One might then construct maps that depict this world from various perspectives. In such a world, even a fairly successful realistic science might well contain individual concepts and relationships inspired by various cultural interests. It is possible, therefore, that our currently acceptable scientific theories embody cultural values and nevertheless possess many genuine representational virtues.

FEMINIST REALISM

There is an unfortunate mismatch in terminology between feminist and general philosophers of science. Within the philosophy of science generally, the distinction between empiricists and realists concerns the sort of epistemic commitment one has toward "unobservable" or "theoretical" entities and properties. Empiricists would restrict our commitments to the observable phenomena; realists make no such restrictions. "Feminist empiricist," on the other hand, characterizes someone who thinks some theories may embody gender biases, but also thinks such biases can be detected using standard scientific methods. Moreover, better theories, which may embody other biases, can be proposed and validated. Feminist empiricism, therefore, is neutral regarding the general debate between empiricists and realists. Of course a feminist empiricist might also be an empiricist in the more general sense, but that would be an *additional* commitment beyond feminist empiricism. More significantly for my purposes, a feminist empiricist could be a *realist* in the more general sense. Thus, feminist realism is not inherently an incoherent doctrine. Given current usage, it turns out, misleadingly, to be a special case of feminist empiricism.

To my knowledge, no feminist philosopher of science has claimed to be a feminist realist. I expect this is because realists have often claimed to know *the truth* about many things, or at least to be *rationally justified* in claiming such knowledge. Feminists, quite naturally, are suspicious of any such claims. From my point of view, this suspicion presupposes the mistaken view that realism must be understood in terms of truth in the standard philosophical sense. Abandoning this presupposition, one is free to adopt a perspectival account of realism which is far more congenial to the interests of feminists. Moreover, adopting perspectival realism does not commit one to any form of special scientific rationality. Perspectival realism is perfectly compatible with a thoroughgoing naturalism which appeals only to the naturally evolved cognitive capacities of human agents together with their historically developed cultural artifacts. It is to be expected that such agents would typically project their cultural values, including gender values, into the models they develop to explain phenomena in the world. And some of these models could be expected to end up part of established science. That is just what feminist philosophers of science have been claiming all along.

Department of Philosophy and Center for Philosophy of Science

NOTES

[1] For an overview and references on this topic see Longino (1990, 106–111).

[2] I have developed these and related themes at greater length in Giere, 1996.

[3] For further elaboration and references on model-based accounts of scientific theories see Giere, 1988.

[4] I find inspiration for both this terminology and the concept in some works of Donna Haraway, particularly her paper, 'Situated Knowledges: The Science Question in Feminism and the Privilege of Partial Perspective', reprinted in Haraway, 1991.

REFERENCES

Giere, R. N.: 1988, *Explaining Science*. University of Chicago Press, Chicago.

Giere, R. N.: 1996, 'From *Wissenschaftliche Philosophie* to Philosophy of Science.' in R. Giere and A. Richardson (Eds.), *Origins of Logical Empiricism*. Minnesota Studies in the Philosophy of Science, Vol. XVI. University of Minnesota Press, Minneapolis.

Gilligan, C.: 1982, *In a Different Voice*. Harvard University Press, Cambridge.

Haraway, D. J.: 1991, *Simians, Cyborgs, and Women*. Routledge, New York.

Harding, S.: 1986, *The Science Question in Feminism*. Cornell University Press, Ithaca, NY.

Jeffrey, R. C.: 1965, *The Logic of Decision*. McGraw-Hill, New York; 2nd edn University of Chicago Press, Chicago, IL, 1983.

Keller, E. F.: 1983, *A Feeling for the Organism*. W. H. Freeman, New York.

Keller, E. F.: 1985, *Reflections on Gender and Science*. Yale University Press, New Haven.

Keller, E. F.: 1992, *Secrets of Life, Secrets of Death: Essays on Language, Gender and Science*. Routledge, New York.

Keller, E. F.: 1995, *Refiguring Life: Metaphors of Twentieth-Century Biology*. Columbia University Press, New York.

Kuhn, T. S.: 1962, *The Structure of Scientific Revolutions*. University of Chicago Press, Chicago, IL; 2nd edn 1970.

Longino, H. E.: 1990, *Science as Social Knowledge*. Princeton University Press, Princeton, NJ.

Popper, K. R.: 1935, *Logic der Forschung: Zur Erkenntnistheorie der Modernen Naturwissenschaft*. Springer Verlag, Wien.

Popper, K. R.: 1959, *The Logic of Scientific Discovery*. Hutchinson, London.

Popper, K. R.: 1974, 'Intellectual Autobiography', in P. A. Schilpp (Ed.), *The Philosophy of Karl Popper*, 2 vols. Open Court, La Salle.

Reichenbach, H.: 1938, *Experience and Prediction*. University of Chicago Press, Chicago, IL.

van Fraassen, B. C.: 1980, *The Scientific Image*. Oxford University Press, Oxford.

van Fraassen, B. C.: 1989, *Laws and Symmetry*. Oxford University Press, Oxford.

van Fraassen, B. C.: 1991, *Quantum Mechanics*. Oxford University Press, Oxford.

NANCY TUANA

REVALUING SCIENCE: STARTING FROM THE PRACTICES OF WOMEN

INTRODUCTION

Work in the social studies of science in the last twenty years has undermined the belief common to positivist models of science that value-neutrality is both a hallmark and goal of scientific knowledge. The ideal of a value-free science was linked to the tenet that neither the individual beliefs or desires of a scientist nor the social values of a scientific community are relevant to the production of knowledge, and models of scientific method were constructed with the goal of factoring out such contaminating influences. The rapid militarization of science in the United States since the 1970s and the current rise of influence of venture capital in charting the direction of scientific research have made it increasingly difficult to draw any clear lines between a "pure," disinterested science, and a goal-oriented, transformative "applied" science. Questions in the philosophy of science have shifted from the "pure" epistemological question "How do we know?" to questions that reflect the locations of science within society and the relationships between power and knowledge: "Why do we know what we know?" "Why don't we know what we don't know?" "Who benefits or is disadvantaged from knowing what we know?" "Who benefits or is disadvantaged from what we don't know?" "Why is science practiced in the way that it is and who is advantaged or disadvantaged by this approach?" "How might the practice of science be different?"

Feminist theorists of science have been active participants in this research program. Our work has added an important dimension to discourses concerning the value-neutrality of science by focusing attention onto the dynamics of gender and oppression in the theories and methods of science.[1] One of the central insights of feminist science studies has been the increased awareness of the ways in which social locations, locations that include political and ethical dimensions, are gendered. Through this attention to gender we have contributed to the transformation of the traditional question "How do we know?" in numerous ways, including investigating whether traditional models of rationality and of the scientific method have been gender biased, that is, have privileged traits viewed as masculine and denigrated those perceived to be feminine; documenting the ways in which scientific theories have reinforced sexist and/or racist biases: delineating the ways in which men in dominant groups have benefited (and been hindered) by the questions asked and avoided in science; and analyzing the impact of the exclusion, as well as the inclusion, of women in science.

L. H. Nelson and J. Nelson (eds), Feminism, Science, and the Philosophy of Science, 17–35.

An important resource for feminist investigations of science has been the practices of women scientists. Many feminist theorists, particularly those who embrace a feminist standpoint epistemology, have argued that the distinctive experiences of women in a gender-stratified society provide an important resource, a resource typically overlooked by nonfeminist theorists, that, in the words of Sandra Harding, enables "feminism to produce empirically more accurate descriptions and theoretically richer explanations than does conventional research" (Harding, 1991, p. 119). One of my goals in this essay is to illustrate the ways in which the experiences of women, particularly women scientists, provide a resource for feminist critiques of the ideal of value-neutrality in science.

Women's differences, both their differences from men and their differences from one another, can highlight overlooked or minimized aspects of the knowledge process in science. I will here limit my analysis to three of these, each of which is relevant to transformations of the traditional epistemological question "How do we know?" and the rejection of the ideal of value-neutrality in science:

(1) replacing the traditional model of the knower as a detached, disinterested individual with the dynamic model of engaged, committed individuals in communities;
(2) recognition of the epistemic value of affective processes;
(3) examination of the role of embodiment in the knowledge process.

INDIVIDUALS IN COMMUNITIES

Descartes envisioned himself alone in this study, attempting to put aside all he had learned from authority and all the beliefs he had unquestioningly inherited from his culture, as well as endeavoring to suppress the needs of his body. Descartes believed that only after he had removed all such influences from his rational processes would he be capable of pursuing his method for gaining true knowledge, alone and unencumbered by others.

Although Descartes was hardly an empiricist, it is the Cartesian subject that is designed to hold the subject position in S-knows-that-p models of knowledge. This is a model of knowledge that aims ideally at removing all individual traces of the knowing subject. Both perception and cognition are assumed to be invariant from knower to knower – at least in the ideal case. All other factors such as personal beliefs, desires, and bodily configurations are deemed irrelevant at best, contaminating at worst. Based on this picture of rationality, much of modern epistemology has been focused on the ways in which variations between knowers could be filtered out.

This model of the knowing subject is in tension with the feminist acknowledgement of the fact that as humans we are always in relations of interdependence and that these relationships are crucial not simply for personal satisfaction, but also for moral, political, and scientific deliberation. In the words of Seyla Benhabib, "the self only becomes an I in a community of other selves who are also I's. Every act of self-reference expresses simultaneously the uniqueness and difference of the self as well as the commonality among selves" (Benhabib, 1987, p. 94).

A careful study of the actual practice of science also discloses a different model of the knowing subject, one that necessitates a rejection of the model of the isolated knower and replaces it with a dynamic model of individuals in communities. An examination of the complexity of the communities relevant to the production of knowledge in science also reveals that the production of good science does not require disinterested, dispassionate scientists. As Sandra Harding has convincingly argued, objectivity does not require neutrality.[2] A scientist's social locations can be epistemically significant to her or his practice of science. The ideal of a pure science, a science uninfluenced by values, and the scientist as a neutral recorder of facts are myths, ones that can be rejected without abandoning objectivity.

Changing the Subject of Evolution

The development of "woman, the gatherer" theories of human evolution has been the subject of much discussion in feminist science literature because this example is an excellent illustration of not only the inescapable fact of value within the construction of scientific theories, but also the potential epistemic significance of the various communities, including political communities, in which the knower participates.[3]

Feminist discussions of the epistemological significance of being part of the feminist community have found the science of primatology to be particularly relevant due to the fact that its stories of human evolution arise out of origin myths, that is, accounts of the origins of the family, of the sexes and their roles, as well as of the divisions between human and nonhuman animals. In other words, accounts of human evolution wear the metaphysics of their authors "on their sleeves" and thus provide clear accounts of the ways participation in alternative communities can be epistemologically significant.

To understand the androcentrism of traditional "man, the hunter" accounts of evolution, we need only attend to the respective roles of women and men. "Man, the hunter" theories of human evolution attribute the evolution of *Homo sapiens* to those activities and behaviors engaged in and exhibited by male ancestors. Males, the explanation goes, having the important and dangerous task of hunting big animals to provide the central food source, invented not only tools but also a social organization, including the development of language, that enabled them to do so most successfully. Hunting behavior is posited as the rudimentary beginnings of social and political organization. "In a very real sense our intellect, interests, emotions, and basic social life – all are evolutionary products of the success of the hunting adaptation. ... The biology, psychology, and customs that separate us from apes – all these we owe to the hunters of time past" (Washburn and Lancaster, 1976, pp. 293, 303).

Such accounts do not omit women, but place them firmly "at home." While men are out hunting, women are taking care of hearth and children, dependent upon the men for sustenance and protection. Note the assumptions embedded in this account. Only male activities are depicted as skilled or socially oriented. Women's actions are represented as biologically oriented and based on "nature." This definition of woman's functions as natural curtails any analysis of them, such as their relation to the physical and social environments or the role they might play in determining other social arrangements. Men are depicted as actively transforming their nature, while women are portrayed as constrained by it.

The alternative origin stories told by feminist primatologists transform women from a passive, sexual resource for males to active agents and creators. The work of Linda Marie Fedigan, Sarah Blaffer Hardy, Lila Leibowitz, Sally Linton Slocum, Barbara Smuts, Shirley Strum, Nancy Tanner, and Adrienne Zihlman, among others, began in the 1970s to transform the complexion of accounts of the nature of woman and man. One key to understanding the explosion of alternative images of women's nature lies in the woman's movement of the 1960s that contested the definitions of woman as the second sex, definitions that simultaneously relegated her role to the private realm of family while designating the public realm of culture and politics as that which makes one fully human.

Feminist attention to perceptions of women's roles and the linkage of woman and nature provided the basis for a rethinking of evolution for a number of scientists. The anthropologist Sally Linton Slocum, for example, in her 1970 essay "Woman the Gatherer: Male Bias in Anthropology" identified ways in which females were being obscured within evolutionary theories by the association of their actions with nature and began to question the assumption that women's actions were unimportant because they were derived from instinct and thus not relevant to the evolutionary process. Slocum's position was in turn developed by the paleoanthropological research of Adrienne Zihlman and Nancy Tanner. This shift of attention was the result not of any biological difference between women and men scientists, but because women scientists were more likely to be affected by and participate in the feminist community – a community that had been actively exposing the history and the impact of the androcentric bias of associating women with nature and men with culture, as well as working to revalue the socially defined work of women, including childcare and housework.[4] This political awareness arising from the influences of the feminist community changed the focus of attention for researchers like Slocum, Tanner, and Zihlman and contributed to the construction of alternative questions.

But it would be inaccurate to see the accounts of these scientists as influenced only by their participation in communities that were redefining woman's nature. These women were also influenced by their membership in scientific communities and the then current theories of evolution. The point is that accounts by women primatologists, particularly feminist primatologists, while marked both by their gender and their politics as they attempt to carve a role for women out of the standard narrative of evolution, nevertheless evolve out of and are influenced by the accepted narratives and standards of evidence of their scientific communities.

Nor should alternative evolutionary accounts such as "woman, the gatherer" be seen simply as feminist "correctives," that is, as an ideological image imposed onto the data. I will argue that this alternative model of evolution arose in response to changes within the scientific community, provided more accurate accounts of the evidence, and was therefore the result of better science. But this is not incompatible with saying that the model emerged from the practice of feminist scientists who, because of the impact of their communities, attended differently to the data. To say that the practice of science is marked by gender and by politics is not the same as claiming that it arises out of wishful thinking or ideological concerns. A scientific theory can provide consistent methods for obtaining reliable knowledge, yet be influenced by certain values or interests. Objectivity and neutrality are not the same thing.

Although it is an error to ignore the influence of the feminist community on Adrienne Zihlman's development of the "woman, the gatherer" account, it was hardly the only or the most important community of which she was a part. Being a student of Washburn, she was influenced by his belief in a human/chimpanzee/gorilla divergence of only 5 million years, rather than the previously accepted 20 million years. This allowed for an alternative perspective of the differences between chimpanzees and humans, one that saw them as less dramatic and emphasized continuities. To this we must also add Jane Goodall's accounts of mother–infant relationships that evinced a heretofore neglected complexity and enlarged perspective of maternal investment, a model Zihlman extended to hominid evolution.[5] There were also numerous primate studies that resulted in modification of previous views concerning the passivity of female roles, including female sexual initiation of copulation and the roles of females in social dynamics. Other primate studies documented the use of tools by both sexes for gathering tubers.

Zihlman's accounts were thus strongly influenced by contemporaneous scientific accounts. Based on chimpanzee studies of diet and tool use, Zihlman denied the evolutionary importance of hunting as inscribed in "man, the hunter" accounts. For similar reasons she rejected a sharp division of labor between females and males, arguing that both sexes shared food gathering activities, and denounced the image of females as basically confined to a "home-base." However, Zihlman focused on the intensification of the mother–infant bond resulting from increased infant dependency as a crucial element in human evolution, arguing that this change mandated both a greater need for regular food supplies to feed the infant and an increase in cooperation amongst individuals, resulting both in the development of increased social skills and in the selection of more socially skilled males by females.

"Woman, the gatherer" accounts of human evolution were not reversals of "man, the hunter" accounts in which women are seen as agents of evolution and men's activities are relegated to the realm of nature. Rather, "woman, the gatherer" theories recast accounts of human evolution such that both female and male roles are seen as important and did so primarily by denying radical differences between them. Previous accounts of human evolution had acknowledged gathering activities by women. What Zihlman's account changed was the evolutionary *importance* of gathering activities. Previous accounts claimed that hunting was something new that accounted for evolutionary change. On Zihlman's account what accounted for evolutionary change was extended gathering by men and women alike. Zihlman depicted gathering activities as skilled and requiring a relatively large body of knowledge concerning which plants are edible. In other words, these activities are constructed as social and skilled rather than individual and unskilled. Hominid evolution arising from the increased importance of gathering is depicted as a central factor in the development of greater social skills, including communication, as well as the development of tools. Zihlman thus depicts gathering as creating the conditions for the evolution of the importance of hunting, a conclusion more consistent with the archaeological evidence for hunting tools.

What this example illustrates is the importance of beliefs acquired from one's participation in various communities for the development of knowledge. Zihlman's creation of an alternative to the androcentric "man, the hunter" theory was made possible by the knowledge she gained from the communities of which she was a member, in this case both scientific and nonscientific. However, it is not a coincidence that the "woman, the gatherer" hypothesis was initiated by the work of Sally Linton Slocum, and developed by Nancy Tanner and Adrienne Zihlman. Being a feminist scientist can affect one's practice of science. In the words of Lynn Hankinson Nelson, "it makes a difference to one's observations, appraisals of theories, and one's own theorizing, if one *recognizes* androcentric and sexist assumptions, categories, or questions and if one questions the inevitability of male dominance and/or the universality of hierarchical dominance relationships. In short, it makes a difference if one is working from a feminist perspective" (Nelson, 1990, p. 224). But Zihlman's participation within particular scientific communities was also a crucial factor in the development of her research. The point is that a scientist is simultaneously a member of a number of different epistemic communities and subcommunities. The values and beliefs of these various communities often interact in complex ways over the course of the knowledge process. Fully understanding the development of knowledge then requires an appreciation of the interactive effects of all relevant communities and an understanding of the underlying presuppositions, metaphysical as well as aesthetic and moral values, of each community's system of beliefs.

ENGAGED KNOWERS

Acknowledging that social values enter into the practice of science problematizes the traditional model of the knower as detached, disinterested, and autonomous. Both the individualism as well as the goal of neutrality posited by traditional accounts of knowledge must be questioned. Many feminist theorists of science contend that women's relative absence from the practice of science is not due simply to institutional barriers such as limited access to advanced science training, but is also an aspect of a model of the scientist that privileges traits that have historically been associated with masculinity (autonomous, detached, disinterested) and suppress those traditionally associated with femininity (dependent, connected, engaged).[6] In other words, despite its professed neutrality, the positivist model of knowledge, like all models, arises out of a tradition and is imprinted with the values of that tradition. Neither science nor our models of science correspond to the neutrality ideal.

Feminist studies of science thus reveal the myopia of traditional individualist accounts of the knowing subject. On the traditional S-knows-that-p model of knowledge, we need have no knowledge of S. Knowers, while envisioned as distinct individuals, are not seen as *distinctive*. Neither the body nor any "subjective" aspect of an individual's mental activity is seen as affecting the proper pursuit of knowledge. This model of knowledge is linked to the belief in a universal faculty of reason common to all potential knowers. Whether it be Descartes' ability to apprehend clear and distinct ideas or the positivist vision of a deductive logic, knowing capacities are invariant (though not all equally developed). S-knows-that-p models thus embrace the vision of the generic "man" – a sameness that removes the threat of allegedly biased or partial perspectives.

Feminist investigations of science are resulting in what Helen Longino labels the strategy of "changing the subject" of knowledge. We are finding that S-knows-that-p models of knowledge are inadequate to the actual practice of science. The conception of the subject of knowledge as "generic" and hence not itself a subject of study does not fit the epistemic importance of differences between subjects. Such a model, for example, does not account for the epistemic role of the complex relationships between agents of knowledge as evidenced in examples like that of "woman, the gatherer" theories of human evolution. Equally problematic, this model overlooks the epistemic significance of subjective aspects of the relationship between scientists and the subject of inquiry, such as the scientist's commitments, desires, and interests. It also ignores the fact and nature of a scientist's embodiment. In this section I examine the role of what has traditionally been labeled "the passions" in the knowledge process in science, reserving the question of the role of the body for the final section of the essay.

A Feeling for the Organism

Evelyn Fox Keller offers a portrait of the geneticist Barbara McClintock that provides a very different image of the scientist than that of the disinterested, detached observer. McClintock describes herself as having developed a close relationship with the objects of her investigation. "I start with the seedling [of maize], I don't want to leave it. I don't feel I really know the story if I don't watch the plant all the way along. So I know every plant in the field. I know them intimately, and I find it a great pleasure to know them" (Keller, 1983, p. 198). McClintock viewed the complexity of nature as being beyond full human comprehension. "Organisms can do all types of things; they do fantastic things. They do everything that we do, and they do it better, more efficiently, more marvelously.... Trying to make everything fit into set dogma won't work.... There's no such thing as a central dogma into which everything will fit" (Keller, 1983, p. 179). In holding to this belief – a metaphysical value – McClintock deviated from the positivist assumption – yet another metaphysical value – that there were underlying regularities of nature, the laws of nature, that were discrete and individually knowable by humans. This difference in basic values contributed to McClintock's commitment of developing a close relationship with the material she was studying, for only by listening carefully can one "let the material tell you." "I feel that much of the work [in science] is done because one wants to impose an answer on it. They have the answer ready, and they [know what they] want the material to tell them. [Anything else it tells them] they don't really recognize as there, or they think it's a mistake and throw it out.... *If you'd only just let the material tell you*" (Keller, 1983, p. 179).

McClintock's respect for the complexity of nature and her desire to listen carefully led her to embrace a close relationship with the maize she studied. Rather than shifting the job of planting and tending the maize to an assistant, McClintock saw these activities as a crucial part of her process of understanding. In fact, she was proud of the fact that she came to know her plants so intimately that she could predict the structure of the cells' nuclei from external traits of the plants. "Before examining the chromosomes, I went through the field and made my guess for every plant as to what kind of rings it would have.... And *I never made a mistake*" (Keller, 1983, p. 102).

McClintock's emphasis on nature's complexity led to her belief that scientists must at all costs "take the time and look" (Keller, 1983, p. 206). At a time when many other scientists abandoned the slow-growing maize for the quickly reproducing *Drosophila*, McClintock's attention to complexity and the related need for time to look, made her conclude that corn's two crops per year was too much. To have time to analyze all that there was to see, McClintock found that one crop a year was all that she could manage.

McClintock's values, particularly her belief in nature's complexity and the corresponding method of taking the time to look, resulted in her being particularly sensitive to difference. Rather than minimizing anomalies, McClintock believed they were to be listened to carefully, for by finding out why they do not fit the model, one is challenged to develop a more complex model which will account for them. This value led her to criticize the emphasis on quantitative analysis in science arguing that the focus on making everything numerical often resulted in overlooking what was different. Her method was to "see one kernel [of corn] that was different, and make that understandable" (Keller, 1983, p. 97). Part of her method of "letting the material tell you" was this attention to difference. "So if the material tells you, 'It may be this,' allow that. Don't turn it aside and call it an exception, an aberration, a contaminant.... That's what's happened all the way along the line with so many good clues" (Keller, 1983, p. 170).

McClintock's description of the process of observation reads very differently from the accounts of the detached, disinterested observer. "I found that the more I worked with them the bigger and bigger [the chromosomes] got, and when I was really working with them I wasn't outside, I was down there. I was part of the system. I was right down there with them, and everything got big. I even was able to see the internal parts of the chromosomes – actually everything was there. It surprised me because I actually felt as if I were right down there and these were my friends" (Keller, 1983, p. 117). In addition to her admonition to take time to listen, to develop an intimate relationship with the subject of study, McClintock's approach was a blend of reason and passion. Explaining why she was so convinced of the accuracy of her theory at a time when other scientists were rejecting it as untenable, McClintock explained, "The logic was compelling. The logic made itself, the logic was it. What's compelling in these cases is that the problem is sharp and clear. The problem is not something that's ordinary, but it fits into the whole picture, and you begin to look at it as a whole.... It isn't just a stage of this, or that. It's what goes on in the whole cycle. So you get a feeling for the whole situation of which this is [only] a component part" (Keller, 1983, p. 67). According to Keller what is significant about McClintock's method and constitutes the wellspring of her powers as a scientist is the intimacy of the relationship that she develops with the object she is studying, a relationship that requires empathy and cultivated attentiveness.

Knowing Other People

Feminist studies of science, particularly the detailed studies of the practices of women scientists, have served as an important resource for feminist epistemologists. Influenced by examples like that of McClintock, many feminists are developing epistemologies that include the tenet that subjectivity is an important and indispensable component of the process of gaining knowledge.[7] But successfully doing so requires offering alternative models of knowledge. Lorraine Code has offered such an alternative, the model of "knowing other people." While S-knows-that-p models of knowledge are based on what Code calls ordinary knowledge of medium-sized objects in the immediate environment – the red book, the open door – Code's model is based on the centrality of our relationships with others. "Developmentally, recognizing other people, learning what can be expected of them, is both one of the first and one of the most essential kinds of knowledge a child acquires" (Code, 1991, p. 37). Code presents this model as an *addition* to the S-knows-that-p epistemologies that perhaps work for simple objects in simple settings. She argues that the latter model is not sufficient for more complex instances in which knowledge requires constant learning, is open to interpretation at various levels, admits of degree, and is not primarily propositional. For such cases, a standard of knowledge modeled after our knowledge of other people would be more accurate.

Code, influenced by examples like that of McClintock, argues for a remapping of the epistemic terrain. A model that posits knowing other people as a paradigmatic kind of knowing challenges the desirability or even possibility of the disinterested and dislocated view from nowhere. Code's model of knowing other people is a dynamic, interactive model. It is a vision of a process of coming to know, "knowing other people in relationships requires constant learning: how to be with them, respond to them, and act toward them" (Code, 1993, p. 33). It is a model of knowledge that admits of degree, that is not fixed or complete, that is not primarily propositional, and is acquired interactively.

Code's model embraces the subjective components of the knowledge process illustrated in McClintock's method. McClintock's desire to develop an intimate relationship with the subject of her study, to take the time to listen, to get right down there, to develop a feeling for the organism, are all aspects of the knowing process accounted for by Code's model. "Rocks, cells, and scientists are located in multiple relations to one another, all of which are open to analysis and critique. Singling out and privileging the asymmetrical observer-observed relation is but one possibility" (Code, 1991, p. 164). Code's alternative model, unlike S-knows-that-p models, embraces McClintock's metaphysical belief that nature, like other people, is far too complex to allow for complete and universal knowledge. For McClintock, our knowledge of nature will always be partial, always changing, always in process – just as is our knowledge of people. This is not a critique or belittlement of our knowledge capacities, but rather a recognition and appreciation of the extraordinary complexity and continual evolution of both nature and of people. Such recognition leads to a model of knowledge that embraces the importance of empathy and imagination as a resource for "letting the material tell you." It is a model that, while acknowledging the importance of categories and theories, does not privilege them over and above the importance of listening attentively and responsibly to the stories told to us – accounting for the differences rather than imposing a model upon the world.

Code's point and one that is shared by many feminist epistemologists is that the traditional image of the dispassionate scientist removed from her or his object of study has blinded us to the complexity of the possible relationships between subjects and objects. Code argues that McClintock gained her knowledge *because* of her engaged relationship with the object of her study. That is, McClintock's fascination with the maize is epistemically significant. She is drawn to it not just to predict the genetic patterns, but because she desires a full understanding of the organism in all its stages. When she refers to her study of maize, she does so with affection – "these were my friends."

Code posits nothing like an essential femininity that entails that all and only women will embrace an engaged style of knowledge production. She argues rather that McClintock's femaleness is one aspect of the complex conjunction of subjective factors at play in her practice of science. Code's goal and the goal of other feminists is to open epistemology and science education and practice to the importance of such subjective features and to argue that S-knows-that-p models of scientific knowledge are inadequate to the full complexity of knowing. Code's intention is to reclaim subjective components of the knowledge process, components often defined as "feminine" and suppressed from traditional accounts. The aim is not to create a "feminine" science, but "to make a space in scientific research for suppressed practices and values that, coincidentally or otherwise, are commonly associated with 'the feminine'" (Code, 1991, p. 152).

EMBODIED KNOWERS

The feminist rejection of the supposedly "generic" knower thus requires that attention be paid to the characteristics and situation of the knower as an important part of the knowledge process. As illustrated in my example of Zihlman's practice of science, the various communities of which one is a part, including one's political beliefs, can be epistemically significant to the knowledge process. As we see with McClintock, a knower's emotive capacities and her or his openness to their relevance to the knowing process, can also be epistemically significant. This is the content of Code's claim that a person's gender can be epistemically significant. In contemporary Western culture, one who is female is more likely than one who is male to be socialized in such a way as to make her more proficient in and accepting of the usefulness of emotions such as empathy and imagination. Just as a feminist is more likely to question the categorization of female activities as "natural" and male activities as "cultural," a woman in contemporary Western culture is more likely to be accepting of and skilled in the employment of emotions in the knowledge process.

But an additional aspect of the knowing subject that is epistemically significant is the fact of and nature of their embodiment. The model of the generic knower has traditionally rejected the relevance of our bodily differences. Attention to the body calls attention to the specificities and partiality of human knowledge, as well as reminding us of the importance of acknowledging the body, and its variations, in the knowledge process. Once we admit the body into our theories of knowledge, we must also recognize its variations; we must, for example, examine the ways in which bodies are "sexed."[8]

Vision

Traditional models of knowledge privilege vision over the other senses. The association of knowledge and vision provides a model of knowledge as disembodied. Vision, perceived as the most detached of the senses, is employed in such a way as to conceal the action of the body. The world appears to my gaze without any apparent movement or action on my part. The action of the body disappears into the background and with it as a model of knowledge, the philosopher places the world at a distance from the observer, thereby dematerializing knowledge. The perceived scene, as well as the perceiver, is to be physically unaffected by the gaze.

There have been many studies that have examined the ways in which this conception of vision has shaped traditional Western conceptions of knowledge.[9] The construction of an image of reason based on metaphors of vision has led to the notion of a "mind's eye" and a conception of knowledge in which the world is separated from the observer who sees it and thereby gains knowledge of it, without in any way contaminating it or being affected by it.

But these disembodied images of vision are possible only by "forgetting" the fact of our embodiment. What we are capable of seeing and what we attend to are part of our location within the world. Let me begin by using a very different case study than those I've so far employed. Let us think about frogs and dogs.[10]

There are many ways to remember the significance of the situatedness of vision and thereby inhibit the tendency to use visual metaphors to construct allegedly generic images of reason. One of these is to reflect upon the significance of the specificities of human vision. A frog's visual cortex is different from ours. Neural response is linked to small objects in rapid, erratic motion. Objects at rest elicit little neural response and large objects evoke a qualitatively different response than small ones. Although this makes sense for frogs, let us imagine, along with Katherine Hayles, that a frog is presented with Newton's law of motion:

> The first law, you recall, says that an object at rest remains so unless acted upon by a force. Encoded into the formulation is the assumption that the object stays the same; the new element is the force. This presupposition, so obvious from a human point of view, would be almost unthinkable from a frog's perspective, since for the frog moving objects are processed in an entirely different way than stationary ones. Newton's first law further states, as a corollary, that an object moving in a straight line continues to move so unless compelled to change by forces acting upon it. The proposition would certainly not follow as a corollary for the frog, for variation of motion rather than continuation counts in his perceptual scheme. Moreover, it ignores the *size* of the object, which from a frog's point of view is crucial to how information about movement is processed (Hayles, 1993, p. 28).

The point is that bodily differences in perceptual organs and neural patterns organize perception in highly specific, in this case species specific, ways. Far from being the neutral receptor or static mirroring of the visual metaphors informing traditional accounts of knowledge, observation is a dynamic process of organization in which our bodily being plays a central role.

The image of disembodied vision is similarly discounted by imagining a walk with one's dog. Haraway reminds us of the lessons that can be learned from such a walk. "I learned in part walking with my dog and wondering how the world looks without a fovea and very few retinal cells for colour vision, but with a huge neural processing and sensory area for smells.... [that] all eyes, including our own organic ones, are active perceptual systems, building in translations and specific *ways* of seeing, that is ways of life" (Haraway, 1991, p. 190).

Although the walk with your dog may remind you of human emphasis on color and shape over a canine attention to smell, it may also remind you, depending on your focus of attention, of the essential and intimate connection of vision with our kinaesthetic sense and our sense of touch. As you walk through a meadow you may meditate upon the way in which your two eyes integrate to produce a unified vision of your dog, and as you reach out to pet her be reminded of the ways in which vision is woven together with motility and touch. Such a walk can impress upon us the realization that the image of vision as disembodied is a circumscribed perspective, and that its emphasis has been the result of complex factors. That is, it is an example of a partial, situated knowledge.

When we consider the human specificities of vision, those mandated by our bodies as well as by the social contexts which shape our experiences of it, we are reminded that the privileging of an image of vision which views it as passive, detached, and disinterested is itself a partial and biased perspective. As any loving parent who looks into the eyes of her or his six month old child or any lover who gazes into the eyes of the person she or he loves know, vision can also be a way in which we actively connect and interact with other people. It can be a way in which we express feelings and negotiate our relationships. Such vision is active, engaged, and reciprocal. An emphasis on vision as passive, detached, and disinterested is a situated vision, one that arises out of particular social situations and values. We are reminded again that vision, as well as objectivity, is not about neutrality, but is embedded in particular and specific embodiments.

A recognition of the epistemic importance of our embodiment requires a conception of knowledge as embodied, in which the emphasis on vision as the primary source of knowledge is replaced by an appreciation of the multiplicity of senses involved in the process of knowledge and an understanding of the ways in which faculties such as empathy, intuition, and reason enter into and interact in this process. I think that Code's model of knowing other people can be fleshed out in actual practice by looking at the science and practice of primary care health providers – nurses, physicians, midwives, etc. – as an alternative to the paradigm of the autonomous, detached, disinterested scientist.[11] Health providers provide a model of an inquirer who is both engaged and objective, whose knowledge is embodied, and whose methods illustrate Code's model of knowing other people. It is important to note that I do not claim that this new image of scientific rationality is gynecentric. In fact I would argue that it is neither androcentric nor gynecentric, but tries instead to more accurately portray the nature of rationality in all its complexity.

Knowledge With and Of the Body

Carefully trained in the latest theories and techniques, the primary care health provider will be successful in her or his diagnoses and treatments only to the extent that she or he can come to fully understand the person she or he is attending. The process of coming to such an understanding is complex. The health provider often begins by looking carefully, even when the source of the malady is not obvious. She or he watches how patients hold themselves, examining their complexion, their eyes, looking for tell-tale signs.

But vision is only one source of information for health providers and often not the most important. Touch enters into the process of diagnosis as well as treatment. The training of health care providers involves an elaborate education of their hands in which they learn not only how to use an instrument or how to stitch a wound, but also how to feel a break in a bone, or an abnormal lump, or the position of the fetus in a woman's uterus. As important as touch is for diagnosis, the health care provider also understands that it serves an important role in both comforting and reassuring the individual, for the process of diagnosis works best when the provider is able to establish a relationship with the patient.[12]

The health care provider also learns from listening. But this listening works best when it can go beyond the stethoscope or monitor and includes dialogue between the provider and the patient. Through trust and empathy, the health care provider develops a relationship with the patient that allows for communication, for she or he views the patient's own experiences as an important source of knowledge. In the words of the physician Chester Keefer, "Listen to the patient. He's giving you the diagnosis" (Keefer, 1994, p. 74). Since successful treatment often involves transformation of life-style choices, health care providers are most successful when they develop relationships with patients that enable patients to become subjects of knowledge – both of the manifestations of their illness as well as how their life-style affects their health.[13] In all these ways the health care provider works with a blend of intellect, imagination, and emotion. The subjective components in this case, far from being a barrier to knowledge and understanding, enhance the process.[14]

My suggestion is that the example of the engaged, embodied scientist we find illustrated so well in the primary care health provider offers a more adequate model of scientific inquiry than the image of the detached, disinterested, autonomous scientist. I personally believe that the actual practice of many contemporary scientists fits this model of the engaged, embodied subject. But I also believe that the legacy of positivism has obscured the variety of ways scientific practice is in fact a blend of subjective and objective factors, and has resulted in a bias in the way in which science is both conceived of and how it is taught to practitioners. McClintock's loving attention to maize is hardly unique among scientists, nor is it limited to women scientists, but this dimension of knowledge has often been overlooked and sometimes explicitly denigrated due to the belief that it was not an appropriate component of the scientific method. Nor has science education acknowledged the epistemic importance of such subjective factors and attended to the ways in which the education of scientists could be enhanced in this regard. Primary care medical education, for all its problems, has long acknowledged, at least in practice, that not all knowledge is propositional, that certain things can be learned only by doing and experiencing, and has stressed the quality of the relationship between the physician and the patient.

Granted, feminists are not the only epistemologists questioning the positivist model of scientific rationality. Michael Polanyi, to cite just one of the many theorists, offered an analysis of the tacit dimension in science that is in many ways compatible with feminist approaches to knowledge. "It is not by looking at things," he explained, "but by dwelling in them, that we understand" (Polanyi, 1966, p. 18). Given this Polanyi concludes that "the ideal of eliminating all personal elements of knowledge, would, in effect, aim at the destruction of all knowledge" (Polanyi, 1966, p. 20). But the feminist concern with gender dynamics and the historical interconnections between power and knowledge has made us particularly attentive to developing alternative models that do not suppress women's cognitive authority. Since the subjective realm has been associated with femininity, given the positivist model, women's choices have been either to attempt to deny their "nature" and/or their socially sanctioned role, or to be excluded from positions of epistemic agency. Feminists, then, are personally invested in the project of transforming traditional accounts of rationality.[15] But again, let us be clear that the alternative being proposed is not gynecentric in either privileging or even equally valuing feminine traits. The model of knowing other people *undercuts* the dichotomy between objectivity and subjectivity and the corresponding gender marking of these terms, denying that they are separate and antithetical.

CONCLUSION

As a final example I would like to quickly mention the ways in which this study, and others like it, serve as case studies of the very model of knowledge I am here professing. Much of feminist scholarship and practice over the last two decades has been devoted to revaluing the importance of interpersonal relationships. Many feminist political theorists have argued that we must revalue the so-called "private realm" of relationships. Psychoanalytic feminists have called attention to the centrality of interpersonal relationships in the development of our personalities, our genders, and our desires. Feminist ethicists have offered and examined an ethics of care in which moral action is intimately linked to our relationships with others. Add to this the fact that women's prescribed social role of primary caretakers of children, of the elderly, and of the ill, contributes to a heightened sensitivity to the fact and importance of our essential relationality and our embodiment, and it should be no surprise that it is feminist philosophers of science and epistemologists who are vociferously rejecting the Cartesian model of the isolated knowing subject and replacing it with models that emphasize the centrality of our relationships with others to the process of knowing.[16]

I and many other feminists came to positions like these *because* of our participation in feminist communities. This obviously was not the only epistemically significant community; we are also philosophers, historians, sociologists, and scientists. Nor does it mean that *only* feminists will hold such views. There are many theorists of science who do not participate in feminist communities who argue for versions of the above tenets. But a difference of feminist analyses is the persistent attention to gender as a variable of analysis. This is how our feminism is epistemologically significant.

What feminist epistemologists have realized is that it is a mistake to ask for a value-free science. As illustrated in the example of McClintock's "feeling for the organism," the development of knowledge, including scientific knowledge, is affectively influenced. And as the example of primatology illustrates, we cannot treat politics as *inherently* distorting the practice of science. Scientific research, as well as all cognitive endeavors, begins with metaphysical and methodological commitments. It arises out of and is conditioned by our participation in various epistemic communities. Each of us, in being part of a community and a number of subcommunities, participates in an evolving conceptual scheme that makes intersubjective experience possible, influences our interests and desires, and also sets the standards of what constitutes evidence.

The acceptance of the essentially relational nature of knowledge and the inseparability of subjective and objective components of knowledge does not result in relativism, though it does require an abandonment of the traditional "view-from-nowhere" conception of objectivity. This alternative notion of objectivity has been the research program of many feminist philosophers of science including a number of those whose work appears in this anthology (see Harding, Longino, Nelson). Although I will refer you to their work for the details of feminist accounts of objectivity, let me call attention to yet another way the development of feminist epistemologies are compatible with the model offered. Although there are significant differences between feminist epistemologies, one common tenet is the emphasis on diversity within the scientific community to ensure objectivity. To cite just one of many possible examples, consider Helen Longino's claim that

> ... because background assumptions can be and most frequently are invisible to the members of the scientific community for which they are background and because unreflective acceptance of such assumptions can come to define what it is to be a member of such a community (thus making criticism impossible), effective criticism of background assumptions requires the presence and expression of alternative points of view. This sort of account allows us to see how social values and interests can become enshrined in otherwise acceptable research programs (i.e., research programs that strive for empirical adequacy and engage in criticism). As long as representatives of alternative points of view are not included in the community, shared values will not be identified as shaping observation or reasoning (Longino, 1993, pp. 111–12).

Once again we see the impact of the politics of feminism upon the development of feminist epistemology, for a central emphasis of feminism has been the importance of inclusion of previously excluded groups and viewpoints. Earlier feminist accounts focused on the impact of including women and attention to gender upon society, scholarly methods, politics, and so on. The last decade has intensified this commitment as feminists have become aware of the differences between women and have acknowledged the ways in which attention to such factors as class, race, and sexuality, as well as gender, reveals previously hidden assumptions and opens up new research programs.

Feminist philosophers of science have thus actively developed research programs consistent with the values and commitments we express in the rest of our lives. In this sense we are creating "feminist sciences," the doing of science from the politics of feminism. We also acknowledge the need for science to be open to diverse groups of individuals and to have these groups engage in what Longino calls "an interactive dialogic community" (Longino, 1993, p. 113). This is not a simple pluralism, but one in which critical interchange between communities is highly valued. This, of course, does not mean that "anything goes." Although scientific standards are not seen as unchanging or unresponsive to such critical interaction, they do provide standards for acceptability. The "woman, the gatherer" model in human evolution studies arises out of a feminist political agenda yet meets the standards set by the field in which it is proposed. And this is important. Only if these alternative models receive a hearing within the scientific community will they ever secure serious attention.

A value implicit in this vision of science is that the best form of science will be that which is the product of the most inclusive scientific community. This suggests that the problem of developing a new science is the problem of creating a new social and political reality.

University of Oregon

NOTES

[1] A relatively recent correction of contemporary feminist theory in general and feminist philosophy of science in particular is that theories that do not attend to the interactions of various forms of oppression, including class, race, and sexuality, distort the nature of gender oppressions. For an important contribution to this discourse in relation to science studies see Harding, 1993.

[2] Harding, 1992. See also Proctor, 1991.

[3] Accounts of "woman, the gatherer" theories can be found in Haraway, 1989; Longino, 1990; and Nelson, 1990. My analysis here is thoroughly influenced by Haraway's *Primate Visions.*

[4] For a more detailed argument in support of this position see Haraway, 1989.

[5] I would argue that the example of Jane Goodall's research, both her methods and values, can be used to support the claim that the sex of the knower can be epistemically significant. However, this is a complicated claim, *not* to be confused with the belief in an essential woman's nature, that must be left for a future essay.

[6] See Keller, 1984 and 1992.

[7] See Code, 1991 and 1993; Jaggar, 1989; Keller, 1984; Longino, 1990; Nelson, 1990; and Rose, 1983 and 1994.

[8] Although I do not have the time to develop this point, I feel it is too important not to mention and to urge readers to explore the work done on this topic in the writings of feminists such as Rosi Braidotti, Elizabeth Grosz, and Luce Irigaray.

[9] See Code, 1991; Keller and Grontkowski, 1983; Jonas, 1966; Leder, 1990; Merleau-Ponty, 1962; and Rorty, 1979.

[10] My account here is influenced by Haraway, 1988 and Hayles, 1993, dogs and frogs respectively.

[11] Just as there are positivist social scientists, so are there positivist primary health care providers. My argument here is that when health care is done well the provider is engaged, embodied, interested, and objective. In other words, objectivity does not require the suspension of subjectivity.

[12] Discussions of doctor–patient relationships in family practice journals often involve the suggestion that the physician touch the patient, even if doing so is not needed for diagnosis. The reason given is that such touch reassures and comforts the patient, thus allowing a relationship of trust to develop between them.

[13] There have been numerous feminist critiques of modern medicine that discuss the ways in which medical institutions reinforce sexism. Vrinda Dalmiya and Linda Alcoff (1993), for example, have argued that modern obstetrics has emphasized propositional knowledge to the exclusion of the practical knowledge of midwifery. I would argue that these accounts, although certainly pointing out problematic areas in contemporary medicine, are not reflective of primary care medicine (as distinct from surgery and other specialty areas). The training, experience, and institutions of primary care physicians are antithetical to a strict division between knowing how and knowing that, and involve an explicit acknowledgement of the importance of the physician/patient relationship both for diagnosis and for treatment. I do not claim that all primary care doctors in fact successfully embody this model, but I do think it is one that the majority of primary care physicians strive for. (But perhaps it is only fair to reveal that my partner is a primary care physician.)

[14] In developing my model of the primary health care provider as a paradigm of scientific practice, I focused on the provider/patient relationship. But given my discussion of the agent of knowledge as individuals in communities, it should be understood that my discussion is only partial. To complete it would require adding the relationships between providers and their research/practitioner communities, and any other epistemically relevant communities.

[15] For a discussion of this point see Lloyd, 1984 and 1993.

[16] I am *not* claiming that feminists are the only theorists developing such a model, but that there is an epistemic link between this model of the subject of knowledge and the politics of feminism.

REFERENCES

Benhabib, Seyla: 1987, 'The Generalized and the Concrete Other: The Kohlberg-Gilligan Controversy and Feminist Theory,' in Seyla Benhabib and Drucilla Cornell (eds), *Feminism as Critique: On the Politics of Gender*. University of Minnesota Press, Minneapolis.

Code, Lorraine: 1991, *What Can She Know? Feminist Theory and the Construction of Knowledge*. Cornell University Press, Ithaca.

Code, Lorraine: 1993, 'Taking Subjectivity into Account,' in Linda Alcoff and Elizabeth Potter (eds), *Feminist Epistemologies*. Routledge, New York, pp. 15–48.

Dalmiya, Vrinda and Linda Alcoff: 1993, 'Are "Old Wives' Tales" Justified?,' in Linda Alcoff and Elizabeth Potter (eds), *Feminist Epistemologies*. Routledge, New York, pp. 217–44.

Haraway, Donna: 1989, *Primate Visions: Gender, Race and Nature in the World of Modern Science*. Routledge, New York.

Haraway, Donna: 1991, 'Situated Knowledges: The Science Question in Feminism and the Privilege of Partial Perspective,' in *Simians, Cyborgs, and Women: The Reinvention of Nature*. Routledge, New York.

Harding, Sandra: 1991, *Whose Science? Whose Knowledge?* Cornell University Press, Ithaca.

Harding, Sandra: 1992, 'After the Neutrality Ideal: Science, Politics, and "Strong Objectivity",' *Social Research*, **59**, 567–87.

Harding, Sandra: 1993, *The "Racial" Economy of Science: Toward a Democratic Future*. Indiana University Press, Bloomington.

Hayles, N. Katherine: 1993, 'Constrained Constructivism: Locating Scientific Inquiry in the Theater of Representation,' in George Levine (ed.), *Realism and Representation*. University of Wisconsin Press, Madison.

Jaggar, Alison: 1989, 'Love and Knowledge: Emotion in Feminist Epistemology,' in Susan Bordo and Alison Jaggar (eds), *Gender/Body/Knowledge*. Rutgers University Press, New Brunswick.

Jonas, Hans: 1966, *The Phenomenon of Life*. Chicago University Press, Chicago.

Keefer, Chester: 1994, 'Skillful Listening,' *Cortland Forum*, June, 74.

Keller, Evelyn Fox: 1983, *A Feeling for the Organism*. W. H. Freeman and Company, New York.

Keller, Evelyn Fox: 1984, *Reflections on Gender and Science*. Yale University Press, New Haven.

Keller, Evelyn Fox: 1992, *Secrets of Life, Secrets of Death: Essays on Language, Gender and Science*. Routledge, New York.

Keller, Eveyln Fox and Christine R. Grontkowski: 1983, 'The Mind's Eye,' in Sandra Harding and Merrill B. Hintikka (eds), *Discovering Reality*. D. Reidel, Dordrecht, pp. 207–224.

Leder, Drew: 1990, *The Absent Body*. University of Chicago Press, Chicago.

Lloyd, Genevieve: 1984, *The Man of Reason: "Male" and "Female" in Western Philosophy*. University of Minnesota Press, Minneapolis.

Lloyd, Genevieve: 1993, 'Maleness, Metaphor, and the "Crisis" of Reason,' in Louise M. Anthony and Charlotte Witt (eds), *A Mind of One's Own: Feminist Essays on Reason and Objectivity*. Westview Press, Boulder.

Longino, Helen: 1990, *Science as Social Knowledge*. Princeton University Press, Princeton.

Longino, Helen: 1993, 'Subjects, Power, and Knowledge: Description and Prescription in Feminist Philosophies of Science,' in Linda Alcoff and Elizabeth Potter (eds), *Feminist Epistemologies*. Routledge, New York, pp. 101–120.

Merleau-Ponty, Maurice: 1962, *Phenomenology of Perception*, trans. Colin Smith. Routledge and Kegan Paul, London.

Nelson, Lynn Hankinson: 1990, *Who Knows: From Quine to a Feminist Empiricism*. Temple University Press, Philadelphia.

Polanyi, Michael: 1966, *The Tacit Dimension*. Doubleday and Co. Garden City, NY.

Proctor, Robert N.: 1991, *Value-Free Science? Purity and Power in Modern Knowledge*. Harvard University Press, Cambridge.

Rose, Hilary: 1983, 'Hand, Brain and Heart: A Feminist Epistemology for the Natural Sciences,' *Signs*, **9**, 73–90.

Rose, Hilary: 1994, *Love, Power, and Knowledge: Towards a Feminist Transformation of the Sciences*. Indiana University Press, Bloomington.

Rorty, Richard: 1979, *Philosophy and the Mirror of Nature*. Princeton University Press, Princeton.

Slocum, Sally Linton: 1971, 'Woman the Gatherer: Male Bias in Anthropology,' in Sue-Ellen Jacobs (ed.), *Women in Perspective: A Guide for Cross Cultural Studies*. University of Illinois Press, Urbana.

Washburn, Sherwood L. and C. S. Lancaster: 1976, 'Evolution of Hunting,' in R. B. Lee and I. DeVore (eds), *Kalahari Hunter–Gatherers*. Harvard University Press, Cambridge.

PART II

FEMINIST AND MAINSTREAM PHILOSOPHY OF SCIENCE: CONTINUITIES AND TENSIONS

HELEN E. LONGINO

COGNITIVE AND NON-COGNITIVE VALUES IN SCIENCE: RETHINKING THE DICHOTOMY[1]

INTRODUCTION

Underdetermination arguments support the conclusion that no amount of empirical data can uniquely determine theory choice. The full content of a theory outreaches those elements of it (the observational elements) that can be shown to be true (or in agreement with actual observations).[2] A number of strategies have been developed to minimize the threat such arguments pose to our aspirations to scientific knowledge. I want to focus on one such strategy: the invocation of additional criteria drawn from a pool of cognitive or theoretical values, such as simplicity or generality, to bolster judgements about the worth of models, theories, and hypotheses. What is the status of such criteria? Larry Laudan, in *Science and Values*, argued that cognitive values could not be treated as self-validating, beyond justification, but are embedded in a three-way reticulational system containing theories, methods, and aims or values, which are involved in mutually supportive relationships (Laudan, 1984). My interest in this paper is not the purportedly self-validating nature of cognitive values, but their cognitive nature. Although Laudan rejects the idea that what he calls cognitive values are exempt from rational criticism and disagreement, he does seem to think that the reticulational system he identifies is independent of non-cognitive considerations. It is this cognitive/non-cognitive distinction that I wish to query in this paper. Let me begin by summarizing those of my own views about inquiry in which this worry about the distinction arises.

CONTEXTUAL EMPIRICISM

I've argued for a view I call contextual empiricism, according to which empirical, that is, observational and experimental, data constitute the least defeasible grounds of theory assessment. This much is the empiricism of the view. But data underdetermine the theories, models, and hypotheses for which they serve as evidence. Theories and hypotheses always overreach available data. More crucially, the content (and language) of data descriptions and of explanatory hypotheses are different. For example, data can consist of correlations while hypotheses assert causal relations among correlated items. Thus, no purely formal relations can be established between them. Evidential relevance of data is secured instead by background assumptions, with the consequence that the same data can in different contexts serve as evidence for different hypotheses. This is the contextualism of the view.

L. H. Nelson and J. Nelson (eds), Feminism, Science, and the Philosophy of Science, 39–58.

Contextual empiricism invites the question what controls background assumptions. If scientific reasoning is so porous to context, what prevents theories from being entirely subjective? My answer, in *Science as Social Knowledge* (Longino, 1990), was that critical interactions among scientists of different points of view were required to mitigate the influence of subjective preferences on background assumptions and hence theory choice.

While intersubjective interaction is a necessary feature of scientific cognition, not just any form of interaction will do. If the point of intersubjective interaction is to transform the subjective into the objective, then those interactions must not simply preserve and distribute one subjectivity over all others, but must constitute genuine and mutual checks. This end can be served by specifying features of the design and constitution of a community that facilitate transformative criticism and enable a consensus to qualify as knowledge. Four such features can be identified.

(1) There must be publicly recognized forums for the criticism of evidence, of methods, and of assumptions and reasoning.

(2) There must be uptake of criticism. The community must not merely tolerate dissent, but its beliefs and theories must change over time in response to the critical discourse taking place within it.

(3) There must be publicly recognized standards by reference to which theories, hypotheses and observational practices are evaluated and by appeal to which criticism is made relevant to the goals of the inquiring community. Such standards serve as ideals regulating normative discourse in a community. That is, by explicitly or implicitly professing adherence to those standards individuals and communities adopt criteria of adequacy by which their cognitive activity may be evaluated. The satisfaction of goals of inquiry is not ascertained privately, but by evaluation with respect to shared values and standards. This evaluation may be performed by anyone, not just by members of the community sharing all standards. Furthermore, standards are not a static set, but may themselves be criticized and transformed, in reference to other standards, goals, or values, held temporarily constant. Indeed, the presupposition of reliance on such standards is that they have survived similar critical scrutiny.

(4) Finally, communities must be characterized by equality of intellectual authority. What consensus exists must be the result not of the exercise of political or economic power, or of the exclusion of dissenting perspectives, but a result of critical dialogue in which all relevant perspectives are represented. This criterion is meant to impose duties of inclusion; it does not require that each individual, no matter what their past record or state of training, should be granted equal authority on every matter.

Discursive interactions reduce the likelihood that the idiosyncratic preferences of individuals will be incorporated in the public body of scientific knowledge. While they cannot eliminate background assumptions altogether, discursive interactions conducted in and among communities satisfying the above conditions not only eliminate the idiosyncratic but insure that no set of assumptions dominates simply by virtue of its commonality or invisibility. The public standards mentioned in condition (3) have two objects. One is to impose obligations on acknowledged members of a knowledge-productive community to attend to criticism that is relevant to their cognitive and practical aims. The other is to limit the sorts of criticisms to which a community must attend to those which affect the satisfaction of its goals. The point of (4) is that such criticism may originate from an indeterminate number of points of view, none of which may be arbitrarily excluded from the community's interactions without cognitive impairment.

I want to focus on the public standards mentioned in the third of these conditions. I originally thought that they contained cognitive values, pragmatic values, and substantive assumptions grounded in either the metaphysical commitments or the social and political commitments of a society, i.e. metaphysical or value-laden substantive assumptions. I argued in *Science as Social Knowledge* that social or non-cognitive values could and did serve as cognitive values. What I did not question, and want to explore more deeply here, is the supposition that the values called cognitive really are cognitive. By "cognitive" here, I mean something like "epistemic", that is, conducing to the truth. There are accounts of knowledge and cognition within which "cognitive" would have a different meaning, but that is not the meaning with which it is used by philosophers promoting cognitive values as solutions to underdetermination.[3]

COGNITIVE VALUES

There have been a variety of proposals as to what count as cognitive values, but there is a great deal of overlap among most of those proposals. In his essay "Objectivity, Values, and Theory Choice", Thomas Kuhn discussed five values that scientists use to guide their judgements in choosing between competing theories (Kuhn, 1977). These are accuracy, simplicity, internal and external consistency, breadth of scope, and fruitfulness. Kuhn had a lot to say about these values and how they functioned; his overall claim was that they constituted objective grounds for theory choice. The elements on Kuhn's list (with the exception of fruitfulness) are just the sorts of consideration that end up in philosophers' lists of what, besides agreement with observational and experimental data, counts for the truth (or acceptability) of a theory or hypothesis. For example, Quine and Ullian, in *The Web of Belief*, list as virtues of a hypothesis conservatism, modesty, simplicity, generality and refutability (Quine and Ullian, 1978). Indeed, when drawing a distinction in *Science as Social Knowledge* between what I called constitutive and contextual values, I used items like empirical adequacy (used interchangeably with accuracy), simplicity, and explanatory power (used interchangeably with breadth of scope) as paradigmatic examples of constitutive values (Longino, 1990).

Although certain items, like simplicity and generality (which overlaps with breadth of scope and explanatory power), consistently recur on philosophers' lists of cognitive virtues, other items are less universally proposed. Laudan, in his paper "Demystifying Underdetermination", proposes internal consistency, the (correct) prediction of surprising results, and variety of evidence (Laudan, 1990). The first is fairly universally accepted, especially if it means nothing more than satisfying some basic logical principles like the principle of non-contradiction. Correct prediction of surprising results comprises two criteria: correct prediction, which we can understand as comparable to accuracy and empirical adequacy, and the surprising or unexpected character of what is predicted. About the former there is general agreement, about the latter, very little.[4] The role and importance of variety of evidence is also disputed.[5] Paul Churchland (1985) identifies simplicity and explanatory power as cognitive or epistemic virtues that enable us to go beyond mere (and for Churchland, highly problematic) empirical adequacy. What seems presupposed in these discussions is that it is possible to identify some properties of theories as cognitively virtuous. This means that judgement guided by these virtues or standards is more likely to lead to true or at least rational belief.

Theoretical virtues can be and are invoked in a variety of contexts: there is theory choice, but also retrospective appraisal, rationalization of commitments made, plausibility assessment, etc., not to mention hallway gossip. Different elements in these lists will be more salient in some contexts than others.[6] My aim here is not, however, to illuminate theory choice specifically or to distribute the cognitive virtues to their appropriate normative contexts. It is instead to cast doubt on the very idea of a cognitive value or virtue, where we mean by that a quality of theories, models, or hypotheses that can serve independently of context as a universally applicable criterion of epistemic worth.[7] For convenience and brevity, I shall focus my argument on the virtues enumerated by Kuhn, beginning with a closer look at what they are.

Accuracy

This virtue is what others might call empirical adequacy, i.e. the observational content of a theory or hypothesis should be in agreement with observational and experimental data. Preference will be given to those models, hypotheses, and theories whose observational elements or consequences are in greater consonance with data as compared with alternatives. As I have argued elsewhere (Longino, 1995), accuracy and empirical adequacy are not as straightforwardly applied in evaluative contexts as might be hoped. And in any case, arguments about the underdetermination of theory by data require that other considerations be brought to bear on theory assessment. That is, given that accuracy or empirical adequacy mean only that the observational content of a theory is in accord with observational data and that the full content of a theory extends in various ways beyond that observational content, other criteria must be invoked in the assessment of the empirical, but non-observational, content. For example, the observational content of elementary particle theory is the predictions of the data various kinds of detector will produce under given circumstances. The particles, their properties, interactions, and disintegrations, are all hypothesized as underlying or causing the manifestations observed in bubble chambers or data tapes. That part of the theory which is about particles cannot be directly assessed since our access to particles is mediated by instruments and theory about those instruments. We cannot assess the truth of statements about particles as we can the truth about the frequencies of certain kinds of signal produced in detectors. Hence the lists of virtues contain additional elements intended to assist in the discrimination among different theories that could be used to account for the same phenomena.

Internal Consistency

The theory or hypothesis contains no contradictions.

External Consistency

The theory or hypothesis is consistent with presently accepted theories in other fields. Quine and Ullian's version of this virtue is conservatism, which they gloss as favoring the hypothesis disconsonant with the fewest number of sentences in the web of belief, including observational ones.

Simplicity

This is a notoriously difficult criterion to pin down. Kuhn says a simple theory brings order to phenomena that would otherwise remain disparate or confused. But this still leaves many ways in which a theory might be simple. Some gloss simplicity as having to do with the order of equations used in a theory or the continuity of the curves used to generate the data points in a graph of the data.[8] Another interpretation, and I think the more common one, is ontological. Any theory (or model, or hypothesis) stipulates an ontology, i.e. it characterizes what is to count as a real or basic or causally effective entity in its domain as well as the kinds of process in which such entities participate. The simpler theory is the one that stipulates fewer entities or fewer processes.[9] Newtonian mechanics, which applies to a universe of bodies characterized by extension, hardness, impenetrability, mobility and inertia, is a prime example of a theory exhibiting the virtue of simplicity, in comparison with its Aristotelian predecessor which held that there were four (sublunary) elements, each with distinctive properties. One might also characterize simplicity in terms of the number of basic or underived principles of a theory, but this shades into the next virtue.

Breadth of Scope

Again there are various ways to characterize this criterion. Kuhn explicated it as requiring that the consequences of the theory extend beyond those the theory was originally developed to explain. Other philosophers talk of explanatory power or generality, by which they mean the diversity of phenomena that can be explained by a single or single set of basic or underived explanatory principles.[10] Newtonian mechanics also exemplifies this virtue, since a variety of hitherto different phenomena (from falling bodies to orbiting planets) were brought under the unifying explanatory umbrella of the three laws of motion.

Fruitfulness

This criterion is unique to Kuhn, but I think he is correct in identifying it as a criterion used by scientists in evaluating theories. A fruitful theory generates new findings or discloses new relationships. Another way of understanding fruitfulness is that by suggesting new hypotheses, it generates problems or questions that the theory can be used to answer. Kuhn, in a footnote, remarks that a young scientist, choosing between two theories knows that the choice will bear on her future research career. She will, if she has any sense, choose the one that promises "the concrete successes for which scientists are normally rewarded" (Kuhn, 1977, p. 322, n. 6). Fruitfulness could be considered a richer version of refutability. I will discuss this suggestion below.

To say that these are values or virtues is to say that they are properties that theories can have or can have to a greater or lesser degree. In practice, no theory can possess all of these properties to the maximum, since some of them are in a certain amount of tension with each other, particularly accuracy and breadth of scope. Thus, an optimum theory exhibits some balance of these desiderata. These traditional virtues could be thought of as explicating what "best" means in inference to the best explanation. The need, however, to trade-off maximum satisfaction of one virtue against another could be interpreted as an argument against these virtues having any epistemic status at all.[11] But their philosophical interest does not thereby melt away. Even if we disallow *inference* to the best explanation, i.e. the inference that because H offers the best explanation of *e* and *e*, therefore H, the virtues could still be thought of as an explication of how a good, better, or best scientific theory is to be characterized, and, hence, of the (internal, scientific) grounds for preferring, in prospect or in retrospect, one theory over another. This, I think, is what Kuhn was suggesting. In the absence of alternatives, and because they are routinely invoked in discussions of scientific values, it is easy to think of them as constitutive values of science (and thus cognitive in some broad sense of cognitive). Let us see, then, whether consideration of some alternatives permits this classification.

SOME FEMINIST THEORETICAL VIRTUES

Feminist writing about the sciences reveals a quite different set of desiderata. Here one finds empirical adequacy, but also novelty, ontological heterogeneity, mutuality of interaction, applicability to human needs, and diffusion or decentralization of power. There are undoubtedly others, but (as Kuhn said about his list) this list is enough to make the points I want to make.

The traits listed are generally invoked singly or in groups of two or three and for the most part become evident as values in the context of their use.[12] No one (to my knowledge) has offered them as a package. But when they appear in feminist work they function, like the more traditional candidates for cognitive or scientific values, as virtues; that is, as qualities of a theory, hypothesis, or model that are regarded as desirable and hence guide judgements about them. I shall refer to them as virtues, values, standards, criteria, continuing to ignore the differences between those concepts for purposes of this discussion. Let me begin by offering some interpretation of the elements of this alternative set based on the contexts in which they've been deployed. Then I shall discuss their relation to the more standard virtues.

Empirical Adequacy

Empirical adequacy is the one item common to both the traditional and the alternative set. A good deal of feminist effort has gone into discrediting research programs that purport to show a biological etiology for differences ascribed on the basis of sex. The (feminist) scientists involved in this effort – scientists such as Ruth Bleier, Anne Fausto Sterling, Richard Lewontin, Ruth Doell – have concentrated on discrediting such research by showing that it fails minimal standards of empirical adequacy, either through faulty research design or improper statistical methodology. I take their appeal to empirical adequacy in the context of their critiques to constitute an implicit endorsement of the standard. Empirical adequacy is valued for, among other things, its power when guiding inquiry to reveal both gender in the phenomena and gender bias in the accounting of them. It is, of course, a standard shared with race- and class-sensitive research communities as well as with most mainstream communities. Failure to meet the standard in a strong sense, i.e. the generation of statements about what will or has been observed that are incompatible with what has actually been observed, is grounds for rejection of the hypothesis or theory in question. In practice, most research communities reserve judgement when one of their central theories is shown to fail the test of empirical adequacy, unless the failure can be made overwhelming and an alternative theory is available to perform much of the same work.

Novelty

By novelty, I understand models or theories that differ in significant ways from presently accepted theories, either by postulating different entities and processes, adopting different principles of explanation, incorporating alternative metaphors, or by attempting to describe and explain phenomena that have not previously been the subject of scientific investigation. Several thinkers have endorsed the novelty of a model or theory as a value. Sandra Harding seems to do so explicitly when she calls both for "successor science" and for "deconstructing the assumptions upon which are grounded anything that resembles the science we know" (Harding, 1986). And one can read Donna Haraway's invocation of the visions of certain science fiction writers as an appeal for or endorsement of a departure from entrenched assumptions, particularly those having to do with the immutability of boundaries between animal and human, organism and machine. Only new frameworks which have eschewed such boundaries, she suggests, will be appropriate for the new circumstances of 21st-century life (Haraway, 1992).

Treating novelty as a virtue reflects a deep skepticism that mainstream theoretical frameworks could be adequate to the problems confronting us, as well as a suspicion of any frameworks developed in the exclusionary context of modern European and American science. Since mainstream traditional frameworks have been used in accounts that either neglect female contributions to processes biological and social, or that treat as natural alleged male superiority in various dimensions, something new will be required to address phenomena in a non-androcentric way and to ensure that invidious distinctions underpinning gender oppression are not persisting in reformed theories and models. Novelty could, of course, have stronger and weaker interpretations. The strong interpretation demands new frameworks and theories to replace current ones in the domains in which they are currently employed. On the weaker interpretation, new frameworks are to be sought in satisfying a demand for scientific understanding of hitherto neglected phenomena.

Ontological Heterogeneity

As mentioned in the earlier discussion of simplicity, any theory posits, implicitly or explicitly, an ontology; that is, it characterizes what is to count as a real or causally effective entity in its domain. A theory characterized by ontological heterogeneity (or ontological diversity) is one that grants parity to different kinds of entities. Ontological homogeneity, or uniformity, by contrast, characterizes theories that posit only one sort of causally efficacious entity, or that treat apparently different entities as versions of a standard or paradigmatic member of the domain, or that treat differences as eliminable through decomposition of entities into a single basic kind. The criterion of heterogeneity is found in two quite different sorts of discussion in the feminist literature on the sciences, which emphasize different aspects of the criterion. One is the respect for particularity and individuality urged by feminists in a variety of research contexts.[13] Feminists writing about biology have urged that we take account of individual differences among the individuals and samples that constitute the objects of study.[14] Although she was not herself a feminist, Barbara McClintock's attention to the individual kernels of a cob of corn (which helped her to recognize an underlying pattern of mutability) has been taken as a paradigm of what a feminist attitude to nature ought to be.[15] Primatologist Jeanne Altmann has insisted on methods of observation that descriptively preserve the differences among the primates and groups of primates that she studies (Altmann, 1974). This methodological focus on individual differences is a form of particularism – an insistence on the priority of particulars to abstractions.

Treating individual differences as important and not to be elided in abstractions or idealizations which smooth out heterogeneity is valuing heterogeneity, taking it as a basic aspect, if not of the natural world, of one's theories of it. One may have a variety of reasons for so valuing models that preserve heterogeneity. One may, for example, think that such a model more accurately captures the diversity of the experienced world.[16] But the reason feminists have embraced this aspect of the work of McClintock, Altmann, and others is connected, I think, to the second discussion I draw on here: the rejection of theories of inferiority. Theories of inferiority are supported in part by an intolerance of heterogeneity. Difference must be ordered, one type chosen as the standard, and all others seen as failed or incomplete versions.[17] Theories of inferiority which take the white middle class male (or the free male citizen) as the standard grant ontological priority to that type. Difference is then treated as a departure from, a failure to fully meet, the standard, rather than simply difference. Ontological heterogeneity permits equal standing for different types, and mandates investigation of the details of such difference. Difference is resource, not failure.

Mutuality of Interaction

While the prior criterion values theories that are pluralist with respect to entities, this criterion values theories that treat relationships between entities and processes as mutual, rather than unidirectional, and as involving multiple rather than single factors. Many feminist scientists have taken complex interaction as a fundamental principle of explanation. Evelyn Keller's (1983) account of the work of Barbara McClintock and her defense of an interactionist perspective in her *Reflections on Gender and Science* (Keller, 1985) may provide the best known examples, but scientists from icons like Ruth Bleier and Anne Fausto Sterling to much less well known practitioners have eschewed single factor causal models for models that incorporate dynamic interaction, models in which no factor can be described as dominant or controlling and that describe processes in which all active factors influence the others. This perspective has been employed in areas ranging from neuroscience to cell biochemistry by scientists self-consciously practicing science as feminists as well as, of course, by non-feminists. It has also been endorsed in texts devoted mainly to reflections about the sciences.

One thing noted by feminist proponents of mutuality is that simple models of single factor control make one party (the dependent as distinct from the independent variable) to an interaction a passive object rather than an agent. This has been the fate of female gametes in accounts of fertilization and of female organisms in accounts of social structure. Asymmetry of agency in the physiological context is used to naturalize asymmetry in the social. These naturalizing arguments are explicit in sociobiological stories attributing the presumed docility of females and activity of males to anisogamy, i.e. the different sizes of the female and male gametes, which involve different kinds of "parental investment". Informally, there is bidirectionality of support: asymmetric models of gametic fusion depend for their plausibility on social ideologies of gender, but their persistence in medical and biological textbooks thereby reinforces the social ideologies. As Emily Martin shows, the personification of egg and sperm in these contexts is one way of effecting the conflation of the natural and the social (Martin, 1991). Replacing simple asymmetric models of single factor control in social contexts with more complex models of social interaction makes visible the role of gender in the structure of social institutions and the role of private, domestic (traditionally, women's) work in maintaining the activity and institutions of the "public" sphere. Similarly replacing models of energetic sperm acting on passive eggs with models of mutual interaction reveals the egg's considerable contribution to the process of gametic fusion.[18]

Applicability to Current Human Needs

This and the next are pragmatic criteria, and more relevant to decisions about what theories or theoretical frameworks to work on than to decisions about plausibility. That is, heterogeneity and mutuality of interaction concern the content of models and theories while applicability and diffusion of power concern the effects of their adoption. This criterion favors research programs that can ultimately generate applicable knowledge. Many, but not all, feminists in the sciences have stressed the potential role of scientific understanding in improving the material conditions of human life, or alleviating some of its misery. Scientific inquiry directed at reducing hunger, promoting health, assisting the infirm, protecting or reversing the destruction of the environment, is valued over knowledge pursued either for political domination, i.e., science for "defense", or for knowledge's sake.[19] As expressed in feminist contexts, this is not just a call for more applied science, such as is heard in the halls of Congress, but for research that can be directed towards meeting the human and social needs traditionally ministered to by women. The applicability criterion could be understood, then, as requiring research into hitherto neglected areas and hence triggering the novelty criterion in its weaker interpretation.

Diffusion of Power

This criterion is the practical version of the fourth criterion, the one favoring models that incorporate mutual rather than dominant–subordinate relationships in explanatory models. This one gives preference to research programs that do not require arcane expertise, expensive equipment, or that otherwise limit access to utilization and participation. This feature has emerged as a value in a number of different contexts. Feminists in engineering and in economics have condemned requirements of mathematical achievement far beyond what is required for successfully engaging in these fields.[20] Other feminists, such as Hilary Rose (1983) and Ruth Ginzberg (1987), have urged a revamping of traditional distinctions to include widely distributed practices such as midwifery as scientific practices. They urge that such practices be used as models for feminist science practice. Feminist health professionals urge a preference for medical practices and procedures that empower the individual woman either to make decisions about her health or to retain control over her own body. And ecofeminists and feminists in developing regions urge the development of technologies that are accessible and that can be locally implemented (Sen and Grown, 1987). Diffusion or decentralization of power interprets the above-cited elements of the applicability criterion as knowledge of soil conservation, intensive small-scale sustainable agriculture, promoting health by preventive measures such as improved hygiene rather than high-tech interventive measures available only to the few, protection of the environment by conservation and widely dispersed renewable energy technologies.

The various proponents of these standards have had different ideas about how they work or ought to work in inquiry. If we treat them as components of a community set of public standards as I am suggesting, we take them as regulative ideals shaping the normative discourse in a scientific community, that is, as criteria invoked in the assessment of theories, models, and hypotheses, guiding their formulation, acceptance and praise, disparagement and rejection, and pursuit or abandonment. As Kuhn noticed of the values he discussed, these alternative virtues require further interpretation to be applied in a given research context, they are not simultaneously maximally satisfiable, and they are not subject to hierarchical ordering or algorithmic application.

Since empirical adequacy is almost universally recognized as a value, and since others of these characteristics have been endorsed as virtues by non-feminists, one might well wonder what about these standards is specifically feminist.[21] Several answers to this question can be discerned in the texts in which these virtues have been endorsed.

One approach holds that these characteristics express a feminine or female orientation to the world, i.e. that women either because of biology or social experience are more likely to understand the world via theories characterized by these traits. This is said primarily of the substantive and pragmatic virtues. Women are said, for example, to be more inclined to perceive mutual influence and interaction than unidirectional single factor control, and to be more interested in research that will improve the conditions of life. What would be feminist, then, would be treating as theoretical virtues characteristics of women's ways of thinking about the natural and social worlds. The problems with this approach are, first, that there's no evidence that women are inclined biologically or culturally to understand the world in these ways; second, that even if they were, we'd still need an argument that these are traits that ought to be valued in theory construction and assessment; and third, that it creates a need to explain the endorsement of these virtues by non-feminists. Of course, if one is antecedently convinced, as some advocates of these virtues are, that the world really is constituted of heterogeneous entities that interact in complex ways, the need for such an argument will be much less apparent than it is to one less certain. But if the world is such as to be more adequately understood via theories exhibiting these virtues, then they ought to be promoted as general theoretical virtues and not just as feminist theoretical virtues.

A second approach suggests that women are more likely to value the characteristics of theories because they are outsiders to mainstream science and so less likely to be acculturated to the values of the mainstream. This avoids the problem why non-feminists would endorse the virtues, but it is an even less plausible candidate for grounding the claim that the virtues would be feminist. Neither female biology nor feminine conditioning, but marginality explains the appeal of these virtues. Marginality, however, is common to any group excluded from the practice of science and so not specifically feminist. Furthermore, while marginal status may alienate or free those marginalized from mainstream values, in some cases preference for alternative values may be the basis of marginalization. And, as is the case for the previous approach, the empirical data supporting the view that marginalized groups are likely to endorse these virtues in particular has yet to be brought forward.

Rather than look to sociological or psychological facts about who uses them, I have suggested that we look to the work these virtues can do for specifically feminist inquiry (Longino, 1994). In the account given above of each of the virtues, I suggested how inquiry guided by them would be thought to reveal gender, either in the form of bias about the phenomena or as a phenomenon in the domain itself, or to reveal the activities of women or females in the domain. Revealing gender means more than mentioning females or even treating males and females as in some relation or other. Revealing gender in a feminist context means revealing an asymmetric power relation that both conceals and suppresses the independent activity of those gendered female. This relation is sustained by social institutions and symbolic practices and is itself made invisible as a relation of power by, among other things, naturalizing models in the life and behavioral sciences of sex and gender relations. The relation of feminist theoretical virtues to the aim of revealing gender is not that gender is always and everywhere revealed, but that if a context is gendered (in the sense of being structured by gendered power asymmetries), inquiry guided by these virtues is more likely to reveal it or less likely to preserve its invisibility than the traditional virtues.

The aim of revealing gender and/or the activities of those gendered female is, I propose, what makes inquiry feminist. Feminist theoretical virtues will be those that serve this aim. Thus, satisfying it is a bottom line requirement on theoretical standards. I should emphasize that I am not arguing here that the virtues I have discussed so far *are* the theoretical virtues feminists should adopt. I think such a claim needs further discussion and argument. What I do propose is that the basis on which such a claim should be argued and disputed is the contribution any proposed virtue can make to furthering feminist goals in inquiry. If the virtues that have been discussed here are feminist, it is because they satisfy this bottom line requirement, and not because of any intrinsic, statistical, or symbolic association with women or cultural femininity.

UNDOING THE DICHOTOMY BETWEEN COGNITIVE AND NON-COGNITIVE

So far, I have described two sets of what I am calling theoretical virtues. An exemplar of a traditional set comprises such items as accuracy or empirical adequacy, internal and external consistency, simplicity, breadth of scope, and fruitfulness. An alternative list contains empirical adequacy, novelty, ontological heterogeneity, mutuality of interaction, applicability to human needs, and diffusion of power. The virtues of the traditional set are usually recommended as cognitive or epistemic virtues or as constitutive values of science, that is, they are taken to conduce to truth or rational belief or they are taken to be characteristic virtues of (good) scientific theories. I argued about the second set that they are neither uniquely nor intrinsically feminist, but that feminists could argue that theories exemplifying them would be more likely to satisfy feminist cognitive aims (which are also socio-political aims) – namely to make women and female-identified phenomena as well as gender relations more visible. Does this mean that we have one set of virtues that are social or political and one set that are cognitive? If that were the case we would have no real question of choosing between them and could dismiss scientific inquiry performed with feminist concerns in view as mere ideology, not science at all. If we examine contrasting pairs from the two sets, however, it doesn't seem that the dichotomy underwriting this dismissal can be sustained. Rather than examine all possibilities, I shall take three: external consistency and novelty; simplicity and ontological heterogeneity; fruitfulness and the feminist pragmatic virtues.[22]

External Consistency or Conservatism vs. Novelty

Kuhn recommended consistency with accepted theories in other domains, Quine and Ullian recommended the theory that least disrupts the web of belief. If we take accepted theories in other domains to be true, then obviously, to the extent it can be determined, consistency or the avoidance of inconsistency with those theories is a good guide to truth. But then, even in its broader, web of belief, version, the value of this criterion is dependent on the truth status of those accepted theories, or sentences in the web, consistency with which is recommended. The novelty criterion recommends theories and models that depart from accepted theories. It recommends disregarding consistency with other theories, and not being hamstrung by conservatism. Different interpretations of the two criteria can produce different articulations of contrast,[23] but what interests me here is their socio-political valence. The socio-political basis for the criterion of novelty is the need for theoretical frameworks other than those that have functioned in gender oppression by making gender invisible. External consistency, in a context in which theories have had that function, perpetuates this invisibility. Those satisfied with the status quo will endorse this criterion, and the effect of its endorsement is to keep from view the ways in which currently accepted theories are implicated in the legitimation of gender oppression.

Donna Haraway (1986) has pointed out, for example, how the retention of a sociobiological framework in Sarah Blaffer Hrdy's feminist primatology replicates problematic moves in liberal feminism, which perpetuates the framing assumptions about individualist and self-regarding human nature of liberal political theory. In both the primatological and the political case, the (liberal) feminist turn is limited to claiming for females what has been reserved for males without challenging the deeper assumptions about human (and animal) nature involved in both the scientific and the political program. And Susan Sperling (1991) develops a similar argument with respect to the functionalist and sociobiological frameworks she identifies in most of the feminist primatology of the last twenty-five years. Her point is that it preserves essentialist and determinist concepts of gender, its feminism being restricted to revaluing the roles of females in primate evolution. The models advanced by these primatologists thus satisfy the mainstream virtue of external consistency. Paying attention to females, making them more central to the analysis, corrects omissions of androcentric field work and does thereby advance what I have termed the central feminist cognitive aim. However, by leaving the theoretical scaffolding in place, these critics argue, the work under discussion fails to challenge the ways in which sociobiological analysis naturalizes the social relations of capitalism. While a few women may benefit in such a system, the vast majority are impoverished. Gender relations under capitalism are intricately entwined with class and race relations. Thus, feminist primatology that utilizes sociobiological analytical tools is only partially revealing of gender by privileging, Haraway and Sperling argue, middle class gender relations. Even though it has been resisted in certain quarters, one reason the feminist primatology has been taken seriously is its conservativism with respect to basic theory. According to Haraway and Sperling, its exemplification of this traditional virtue is also a cause of its political regressiveness.

Endorsing novelty is not claiming license to depart from the standard of empirical adequacy. The feminist critic can argue that new theories would (or might) produce new observational content about qualitatively different but nevertheless observable phenomena. The empirical data associated with the more standard theories might just lose their salience or even dissolve in the context of an alternative model.[24]

Simplicity and Ontological Heterogeneity

Pursuing another contrast, we can see how certain interpretations of the simplicity criterion are laden with socio-political values. The interpretation that contrasts with ontological heterogeneity is an ontological one: the simpler theory is the one positing the fewest different kinds of fundamental entity (or of causally effective entity). This encourages us to find ways of treating putative entities which are not members of the privileged class either as epiphenomena, as constructions that can be disassembled into collections of entities of the privileged class (cells into molecules, molecules into atoms, etc.), as parts of members of the privileged class, or as variants whose deviations from the standard can be disregarded. To suppose the social world is composed of just one or a few kinds of basic entity (e.g. rational self-interested individuals in neoclassical economic theory) erases the differences among persons, including their social positions, that are fundamental to how they act. Economics, for example, treats the head of household as the main economic actor – assuming its (his) dominance in the household – and assuming that the interests of other members of the household – spouse, partner, children, elderly parent – are identical with those of the head. By erasing the independent interests of other household members from theoretical view, these models prop up an oppressive family structure (one person – "the benevolent patriarch" – is *supposed* to make the decisions) and indirectly legitimate the assumption by welfare policy makers, family policy makers, etc. that this structure is the primary and appropriate family structure in our society.[25]

This treatment of the household preserves the uniformity of the effective entities in economic theory. Thus, it satisfies the virtue of simplicity. Now, suppose we have some alternative theories that, say, treat the household as an internally heterogeneous zone, structured by gender. Such feminist theories of the household disrupt the uniformity of effective entities posited in the more standard neoclassical theories. The heterogeneity of ontology tolerated in such theories helps to make visible gender relations and the activity of those gendered female in the household and its interactions with the larger economic context. If we suppose that we have equally empirically adequate models, can the virtue of simplicity be used to rule out this alternative theory? Only if simplicity could really be shown to be a criterion of truth or likelihood. For one committed to a metaphysical view about the simplicity of the universe, the greater parsimony in postulation of entities might be indicative of the greater likelihood of truth of the simpler theory, but this is now relative to the truth of the metaphysical view. But if one has no grounds for this metaphysical view, metaphysics *and* method are operating in the service of politics.

Fruitfulness and the Feminist Pragmatic Virtues

Fruitfulness, for Kuhn, referred to the capacity of a theory to generate problems or puzzles demanding solutions and to provide the resources with which to solve them. This, of course, means more opportunities to articulate connections between the theory and putatively established phenomena as well as other theories. While fruitfulness might be interpreted in more pragmatic ways, one might also see fruitfulness as a kin to refutability: a theory that generates more problems for solution than another is a theory that offers more opportunities for its confirmation and disconfirmation than that other.[26] The generation of problems, however, is not purely internal to a theory but depends on its relations with other theories and the state of instrumentation and experimental sophistication available at any given time. A theory might be fruitful in one context, but not in another.

The same might be said for the feminist practical virtues – a theory might exemplify them (or we could attribute them to a theory or model) in one context but not in others. But the contrast lies in the following: the feminist practical virtues favor theories and models that can be used to improve living conditions in a way that reduces inequalities of power. Taking them seriously requires looking beyond the immediate (internal) context of research to the ways in which that research might or might not be developed. This in turn requires taking stock of the social, political, and economic context in which development might take place. Fruitfulness is by contrast conservative in that it is inward looking. Understood as the capacity to generate either puzzles or predictions whose non-fulfillment will count as refutations, it directs attention away from the social and technological applications of research, whether they be beneficial or harmful. There is a further dimension to this contrast. One of the consequences of incorporating these quite specific and politically informed pragmatic values into a set of community standards of inquiry, is that the thesis of the political neutrality of science becomes itself a political rather than a methodological or epistemological position.

CONCLUDING REFLECTIONS

I've argued that by identifying values of a scientific community other than the traditional ones we can get insight into important features of the latter. In particular, I've tried to give some reasons for thinking that those traditional values are not purely cognitive (if at all), and that their use in certain contexts of scientific judgement imports significant socio-political values into those contexts. That is, I've argued, by comparing them with contrary theoretical virtues, that in specific research contexts the traditional virtues have a demonstrably political valence. I don't want to say the traditional virtues are always politically regressive, but that the fact that they sometimes are means that we cannot treat them as value-neutral grounds of judgement. There may be reasons for relying on them in a given context, just as there are reasons for relying on the alternatives in certain contexts. But the arguments we can give for them will be context-limited in their validity.

I do not, therefore, want to claim that the virtues or criteria I've discussed have fixed and absolute socio-political meanings. Furthermore, whatever valence they have in a given situation will be modified by their interaction with whatever other values are brought to bear, the relative priorities assigned to these values, and the reasons for which they are being endorsed in that situation. And the social context in which they are used will also make a difference. Thus, it is not clear that treating simplicity as a theoretical virtue would have the same socio political resonance in a socio-political context which values diversity and equality. But in *our* context, in which diversity and equality are granted lip service but made to defer to more important social values like order and economic competitiveness, and in which the physical and life sciences possess a greater cognitive authority than other intellectual sources of value, it does serve anti-progressive ends. Similarly, heterogeneity could, in a context other than our own, fail to be a theoretical virtue with a liberatory potential.

If the cognitive virtues, that is, the standards that regulate discursive interactions in a scientific community, lose their context-independent, universalist, status, as I have been advocating, then what is left to adjudicate scientific disputes? If underdetermination undermines even empirical adequacy's ability to put a definitive, uninterested, end to disputes, are we not faced with either anarchy or the rule of the powerful – a tyranny of the majority? I think these worries are pressing against the background of certain conceptions of scientific inquiry and of scientific knowledge, and perhaps against the background of a hope of a truth that could adjudicate the hurly-burly of the political. Without fully addressing here what I take to be this background, let me offer the following.[27]

To the extent cognitive anarchy does emerge as a consequence of the view outlined here, it is a global, rather than local anarchy. Locally, communities will and must adopt standards that express their aspirations. It's just that these standards – like the aspirations that ground them – are provisional and subject to modification as a consequence of interaction with other communities as well as with the world a community seeks to know. This is why I would describe the view as pluralist rather than anarchist. In any case, there's hardly enough diversity now – a little more could improve things significantly in some of the sciences.

As for the danger of tyranny of the majority, I admit that the criteria of community interaction outlined in the opening pages of this essay, especially the fourth, are idealistic. But they function as, among other things, grounds for the critique of actual practices. While idealistic, they are no less powerful than the truth is in the face of brute force; that is, they are practically ineffective, but serve nevertheless to animate critical reflection and, where appropriate, resistance. The fourth criterion, requiring equality of intellectual authority, invalidates consensus that is achieved by means other than free and open critical discourse, by, for example, tyranny of the majority. Tyranny of a minority, one could say, is what we have now, and its problematic nature is only revealed by the kind of critique advanced here and elsewhere by like-minded analysts. We should worry more about the concealing of political agendas behind the mantle of scientific neutrality than about the consequences of abandoning the illusion of neutral arbiters of our cognitive practices.

One can see the claim I have been defending – that the traditional virtues have a political valence – as leading to a dilemma: either the traditional virtues cannot be taken as constitutive of "best explanation" or of "science" in some social-value neutral sense, that is, we cannot maintain the dichotomy between cognitive and non-cognitive values, *or* the doom sayers are right and science is just a vehicle for the maintenance of political control: "science is politics by other means", to generalize Haraway's (1986) paraphrase of Clausewitz. We can reject this second leg of the dilemma only if we can be satisfied with at best local, sometimes politically grounded, and always negotiated, vindications of virtues and the pluralism this entails.

Center for Philosophy of Science and Department of Women's Studies, University of Minnesota, Twin Cities

NOTES

[1] This is an expanded and revised version of the essay "Gender, Politics, and the Theoretical Virtues" (Longino, 1995). I am grateful to members of the Philosophy Departments at Carleton College, St. Olaf College, the University of Toronto, the History and Philosophy of Science Department at Indiana

University and the Committee on the Conceptual Foundations of Science at the University of Chicago for their comments, and to Marta Gonzalez-Garcia and Lynn Hankinson Nelson for their instructive readings of earlier drafts of the essay.

[2] There are a variety of ways of formulating and demonstrating underdetermination. This formulation is supported by arguments in Longino (1990, Ch. 3) and is consonant with Bas van Fraassen's (1980) distinction between the truth and the empirical adequacy of a theory. See Elizabeth Potter's essay in this volume for discussion of other versions of the underdetermination thesis.

[3] I suspect, in fact, that "cognitive" may be used in preference to "epistemic" because it has a certain vagueness that "epistemic" does not, and it should be noted that not all advocates of this solution use it. McMullin (1983) speaks, for example, of epistemic and non-epistemic values. These are issues to be pursued in another paper.

[4] See the Symposium on 'Do Explanations or Predictions (or Neither) Provide More Evidential Support for Scientific Theories?', (Brush, Achinstein, and Shimony, 1995).

[5] See van Fraassen's (1983) comments.

[6] I owe this point to a comment by Ian Hacking.

[7] I am not addressing the question whether there might not be traits of *cognizers* that could be termed virtuous. These would, of necessity, be different sorts of traits than the traits of models, theories, or hypotheses under discussion here.

[8] Some have understood simplicity as having to do with the character of the calculations needed for derivations in the theory, but this is to treat simplicity as a matter of ease of use, rather than as a property of the theory or its principles.

[9] Of course, there may be a trade-off, as a model using fewer kinds of entities may require more kinds of processes to account for a given range of phenomena than a model which uses more kinds of entities. That is, there may be an inescapable degree of complexity that must be built in somewhere.

[10] Philip Kitcher (1993) has argued for unification as a scientific desideratum. This, for present purposes, can be considered as a variation on breadth of scope. Kitcher, it might be argued, has a more precise measure in mind than the notion of breadth of scope allows.

[11] This is the position taken by van Fraassen, excepting, of course, empirical adequacy (1989, pp. 40–64; 131–150). It has much to recommend it, from a God's-eye point of view. But bundling and dismissing the so-called cognitive virtues saves them from the political critique to which I wish to subject them.

[12] I first discussed what I have called the feminist theoretical virtues in Longino, 1993b. I used them again in a discussion of the possibility of feminist epistemology in Longino, 1994. The exposition of the next several pages borrows from those earlier publications.

[13] Stephen Kellert suggests that the virtue at play is particularism, rather than heterogeneity. This is a suggestion worth exploring in a fuller treatment of the very idea of feminist virtues.

[14] See Bleier (1983), Keller (1985), Fausto-Sterling (1985).

[15] McClintock's embrace of heterogeneity in the phenomena is carried through to her explanation of those phenomena which involves the invocation of different kinds of causal factor and the resistance to subsuming one to the othe (Keller, 1983, 1985).

[16] Philosopher Nancy Cartwright clearly wants our (interpretations of) scientific theories to allow that the world is constituted of highly diverse entities and seems herself committed to a metaphysics of heterogeneity (Cartwright, 1987, 1995). It is less clear that she would want our theoretical (as distinct from our phenomenological) ontologies to exhibit heterogeneity.

[17] Evelyn Keller detects what I would describe as a commitment to ontological homogeneity in the Human Genome Project's ambition to map the genetic complement of a "normal" human being. Who determines what will count as normal? she asks (Keller, 1992). Elisabeth Lloyd raises similar issues in her essay "Normality and Variation," stressing the variability among humans and the value-laden character of judgements about normality (Lloyd, 1994).

[18] To the best of my knowledge, although Margolis and Sagan (1986) gloss "fertilization" as nucleic fusion, no one else has used this obviously superior alternative expression to refer to the process generally referred to by "fertilization." The latter term, conveying action upon something, facilitates asymmetric thinking where "fusion" does not.

[19] Mary Tiles's essay "A Science of Mars or a Science of Venus?" argues for the inevitability of such social choices in the pursuit of scientific knowledge (Tiles, 1987). Kristina Rolin argues that the search for knowledge for "knowledge's sake" underdetermines the direction of inquiry. Particular kinds of knowledge are sought (Rolin, unpublished ms).

[20] This is not a rejection of mathematics but of requiring mathematical knowledge that is not necessary for the discipline. Such requirements restrict who will be able to engage in engineering or in economic analysis to those with certain intellectual skills (which are not actually employed in those fields) and not others, and thus shape the knowledge and products of those fields.

[21] For example, Levins and Lewontin (1985) embrace both heterogeneity and a strong form of interaction they label "dialectical" as features of dialectical biology. Literary scientists Stephen J. Gould and Lewis Thomas endorse interaction as a principle of explanation. Whatever sympathies with feminism they may have, it is not feminism that leads them to heterogeneity or interaction. Indeed Gould (1986) explicitly says that gender or feminism have nothing to do with it. It's just a matter of good science. Finally, Noretta Koertge, upon hearing an early version of these arguments at Indiana University, argued that these (at least heterogeneity) should not be taken as feminist virtues at all.

[22] There is discussion of additional contrasts, including a discussion of problems with the concept of empirical adequacy, in the earlier version of this essay (Longino, 1995).

[23] For example, novelty at least within some limits is required to earn the highest accolades within mainstream science. Only when such novelty does not challenge accepted theory (as in the case of the discovery of the top quark) or when the web of belief has been sufficiently repaired (as was the case for Barbara McClintock) are those accolades extended. On the other hand, the ideal of unified science requires, in the end, a set of theories that are not only mutually consistent, but all equally consequences of a set of basic principles. To the extent novelty licenses fundamentally different explanatory principles for different phenomena, it is contrary to the ideal of unified science.

[24] See Longino (1990, Ch. 7; and 1987) for examples.

[25] For discussion, see England (1993) and Strassman (1993).

[26] This was suggested to me by David MacCallum.

[27] A question from Lorraine Daston persuaded me that it would be important to address this issue. I have discussed what I take to be part of the background in Longino, 1993a.

REFERENCES

Altmann, Jeanne: 1974, 'Observational Study of Behavior: Sampling Methods', *Behavior*, **49**, 227–67.

Bleier, Ruth: 1983, *Science and Gender*. Elmsford, NY: Pergamon Press.

Brush, Stephen, Peter Achinstein, and Abner Shimony: 1995, Symposium: 'Do Explanations or Predictions (or Neither) Provide More Evidential Support for Scientific Theories?' David Hull, Mickey Forbes and Richard Burian (eds), *PSA 1994, Vol. 2*. East Lansing, MI: Philosophy of Science Association.

Cartwright, Nancy: 1987, *How the Laws of Physics Lie*. New York: Oxford University Press.

——: 1995, 'The Dappled World', in Mickey Forbes and Richard Burian (eds), *PSA 1994, Vol. 2*. East Lansing, MI: Philosophy of Science Association.

Churchland, Paul: 1985, 'The Ontological Status of Observables: In Praise of the Superempirical Virtues', in Paul Churchland and Clifford Hooker (eds), *Images of Science*. Chicago, IL: University of Chicago Press.

England, Paula: 1993, 'The Separative Self: Androcentric Bias in Neo-Classical Assumptions', in Marianne Ferber and Julie Nelson (eds), *Beyond Economic Man*. Chicago, IL: University of Chicago Press.

Fausto-Sterling, Anne: 1985, *Myths of Gender*. New York: Basic Books.

Ginzberg, Ruth: 1987, 'Uncovering Gynecentric Science', *Hypatia*, **2.3**, 89–106.

Gould, Stephen: 1986, 'Review of Bleier *Science and Gender*'. *New York Review of Books*.

Haraway, Donna: 1986, 'Primatology is Politics by Other Means', in Ruth Bleier (ed.), *Feminist Approaches to Science*. Elmsford, NY: Pergamon Press.

——: 1991, *Simians, Cyborgs, and Women*. New York: Routledge.

____: 1992, 'The Promises of Monsters: A Regenerative Politics for Innappropriate/d Others', in L. Grossberg, C. Nelson, and P. Treichler (eds), *Cultural Studies*. New York: Routledge.

Harding, Sandra: 1986, *The Science Question in Feminism*. Ithaca, NY: Cornell University Press.

Keller, Evelyn F.: 1983, *A Feeling for the Organism*. San Francisco, CA: W. H. Freeman and Company.

——: 1985, *Reflections on Gender and Science*. New Haven, CT: Yale University Press.

——: 1992, 'Nature, Nurture, and the Human Genome Project', in Daniel Kevles and Leroy Hood (eds), *The Code of Codes: Scientific and Legal Issues in the Human Genome Project*. Cambridge, MA: Harvard University Press.

Kitcher, Philip: 1993, *The Advancement of Science*. New York: Oxford University Press.

Kuhn, Thomas: 1977, *The Essential Tension*. Chicago, IL: University of Chicago Press.

Laudan, Larry: 1984, *Science and Values*. Berkeley, CA: University of California Press.

——: 1990, 'Demystifying Underdetermination', in C. Wade Savage (ed.), *Minnesota Studies in the Philosophy of Science, Vol. XIV: Scientific Theories*. Minneapolis, MN: University of Minnesota Press.

Levins, Richard and Lewontin, Richard: 1985, *The Dialectical Biologist*. Cambridge, MA: Harvard University Press.

Lloyd, Elisabeth: 1994, 'Normality and Variation', in Carl Cranor, (ed.), *Are Genes Us? The Social Consequences of the Human Genome Project*. Berkeley, CA: University of California Press.

Longino, Helen E.: 1987, 'What's Really Wrong with Quantitative Risk Analysis', in Arthur Fine and Peter Machamer (eds), *PSA 1986, Vol. 2*. East Lansing, MI: Philosophy of Science Association.

——: 1990, *Science as Social Knowledge*. Princeton, NJ: Princeton University Press.

——: 1993a, 'Subjects, Power, and Knowledge: Description and Prescription in Feminist Philosophies of Science', in Linda Alcoff and Elizabeth Potter (eds), *Feminist Epistemologies*. New York; Routledge.

——: 1993b, 'The Relevance of Gender to the Philosophy of Science', in M. Forbes, A. Fine and K. Okruhlik (eds), *PSA 1992, Vol. 2*. East Lansing, MI: Philosophy of Science Association.

——: 1994, 'In Search of Feminist Epistemology', *The Monist*, **77**, no. 4.

——: 1995, 'Gender, Politics, and the Theoretical Virtues', *Synthese* (October).

Margolis, Lynn and Dorion Sagan: 1986, *Origins of Sex: Three Billion Years of Genetic Recombination*. New Haven, CT: Yale University Press.

Martin, Emily: 1991, 'Egg and Sperm', *Signs: Journal of Women in Culture and Society*, **16**, 3.

McMullin, Ernan: 1983, 'Values in Science', in P. D. Asquith and T. Nickles (eds), *PSA 1982, Vol. 2*. East Lansing, MI: Philosophy of Science Association.

Quine, Willard. V. and J. S. Ullian: 1978, *The Web of Belief*, 2nd edn. New York: Random House.

Rolin, Kristina (unpublished ms) 'Gender and Scientific Knowledge in High Energy Physics'.

Rose, Hillary: 1983, 'Hand, Brain, and Heart: A Feminist Epistemology for the Natural Sciences', *Signs*, **9**, no. 1, 73–90.

Sen, Gita and Caren, Grown: 1987, *Development, Crisis, and Alternative Visions*. New York: Monthly Review Press.

Sperling, Susan: 1991, 'Baboons with Briefcases: Feminism, Functionalism and Sociobiology in the Evolution of Primate Gender', *Signs*, **17**, no. 1.

Strassman, Diana: 1993, 'Not A Free Market: The Rhetoric of Disciplinary Authority in Economics', in Marianne Ferber and Julie Nelson (eds), *Beyond Economic Man*. Chicago, IL: University of Chicago Press.

Tiles, Mary: 1987, 'A Science of Mars or a Science of Venus?', *Philosophy*, **62**.

van Fraassen, Bas: 1980, *The Scientific Image*. New York: Oxford University Press.

——: 1983, 'Theory Comparison and Relevant Evidence', in John Earman (ed.), *Minnesota Studies in the Philosophy of Science, Vol. X: Testing Scientific Theories*. Minneapolis, MN: University of Minnesota Press.

——: 1989, *Laws and Symmetry*. New York: Oxford University Press.

JACK NELSON

THE LAST DOGMA OF EMPIRICISM?[1]

INTRODUCTION

This is, in the jargon of 25 years ago, a paper about the fact/value distinction, or, better, the science/values distinction. The question mark in the title concerns whether the science/value distinction is a dogma or a defensible distinction to be drawn within holism. I begin from the position of a Quinean holist and argue that value claims should be seen as part of our holistic world theory that is, as a whole, supported by experience. However, I do conclude that there may still be a distinction, though perhaps one of degree, to be drawn between values and value laden claims and other parts of our world theory. The moral of this paper will be that whether there is a science/values distinction of some sort is far less important than is the recognition of two "facts" about "values", that values frequently do influence the course of science and that we cannot, for this very reason, afford to treat values as matters of personal preference or as subjective or as in any other way wholly or even largely exempt from the standards of evidence and evaluation that apply to science itself.

STARTING POINTS

It will be useful to enumerate at the start what I take to be some fairly non-controversial boundary conditions for any view about the relation between facts and values and about the nature of science. I assume that on any acceptable view of science and values there must be room for social, political, and moral theory as well as for physics, chemistry, and biology, that there must be room within all of these areas of inquiry for arguments and for evidence. Within the sciences, however broadly construed, there must be a place for testability and empirical confirmation. I am a holist, and as such believe that none of our currently accepted theories, either in part or in whole, are immune from possible revision and even abandonment. I think it is a consequence of holism, though not one Quine fully recognizes, that it is we, and not I, who know, in more than the editorial sense (see Nelson, 1990). As a holist, I believe that no theory or thesis that we are now in a position to put forward that challenges the basic success, accuracy, or objectivity of science, or large parts thereof, will be viable. We are too close to current science, our standards of evidence and reasonableness are too much part and parcel of our science, to be in a position to reject that science in total. And this is as it should be. But it does not follow that all intelligent creatures, everywhere, and for all time, will be subscribers to science as we now know it. There is no paradox here.[2]

L. H. Nelson and J. Nelson (eds), Feminism, Science, and the Philosophy of Science, 59–78.

Like most philosophers of science, I hold both that science does not rest on *a priori* truths that it is the business of philosophy to lay bare and, where necessary, explain, and that the philosophy of science should be equated neither with the history of science nor with the sociological description of the activity of scientists. Rather, philosophy of science is, despite the ill repute into which the phrase has fallen, the identification and *rational reconstruction* of good science. As such it presumes that it is possible to distinguish good from bad science, and to say what it is about each that makes it so. But there is no presumption here that there are any standards other than those of science itself, broadly construed, for so doing.

One of the great failings of the naive empiricism of logical positivism is that the story it told about scientific investigation failed, almost completely, to match the actual practices of scientists. Logical empiricism (or immediate post-logical positivist philosophy of science[3]) adopted, as a proposed remedy for this failing, the context of discovery/context of justification distinction, and, at least apparently, abandoned all pretext of giving an account of the practice of scientists in favor of a way of distinguishing good theories, however arrived at, from bad theories. This is not, as most philosophers of science now recognize, a viable position, for to suggest that there is no discoverable connection between how science is done and whether viable theories are produced is to assign a substantial part of the process of science to the realm of the irrational, or at least the non-rational.

Any viable philosophy of science must recognize the indisputable success of science. But what does this mean? At least this: we are able to build automobiles and nuclear reactors, and to explain why both run and, normally, neither explodes. We understand something about how diseases spread and can, sometimes, intervene to cure or control them. We can relate the gross structural properties of middle sized objects to properties attributed to molecular, atomic, and sub-atomic particles. For a wide range of events and actions, we can predict at least some ensuing events and actions. We have organized our knowledge into broad fields, and know something of the connections between and across those fields. And so on.

The success of science is due, in large part, to the stress it places on empirical confirmation. The traditional view has been that the questions science deals with can, in principle, be settled by the normal techniques of science, including empirical confirmation, whereas questions of value cannot. Hence, if the norms of objectivity, accuracy, and the goal of truthfully representing nature are to be accomplished, values must be excluded from science. Since anything goes in the realm of values, the success of science can only be accomplished by building and maintaining an absolute wall between science and values.

For these reasons, scientists are trained to "leave their values at the laboratory door". And, by and large, good scientists do leave at least their most obviously value laden views at the laboratory door. That is, for the most part the various sciences have been fairly successful in developing and enforcing canons of intellectual responsibility that have either prevented, or exposed and condemned, the blatant intrusion of values into research. However, over the past 20 years research into the theory and practice of science, including research by feminist science critics and by practicing scientists – some of which is presented in papers in this volume – has shown that values do influence science more frequently and in more subtle ways than positivist and post-positivist philosophy of science would have us believe. Personal and societal norms and values do influence, for better and for worse, the selection of research agendas, the way research is carried out, what is counted as "data" and as "results", and when a hypothesis or theory is counted as confirmed or disconfirmed. The norms and values that affect science include not only what are taken to be constitutive values of science (e.g., simplicity, explanatory power, fecundity ...) but also traditional moral values of various sorts, broad and usually unexamined views and presuppositions about what research agendas are important and worth pursuing, about what is and is not relevant to those agendas, about the standards for what constitutes good scientific practice and good scientific theories, about what the relation of science to society and social values is, and about what is and is not obvious. Often cited examples of social values influencing science include the long unchallenged dominance of "man-the-hunter" theories in anthropology, the unwarranted and widespread acceptance of Wilson's Sociobiology, the emphasis in medical research on diseases and ailments primarily affecting males, and the continuing currency of the view that women and minorities "can't do science and math".

Reactions to the litany of cases in which values seem to have influenced science range from attempts to deny, or dramatically minimize, the effects of such influence, to pronouncements that these are examples of bad science and that the scientific community must hereafter be more vigilant about keeping values out of science (Haack, 1993), to suggestions that the problem is not that values influence science, but rather that *the wrong values* have influenced science (along with suggestions as to what the right values might be) (Longino, 1990), to proposals to reconstruct science with values but without empiricism (Harding, 1986). A few have argued that the notion of evidence can reasonably be construed so as to include value-laden assumptions broadly held within the relevant research community without transforming science into a radically different enterprise (Nelson, 1990).

I end this section with what may seem a digression. Many years ago I was discussing meta-ethics with a colleague. I was maintaining that, despite all of the problems with utilitarianism, it is still the only theory that makes sense, in the end, in a world in which determinism or something very like it holds. After a lot of verbiage, my colleague, who was defending a deontological view, said, in exasperation, "Well, *everyone* is a utilitarian in your sense", by which he meant that everyone, or everyone who has thought seriously about the matter, agrees that in the end what is of concern is maximizing the welfare and happiness of individuals. He thought the only interesting disputes are about what sort of a society is likely to best address those concerns, his own view being that the notion of justice or fairness must play a pivotal role in any such society.

In recent years it has become fashionable to announce that empiricism is dead and that we must move beyond that naive approach to science. The context within which such claims have been made is not that of a discussion of philosophical empiricism versus philosophical rationalism, both of which grant experience an important role in theory formation and confirmation, but differ on whether, in epistemic activities, the mind is "active" or "passive". That is, the empiricism under attack is that which is assumed by both philosophical empiricism and philosophical rationalism, and is the touchstone of modern science. It is the view that our knowledge of the natural world is built on sensory experience and is intended to predict, explain, and integrate that experience, present, future, and past. I feel about pronouncements of the death of this minimal kind of empiricism, call it science empiricism, much as I do about the professed demise of utilitarianism – that people who reject either utilitarianism or science empiricism are being disingenuous. What they are really saying is that certain unduly restrictive or naive formulations of utilitarianism or of science empiricism must be rejected. This is of course true. But in the end we are all utilitarians, and, I think, science empiricists. Those who deny a connection between knowledge and experience, and are not disingenuous in doing so, are unlikely to successfully cross a street on which there is even a moderate amount of vehicular traffic.

The point of this disgression is that some recent attacks on science, and more specifically on empiricism, including those motivated by the revelation that values have influenced various sciences, are either disingenuous or very badly stated, mistaking science empiricism for logical positivism or for a naive inductivist view of the growth of scientific knowledge by accretion, or for the post-positivist distinction between the context of discovery and the context of justification and the deductive nomological model of explanation developed in tandem with that distinction (see, for example, Harding, 1986 and 1993). But science empiricism is not any of the above positions. The core of science empiricism is simply that experience matters, is the touchstone of knowledge claims and theory formation.

THE QUINEAN CONTEXT

Quine's legacy will surely include the lesson that what we have come to think of as natural, intuitive, and structure-yielding distinctions and boundaries are frequently anything but natural, seem intuitive only because we have learned to see philosophy, science, and the world through their blinders, and that the structures they impose are not always the most useful ones. There is no analytic/synthetic distinction because there are no analytic truths. There is at least no sharp theory/observation distinction because all, or almost all,[4] sentences are theory-laden. No sentences of our accepted theories, including definitions and observation sentences, are immune from possible revision or abandonment in the light of new experience or of reconsidered past experience. Truth can be defined, *a la* Tarski, for a language and hence for a theory, but the old notion of a naive correspondence theory of truth, with its simplistic dichotomy between words and world, goes by the way, as, in the end, do reference and meaning, at least in the sense of intension. Confirmation and testability are viable and useful notions, but it is theories as wholes that face the tribunal of experience, not individual sentences, be they putative laws, observation sentences (except as construed holophrastically), or proposed definitions.

Science itself is "a conceptual bridge of our own making, linking sensory stimulation to sensory stimulation" (Quine, 1981d, p. 2). We, as a species and as individuals, are as much posits of our theories as are quarks, atoms, tables, and chairs. Hence, we are as much creatures of our theories as they are of us. And once we have posited *us*, *Homo sapiens*, the data that science must accommodate come to include our reactions to the firings of our sensory receptors.

Some have concluded that for all of these reasons we are trapped by our own conceptual scheme, powerless prisoners of our past, and have turned to Kuhn's account of scientific revolutions for an explanation of theory change. But the proper metaphor is not that of being trapped or constrained by our own theories. The proper metaphor is rather that of being enabled and empowered by our best going theories, for these theories not only structure our view of the world and our place in it but also themselves provide us with the tools whereby we can reshape those very theories, though not, of course, "all at once". The history of science is the history of our doing so, with varying degrees of competency and success.

Given that Quine has challenged so many traditional distinctions and dogmas, one might expect him to reject the science/value distinction as well.[5] But Quine has a foot firmly in the tradition of Carnap and the empiricism of the first half of this century. He would certainly defend many views and policies that are taken to follow from the science/value distinction, for example that religious, political, social, and moral values have no place in the laboratory, that social and political policies are not logical consequences of scientific discoveries, and that "good" science (i.e., technically well done research and the theories resulting from it) can lead to, or at least make possible, horrendous social and political acts. And in a relatively late piece, 'On the Nature of Moral Values', in *Theories and Things*, he does explicitly defend a distinction between values and science.

There are a number of real, or presumed, difficulties about value sentences that are not, I suggest, the basis for Quine's insistence upon a science/values distinction. The truth conditions for value claims are often seen as problematic, but this is not why Quine excludes such claims from science. In 'On Austin's Method' Quine notes that "The paradigm [Tarski's theory of truth] works for evaluations ... as well as for statements of fact ... 'Slander is evil' is true if and only if slander is evil. ..." (Quine, 1981b, p. 90). There is a difficulty in accommodating value claims to the extensional logic of first order quantification theory, but this is not, I think, Quine's reason for excluding values from science, for the same concern holds for a vast range of sentences that we are not now in a position to treat within first order quantification theory but are also not prepared to abandon, including, for a start, the whole range of sentences ascribing intensional states to agents, many principles and claims of the social sciences, and much of psychology. Nor does Quine exclude values from science because he is a reductionist and value sentences are not reducible to sentences of physics. Whether Quine is a reductionist, and if so of what sort, need not be settled here, for there are, again, a whole host of sentences that are not obviously reducible to sentences of physics that Quine does include within science, including most of the substance of the life sciences and the social sciences.[6]

Finally, Quine does not exclude value sentences from science because there is *no* evidence for value claims. In 'On the Nature of Moral Values' Quine allows that holism does place constraints on the values we posit as well as on the scientific theories we construct. He also recognizes the constraint imposed by the survival value to societies of moral, social, and political systems that serve to promote at least minimal cooperation and to prevent behavior that works against the smooth functioning of society. "[W]e can expect a common core, since the most basic problems of society are bound to run to type. Morality touches the common lot of mankind. ..." (Quine, 1981c, p. 62). Of course, sentences that *ascribe* likes and dislikes, values, or value systems to individuals or societies *are* part of science, for they do support predictions. People who like raw oysters behave differently from those who do not, at least in the presence of raw oysters and with all other things being equal. So too, there are situations in which racists and non-racists behave, predictably, differently.

The primary reason Quine excludes values from science is, I think, that in his view "empirical controls" work only on science: "Science, thanks to its links with observation, retains some title to a correspondence theory of truth; but a coherence theory is evidently the lot of ethics" (Quine, 1981c, p. 63). The empirical controls on science function via observation conditionals. "A reasonably inclusive body of scientific theory, taken as a whole, will ... imply a lot of observational conditionals, as I call them, each of which says that if certain observable conditions are met then a certain observable event will occur" (Quine, 1981a, p. 70). But no set of non-descriptive value sentences will, Quine apparently believes, imply any observational conditionals not derivable without the value sentences.[7] There is also, for Quine, the stumbling block of modal logic. Quine believes that all of modal logic rests on a confusion, and the "logic" of values (deontic logic) seems to be a variety of modal logic.

ARGUMENTS AGAINST A SCIENCE/VALUES DISTINCTION

A compelling reason for seeking to accommodate value claims within science is the hope of bringing something of the "objectivity" and openness that are characteristic of science to bear on value claims. The hope here is that rejecting the science/value distinction would result not, as many scientists and traditional epistemologists and philosophers of science assume, in relativism, in everything becoming a matter of personal or group preference, but rather in all claims, including value claims, becoming subject, more or less, to the same canons of evidence and evaluation. Values, that is, might become more like facts than facts become like values. And were this to happen, the fact that values do, to whatever extent, influence science would be less of a concern than it is now.

I find two interlinked themes within Quine's work that argue against a complete science/value distinction. The first is Quinean holism, the second Quine's notion of posits and reality. I begin with holism.

Quinean empiricism, as I interpret it, is the view that experience is the firings of sensory receptors, and that science is a set of theories whose aim is to explain, predict, and systematize experience. Quinean holism is the story of how this is accomplished. (I take Quinean empiricism to be a theory of evidence, not a theory of meaning, although Quine's talk of words and sentences as having meaning only within the context of larger theories sometimes makes it seem as though he were offering a theory of meaning.[8]) Quine's earliest, and perhaps clearest, statement of holism is in 'Two Dogmas of Empiricism':

> The totality of our so-called knowledge or beliefs, from the most casual matters of geography and history to the profoundest laws of atomic physics or even of pure mathematics and logic, is a man-made [*sic*] fabric which impinges on experience only along the edges. Or, to change the figure, total science is like a field of force whose boundary conditions are experience. A conflict with experience at the periphery occasions readjustments in the interior of the field. Truth values have to be redistributed over some of our statements. Reevaluation of some statements entails reevaluation of others, because of their logical interconnections – the logical laws being in turn simply certain further statements of the system, certain further elements of the field ... the total field is so underdetermined by its boundary conditions, experience, that there is much latitude of choice as to what statements to reevaluate in the light of any single contrary experience. No particular experiences are linked with any particular statements in the interior of the field, except indirectly through considerations of equilibrium affecting the field as a whole (Quine, 1963, pp. 42–43).

What counts for or against a claim or theory is how it is integrated in a larger theory or set of theories, and how well these interconnected claims and theories collectively predict, explain, and integrate the firings of our sensory receptors. All of this seems to invite, or at least allow, the inclusion of value claims, or value laden claims, within science – or at least within the holistic web consisting of "[t]he totality of our so-called knowledge or beliefs. ..."

Consider now Quine's view of posits, reality, and conservatism. Quine is neither a realist nor an instrumentalist. Theories are sets of sentences. Our ontic commitment is to those things that we must posit to make the sentences of our theories come out true. To ask what there is is to ask what posits we must make for our theories to come out true. Among the posits of the going theories of the natural and social sciences are atomic and subatomic particles and middle sized bits of animate and inanimate stuff, including us, *Homo sapiens*. While we do, at times and for certain purposes, make use of theories whose posits include only esoteric objects, e.g., chunks of space-time, we are not prepared, at least yet, to dispense with theories that do include more standard objects, including *us*. Indeed, the theories within which we do most of our applied and theoretical work do posit the mundane objects of ordinary life.

To the extent that we are concerned to understand us, *Homo sapiens*, our theories also posit social structures and individual and social goods and rights. For good and not so good reasons we have posited racial and sexual categories, and have constructed notions of sexuality, most of which are value infused. We collectively construct a world in which pain, suffering, happiness, joy, and prejudice are as real as trees, cars, political entities, and quarks. The social and the physical worlds are both posits, and the very distinction between these is itself a boundary of our own making and open to revision. If this is indeed the boat of Neurath, within which we collectively find ourselves, then values and value theory are important parts of the holistic web that, as a whole, confronts experience. And the evidence of holism is available as much for value claims as it is for claims of logic and number theory. In these as in all cases we appeal to how the claims or theories in question help to integrate our experience, including our social experience, how they connect with other going theories, and how they together help us predict future experience, including our own and others' reactions to the violence, kindness, rudeness, and considerateness we encounter.

A MORE MODEST HOLISM

But there is good reason to think the foregoing view is not Quine's view, or at least not Quine's most recent view. The last sentence of Quine's statement of holism in 'Two Dogmas' (quoted above) is "The unit of empirical significance is the whole of science". I have always been inclined to read Quine, in passages such as this, as including within science all beliefs worth holding. So, since beliefs concerning values are worth holding, they must be included within that holistic system. But I now think this is a misreading of Quine. Even in 'Two Dogmas' Quine probably includes within science only the natural and social sciences (he explicitly includes geography and history), not all of our seriously held beliefs. In 'The Nature of Moral Values' Quine does, as noted above, explicitly advocate a science/values distinction. And Quine's views have evolved. In his conversation with Giovanna Borradori in *The American Philosopher* Quine says:

> I have actually tempered the extreme holism of my first writings. ... The way I look at things today is that there isn't only one science, but a big enough bundle of laws not to be comprehended as a single hypothesis. The big enough bundle implies logically some observational conditions,[9] namely some categories defining observable situations ... holism is needed to the extent that you have a big enough combination to apply some of these testable categories (Borradori, 1994, p. 36).

The theme of a more modest holism is also to be found in 'Five Milestones of Empiricism', where Quine writes:

> When we look thus to a whole theory or system of sentences as the vehicle of empirical meaning, how inclusive should we take this system to be? Should it be the whole of science? or the whole of *a* science, a branch of science? This should be seen as a matter of degree, and of diminishing returns. All sciences interlock to some extent. ... It is an uninteresting legalism, however, to think of our scientific system of the world as involved *en bloc* in every prediction. More modest chunks suffice, and so may be ascribed their independent empirical meaning, nearly enough. ...
>
> Thus the holism that the third move brings should be seen only as a moderate or relative holism (Quine, 1981a, p. 71).

Finally, there is a clear difference between Quine's use of Neurath's metaphor of sailors rebuilding a ship while on the open sea in *Word and Object* and in 'Five

Milestones of Empiricism'. In the former, the metaphor is parsed as follows: "Neurath has likened science to a boat which, if we are to rebuild it, we must rebuild plank by plank while staying afloat in it" (Quine, 1960, p. 3). In 'Five Milestones of Empiricism' the crew of Neurath's boat is not the inclusive 'we' but rather "the naturalistic philosopher":

> The naturalistic philosopher begins his [*sic*] reasoning within the inherited world theory as a going concern. He tentatively believes all of it, but believes also that some unidentified portions are wrong. He tries to improve, clarify, and understand the system from within. He is the busy sailor adrift on Neurath's boat (Quine, 1981a, p. 72).

Here the holistic story is the story of the *naturalistic* philosopher, the scientist. The story is that she or he begins with the sum of the sciences of her or his time (presumably including the social sciences, for though Quine uses the expression 'naturalistic philosopher' he also talks of the "inherited world theory as a going concern") and goes on to refine and expand that body of knowledge, or a chunk thereof.

Should we restrict the web or network of sentences and theories that constitutes Quinean holism to sentences and theories of the sciences, and retreat from the view that it is the totality of those sentences and theories that confronts experience? Doing so would certainly make it easier to maintain the science/values distinction. Here it will help to remind ourselves of how an individual acquires "a world theory as a going concern".

As individuals we begin neither with a manageable chunk of science, nor with all of science, narrowly construed. The body of beliefs we inherit from our parents and our culture does not come to us separated into science/non-science components. We do not make the move from infancy to childhood to adolescence to adulthood by acquiring only, or in clearly delineated categories, a rudimentary knowledge of physics, chemistry, mathematics, logic, and other disciplines Quine includes within science. What we acquire is a much more comprehensive and frequently inconsistent theory of physical objects, including human beings, of their patterns of behavior, and of our and others' reactions to that behavior, along, in some cases, with some ill-defined views about gods, extrasensory perception, magic, and whatever. Mixed in with all of this, not always in clearly distinguishable ways, are our values, what we like and dislike, what is worthwhile and what isn't, what is right and wrong. As we mature we refine our theory, throwing out gratuitous and distracting talk of extrasensory perception, magic, and, with any luck, of gods and the efficacy of prayer. Our theory comes to include more and clearer sentences about physical objects and physics, logic and mathematics, and about psychology, history, geography, morality, social theory, our likes and dislikes, and perhaps the meaning of life, or at least about what is worth doing. The educational system, with its traditional disciplinary distinctions, works its way and we come to see the world and our own beliefs largely in disciplinary terms, with values largely relegated to a separate domain. Some beliefs do persist that, when pressed, we would be hard put to label as purely factual or purely valuational, e.g., beliefs about what is 'natural', medical beliefs about what constitutes health, disease, successful treatment, and beliefs about social and political structures and practices.

The evolution in our beliefs is in part the evolution from 'I' to 'we'. The beliefs we retain are the beliefs we find others in our communities holding or amenable to holding. We learn that most of our beliefs are shared with some, but not all, and not always the same, members of society at large. (Those that are shared with no others tend to wither away.) That is, we become members of distinct but overlapping communities each with an associated set of beliefs. Some communities are based on loyalty to these particular kith and kin, deriving from a shared and idiosyncratic family upbringing, others on broader beliefs about the proper organization of society and socially acceptable behavior (deriving from cross-family training and experience), others on shared religious beliefs derived from a community broader than the family but narrower than one's political and economic community, others on one's exposure to or inclusion in the natural science community, with its commitment to standards of evidence and testability. Experience shows, by and large, that we cannot expect all others to share our views about our own kith and kin, our religion or lack thereof, our views of government and social organization. We are, in fact, taught to tolerate if not encourage differences in beliefs about religion, politics, social structures, and appropriate behavior (within limits). But we do think everyone will share, or can be brought to share, the claims that are constitutive of the sciences. We are taught that differences in belief within the natural sciences must be resolvable. Given this view, holism, the view that every claim is connected, however remotely and tenuously, with every other claim, can work only within the natural sciences, for elsewhere we are prepared to tolerate conflicting claims. And the only posits that deserve to be thought of as constitutive of reality will be those required to make the sentences constitutive of the holistic web of the natural sciences true. An implication of this view is, of course, that rationally based agreement is obtainable on all and only the claims of the traditional sciences.

Two untoward results are likely to follow: first, we are likely to mistake broad agreement concerning a belief with that belief's being part of science (for we expect broad agreement only within science), and, second, we may be too ready to tolerate divergence of belief in areas outside the traditional sciences.

A BROADER HOLISM

A preferable view is surely that the crew of Neurath's boat is indeed 'we' and not 'I', and that the boat itself consists of all our seriously held beliefs, our whole world view including beliefs about values in so far as we expect to be able to achieve community-wide consensus about those beliefs. This is our "world theory as a going concern". To Quine's remark, quoted earlier, that "Science, thanks to its links with observation, retains some title to a correspondence theory of truth; but a coherence theory is evidently the lot of ethics" we can reply that, on Quinean grounds, neither a correspondence nor a coherence theory of truth is viable. The only viable theory of truth is Tarski's semantic theory, and that theory accords well enough, as Quine recognizes, with value claims.

The broader holism I am urging will grant that some beliefs included in our world theory will express values or be value laden. For example, until recently the inherited world theory of most scientists included the view that women are less intelligent and less capable than men, that the activities of men (and not those of women) are at the heart of the explanation of the development of civilizations and culture, and that there are racial differences in intellectual abilities. On the view I am urging, once these and other views are seen to be value laden they will not be for that reason expelled from our going world theory (and thus exempted from the demand for evidence and justification), but scrutinized and adjusted as the (holistic) evidence warrants.

There are, I think, two ways in which we might see the science/value distinction given the broader holism I am urging. Both involve including value and value laden beliefs within our holistic world view. The first would be analogous to the way Quine thinks the work of the analytic part of the analytic/synthetic distinction is diffused throughout science. Recall that the analytic/synthetic distinction presupposed two grounds for truth, linguistic convention and correspondence to the world. In abandoning the distinction we do not abandon the view that world and language both contribute to the truth of our claims:

> Taken collectively, science has its double dependence upon language and experience; but this duality is not significantly traceable into the statements of science taken one by one (Quine, 1963, p. 42).

And in 'Five Milestones of Empiricism' Quine says:

> Holism blurs the supposed contrast between the synthetic sentence, with its empirical content, and the analytic sentence, with its null content. The organizing role that was supposedly the role of analytic sentences is now seen as shared by sentences generally, and the empirical content that was supposedly peculiar to synthetic sentences is now seen as diffuse through the system (Quine, 1981a, pp. 71–72).

Analogously, perhaps, we might see views about *what is* and *what is of value* as jointly contributing to our understanding of the world, but with these contributions also not being "traceable into the statements" of our world view taken one by one. On this suggestion we would see our valuational concerns as "shared by sentences generally", and both factual concerns and value concerns as "diffused through the system".

The work of "analytic" sentences is spread across the linguistic net in the sense that the "meaning" of a term is given, not by a set of analytic truths containing that term, but by all the sentences containing that term, embedded as they are in a wider theory from which observation conditionals can be derived.[10] Similarly, the "meaning" of a value term is given by the totality of sentences in our theory in which that term occurs. The question is what the connection or overlap is between our commitment to sentences containing value terms and the rest of our world theory. Are all the observation conditionals derivable from our world theory also derivable when all sentences containing value terms are excluded? I do not know what the answer to this question is, in part because I am not sure we are clear about the distinction between "value term" and "non-value term". However, if pressed to supply evidence for the minimal utilitarianism to which I subscribe, I would be tempted to say that the evidence is diffused throughout the world view to which I subscribe, flows from the way that view pictures individuals and their place in the world and their relations to other individuals. If this is so, it suggests that the value content of utilitarianism is spread through at least large parts of our world theory.

The second way in which we might bring values within our holistic world theory is by simply extending that theory so as to include as disciplines within it moral theory and social and political theory. We would then come to see our world theory as a continuum, with values and value issues being largely if not entirely excluded from the physical sciences, having some substantive connection with the biological and biobehavioral sciences, more with the traditional social sciences, and of paramount importance in social and political theory and moral theory. This would retain something of the science/value distinction, but as a distinction drawn as a matter of degree, and within the holistic world theory. The overall goal would remain that of delineating, explaining, and predicting experience, the firings of our sensory receptors, and our reactions thereto. The degree to which theories and parts of theories are removed from the experiential edge of this web will, of course, vary, and it is not clear that all parts of the physical sciences, e.g., high energy physics, will be closer to the experiential edge than all parts having a clear value component, e.g., social theory.

Does the foregoing view place science and values on the same footing, obviating any need to disentangle the two wherever they are intertwined? Does it address Quine's concern for empirical adequacy, for the requirement that to be part of a serious theory of nature beliefs must be constituents of some block of theory that does imply observation conditionals? Is it enough that the world theory as a whole does imply observation conditionals? We might note here that pure logic and number theory do not themselves imply observation conditionals. They are, however, infused in chunks of our world theory, e.g., physics, that do imply observation conditionals. Are value claims similarly infused in parts of social and political theory? Might it be the case that values and value theory serve to enhance the explanatory power of our broader biological, biobehavioral, and social theories, not by implying observation conditionals, but by shaping the way in which we look at the world, by highlighting certain aspects of experience. Might not value claims and value theory, including a commitment to a minimal utilitarianism, help us make sense of and integrate the firings of our sensory receptors, past, present, and future, and *our reactions thereto.*[11] For example, might value-laden claims help us make sense of our and others' reactions to natural and human-caused disasters? Does positing that human suffering is evil perhaps make our reactions to suffering more comprehensible than it would otherwise be?

I am not sure what the answers to the foregoing questions are. Nor do I think answering those questions is as important as is recognizing that values and value theory are not everywhere a matter of personal taste or preference, that a minimal commitment to utilitarianism and a recognition of the interconnections among the biological, the biobehavioral, and the social sciences, and value theory yields a more viable and valuable world theory than does any sharp science/values distinction that excludes values and value theory from the realm of evidence and evaluation.

For example, the work of feminist science critics of the last 20 years suggests that the history of science is replete with examples of value-laden beliefs, social, political, and moral, influencing what research project is undertaken and what hypotheses entertained. If this is so, an alternative to striving even harder to exclude those beliefs would be to work harder to make sure such beliefs are reasonable beliefs, are congruent with our collective wider network of beliefs. This will require dropping the assumption that in all value-laden areas agreement is not to be expected, that wildly differing views are to be tolerated, if not encouraged, but excluded from science. Given a commitment to a minimal utilitarianism, it does not seem crazy to think we might obtain agreement about such issues as social and political policies, as well as about standard moral issues.

The attempt to isolate science from values can be traced back a long way, and has in some clear ways contributed to the advancement of science. One can, for example, read Descartes' *Meditations* as a subversive tract designed to show that god is unnecessary (the epistemic arguments of the *Meditations* work as well without any theistic assumptions) in ways that clerics would not understand, and, more to the point here, to get the Church off the back of science. That is, one can read Descartes as providing a pretext for the Church's giving free reign to science, since the latter deals only with the material world, a world, as Descartes showed, of little or no importance compared to the world of the mental.[12] One can also see Descartes' gambit as having worked. Science prospered. But this gambit, which might be termed "Descartes' Compromise", had a price: it left religion and ethics within the domain of the Church and, wittingly or not, laid the groundwork for a strong science/value distinction and the ensuing view that there is no settling issues involving values (except perhaps by appeal to religion).

Had this dichotomy not been established, whether by Descartes or the larger intellectual tradition, science and the industrial revolution might not have progressed as quickly. On the other hand, had it not been put in place we might now have a more sophisticated view of value issues and of the interplay of values and what became traditional science. While allowing that "anything goes" in the areas of values and religion mitigates against holy and ideological wars (if anything goes, then whatever my views are, they are not worth fighting and dying over), it also mitigates against the development of a broader perspective in which it becomes apparent how "facts" and "values" intertwine.

HOLISM WITH DISTINCTIONS

Without assuming that the issue of the coherence and usefulness of the science/value distinction has been settled, I do want to explore the consequences of a science that, minimally, makes less of that distinction. In no sense would it follow from such a down-playing of the science/value distinction that "anything goes" in theory construction and testing, be it in evolutionary biology, social theory, or theories of the hormonal determination of sex-linked differences in cognitive ability. There are constraints on all, constraints traceable to our best going theories of the world and its contents, and to experience, upon which those theories collectively impinge. If well done, none of our theories will be arbitrary or capricious. Accordingly, it would be a mistake to conclude that if we allow both science and values into our going world theory we will, as a result, everywhere mix values and science and that as a result everyone will be free to bring her or his own preferred values into science.

At any given point in the history of research our world theory will be a collection of more modest theories, some well advanced, others in their infancy, still others ripe for abandonment, and together almost certainly inconsistent. There are, therefore, practical limits to applying the results of one field to another, even where they are related. For example, E. O. Wilson's own research, limited as it has been in the main to ants and termites, seems obviously not at a stage where it can serve as a model for Sociobiology applied to humans, even if Sociobiology ultimately proves to be a viable field. The second reality is that researchers are often not broadly educated, their expertise being limited to the narrowly defined segment of the research program for which they are responsible. Finally, researchers holding almost every imaginable combination of political, religious, ethical, and social views do participate in science.

Traditional science accommodates these realities by denying holism – not all of our seriously held beliefs are relevant to all other beliefs, and by denying that values are relevant to any part of science. One advantage of so doing is that it makes the qualifications for doing science very modest indeed. When taught the scientific method and accepted research procedures (which has generally come to mean the method and procedures of one rather narrowly defined field), those of all political, religious, ethical, and aesthetic persuasions, no matter how ignorant of virtually everything except what is narrowly involved in the research project in question, can and do pursue science. The fact that researchers believe the silliest things, and engage in the silliest, or most heinous, practices outside their laboratories, or are woefully ignorant outside their own field, is taken to be no indicator whatsoever of the quality of their scientific research. All traditional science asks is that when doing science researchers neither appeal to nor let their research be influenced by their non-science views and opinions.

To what extent can, and should, a holism that allows values within our world theory seek to accommodate the above realities of scientific practice? As a first step, we must follow Quine in conceding that a more manageable holism, a holism that most often works in terms of chunks of particular theories of particular disciplines, will generally suffice for the *bench scientist*. The theorist will have to deal with larger chunks or whole disciplines, and sometimes, when disciplines are in conflict, with our entire world theory. But the latter will be a very rare exception, not the rule. We can have, that is to say, holism *with distinctions*. We can allow for differential progress across the holistic world theory and can demand far less of bench researchers than a mastery of all of science. In the present taxonomy of disciplines, physics is not political theory, biology is not psychology, and population genetics is not social theory, though in each case the first may have important implications for the second, and in the last case at least, the second for the first.

At the same time we can hope that a broader holism of the sort I am advocating will bring with it a greater willingness to reexamine and redraw disciplinary boundaries as needed and to recognize connections between disciplines, perhaps thereby lessening the disparity in rates of progress among related disciplines, including areas of value theory. Disciplinary boundaries are, after all, of our own making. They evolve as our ongoing theories evolve. They are artificial but not imaginary.[13]

The acceptance of holism may also encourage individuals and society to take research in moral theory and in social and political theory more seriously. This, in turn, might mitigate against both individuals and groups holding and espousing ill-thought-out views about all manner of topics, including race, gender, and class. We can grant individuals the right to hold whatever views they want, while insisting that they are responsible for the views they hold. While we cannot expect individuals to examine, *à la* Descartes, every view they hold, we can expect them to examine views that are called into question by new or reconsidered experience or by other views they or others hold. And holism will reveal that there are more connections between seemingly disparate views than previously thought, e.g., connections between views about race, class, and gender and specific research projects concerning cognitive abilities and sexual preferences.

Suppose we do, as individuals and as communities, take values more seriously and countenance the kinds of evidence holism affords for and against value and value-laden claims. When this is so, should researchers be debarred from doing research when their views on value-infused issues are not those of the larger research community or society at large? No, of course not. But the larger research community and society at large will appropriately view the results of the research in question as in part a function of views and assumptions that they do not share with the researcher. Should funding agencies refuse to fund researchers whose value-infused views about the proposed research are other than those of the larger research community? Perhaps, it depends on how divergent the views are, whether the researcher can give good (even if not compelling) arguments for those views, and the availability of funding and nature of the research project. The openness of science allows for a renewed defense of the Ptolemaic system. But most of us have a very hard time imagining the NSF funding a research proposal to do so.

To take an extreme case, can a researcher who believes that both euthanasia and infanticide should have a place within socially sanctioned practices usefully and appropriately engage in social policy development? There is certainly no reason to prevent such a person from proposing a social policy that sanctions both euthanasia and infanticide. The proposal will be evaluated on its merits by the larger community. Is there a reason to fund such research? Perhaps, though the reality of the current political climate is surely that it will not be funded. Can such a researcher reasonably work on the development of a social policy as part of a larger research group, a group that may have public funding? I think the answer is 'yes'. If the working assumptions of the research group include the view that infanticide and euthanasia are not to be tolerated, then our more radical researcher will either have to challenge those assumptions or agree to work within their confines. That is, if our researcher wants to work with the group, without challenging, or after unsuccessfully challenging, their assumptions, then she or he will have to leave all views about the merits of infanticide and euthanasia at the laboratory door. And this may be a perfectly reasonable accommodation, for there may be large areas of social policy where our researcher and the larger group are in agreement.

In general, then, while the inclusion of values and value theory within the web of seriously held beliefs challenges the view that values have nothing to do with research, it remains appropriate to require that researchers who are engaged in research whose assumptions are at odds with their own views must either challenge those assumptions or set their own conflicting views aside while engaged in that research.

We must also remember that the value-infused views and assumptions that influence research are not always consciously held or made. They are equally or more frequently non-consciously held biases, prejudices, or general but unwarranted or at least untested assumptions about, e.g., race, class, and gender. They also include naive views about the nature of science, about the theory/observation distinction, about naive induction and theory formation, and about the relation of science to society. They include, finally, unspoken assumptions about the privileged status of the experience and views of white middle class males and of women and minorities who have been made into surrogates for white middle class males. There is no panacea for this problem. We can only work to make individuals more conscious of their own views, more aware that there may be connections between those views and what previously seemed to be wholly unrelated views. Making values and value theory a legitimate area of investigation, where rational discussion, argumentation, and connections with various of the traditional sciences are expected, is a good step in this direction.

The holism I am advocating also suggests we need to reconceptualize the way we train scientists. A reasonable (and not a rationalized) knowledge of the history of science should be required of all science majors. Ph.D. programs should place more emphasis on understanding the broad field and its presuppositions, including its relation to society and social issues, and less on purely technical skills. While mastering those technical skills allows Ph.D. students to replace traditional laboratory assistants in the research programs of their mentors (frequently without understanding the significance of those programs), it too often prepares those students only for specialized laboratory work, the demand for which has often disappeared by the time they graduate.[14] We need, in short, a better and more broadly educated, as well as a more diverse, research community.[15] We need to challenge the misconception that "anything goes" in the realm of religious, social, political, and personal views without inciting holy or ideological wars and while preserving an open society. We need to allow members of society to believe the silliest of things, but we need not deny that they are silly.

Will we ever, however we reform science and research, be in a position to be sure that some unseen prejudice or unwarranted assumption has not biased the results of a particular piece or whole area of research? The answer here is clearly 'no'. We can be more careful, we can make the science community more diverse, we can urge that problems and research agendas be viewed from various perspectives, but we cannot know that nothing has biased our research. We can only do our best and accept our results tentatively. But what else could we ask for?

POSTSCRIPT

One commentator has suggested, in less bald terms than these, that my thesis is muddled. When unmuddled it comes either to the view that value-laden beliefs are, at times, actually nodes in the theories and reasoning that lead ultimately to observation conditionals – in which case it is false, or to the view that values do sometimes influence the formation of research agendas and of perspectives taken, the range of experiments deemed relevant, and the use to which research results are put, but *not* the actual findings of science, in which case it is obvious (and uninteresting). In many ways, I am content to accept the interpretation that makes my position obvious and uninteresting, for I think that, in the end, the business of philosophy is in large part to make the obvious obvious. But I also think that what comes to be seen as obvious is not always uninteresting and unimportant. If we can establish the principle that not anything goes in the realm of values and value theory, then the influence values have on science, on the selection and delineation of research agendas, on the range of experiments deemed relevant, and on the use that is made of the findings, will become less pernicious.

It also seems to be the case that when one explores actual research programs that have been influenced by values, it becomes very hard to decide whether the best description of that influence is that of value-laden views that are clearly separate or separable from science influencing and shaping research agenda, or that of value-laden views becoming nodes in the theory and reasoning within which the research takes place. Consider, for example, research in neuroendocrinology in fetal development. The work of biologist Anne Fausto-Sterling and others makes it clear that this research presumes that it is only male development that is to be explained, and that it is only male hormones that are seen as having "organizing effects" on development (Fausto-Sterling, 1991). Fausto-Sterling notes that textbook accounts of "fetal sexual development" are accounts of male sexual development. One such text includes the assertion that it is "the natural tendency of the body to develop along female lines in the absence of other modifying influences" (Carlson, 1981 as quoted in Fausto-Sterling, 1991). As Fausto-Sterling notes, the question "How does [the female direction of development] *happen*, what are its mechanisms?" is, by and large, not asked. So too, it is only the effects of androgens, not those of estrogens, that have been widely studied, including their effects on the morphology of rat brains and subsequent changes in maze-negotiating skills (with some taking the results to be relevant to an explanation of why males are purportedly better than females at science and mathematics). This assumption of what needs to be explained and what the explanatory mechanisms might be is clearly value-laden and seems to be central to the research conducted. But it is not clear that the research in question is bad research; in fact Fausto-Sterling takes it to be commensurate with and supported by accepted theories (see also Nelson and Nelson, 1995).

Feminists have seen a broader assumption at work, the assumption that it is males and male activity that are the appropriate focus of explanations in the social and biobehavioral sciences, as shaping and limiting, if not distorting, work not only in neuroendocrinology but also in evolutionary theory, primatology, and anthropology. Whether the narrow assumption about what is to be explained in fetal development, and the broader assumption about the centrality of male and maleness are seen as value-laden cultural beliefs that have influenced science or nodes within the theories of science matters less, as I suggested in my opening remarks, than does challenging those assumptions and modifying them as the evidence warrants, and this requires serious discussion and evaluation, not relegation to somewhere outside our world theory where evidence is of no consequence.

ACKNOWLEDGEMENTS

I thank Steve Cohen, Michael Shepanski, and Richmond Campbell for helpful comments on several earlier versions of this paper. I thank the members of the Institutionen för filosofi och vetenskapsteori at the University of Umeä, and in particular Sten Lindström, for enduring the reading of a much longer and denser version of this paper, and for convincing me that major changes were required. Finally, I thank Lynn Hankinson Nelson for valuable comments on many versions, and the moose and loons of Maine, who provided welcome interruptions to two summers of work on this paper.

Temple University

NOTES

[1] There are two plausible claimants to this title, epistemic individualism, which has already been undermined by Lynn Hankinson Nelson and others, and the science/value distinction. This is a paper about the latter.

[2] The view just expressed is, however, controversial. It is not held, for example, by Sheldon Glashow, who made the following remarks at a Nobel conference in 1989:

> We [scientists] believe that the world is knowable, that there are simple rules governing the behavior of matter and the evolution of the universe. We affirm that there are eternal, objective, extrahistorical, socially neutral, external and universal truths, and that the assemblage of these truths is what we call physical science. Natural laws can be discovered that are universal, invariable, inviolate, genderless and verifiable.
>
> They may be found by men or by women or by mixed collaborations of any obscene proportions. Any intelligent alien anywhere would have come upon the same logical system as we have to explain the structure of protons and the nature of supernovae.
>
> This statement I cannot prove. This statement I cannot justify. This is my faith (Glashow, 1989).

[3] I include as post-positivist philosophers of science Carnap, Hempel, Nagel, Popper and others who recognized the failure of logical positivism but still sought to retain the analytic/synthetic distinction and the theory/observation distinction.
[4] Quine does exempt occasion sentences viewed holophrastically from the charge of being theory-laden (see Quine, 1990).
[5] In one sense, of course, whether there is a science/value distinction is an issue that cannot arise at all for Quine, for his most serious ontology presumably includes neither facts nor values. But there are sentences that include value terms, and whether the evidence that bears on these sentences is substantially different from the evidence that bears on sentences not containing value terms is an issue that can and does arise for Quine.
[6] In fact, I do not think Quine is a reductionist. That is, I do not think he holds the view that the relation between our most serious theory, whatever that turns out to be, and the multiple theories we use in various contexts, is one of reduction. To the extent and for the period that it suits our purposes, we may talk within a theory that countenances Tom Sawyer, Lady Macbeth, and Calvin and Hobbes, not as fictional entities with some kind of funny ontological status, but as straightforward members of our current theory's domain. When we move or retreat to a theory whose domain includes only entities to whose existence we are more seriously committed, the move is not one of reduction, but rather one of shifting from one theory to another. So too, it may be that at some point Quine shifts from a theory that includes people, table, chairs, and other middle sized bits of dry goods to one that includes only chunks of space time. At that point, at the point where we give up talk of people, it may be that we also give up talk of values.
[7] Michael Shepanski has pointed out that non-value claims are readily derivable from value claims. His example is this: "All blue-eyed people are evil" and "Jones is not evil" (both value claims) logically imply "Jones is not blue-eyed", a non-value claim. Shepanski also suggests, and I think rightly, that the view Quine holds or would agree to is that "evaluative claims generally never 'add' to the total observable consequences of our theory", that is, whatever observable claim, or even observation conditional, we derive from value claims we will always be in a position to derive without those value claims.
[8] To the extent that Quine has a theory of meaning it might be termed the disappearance theory, for in the end meaning does disappear, except in the sense that the meaning of holophrastically construed observation conditionals can be specified, in terms of their verification conditions, and in the sense that we can still say, e.g., that 'Maine' denotes Maine.
[9] One suspects that the phrase should be 'observation conditionals' – Quine's knowledge of Italian is almost certainly impeccable, but perhaps Borradori's knowledge of English is not.
[10] I am here indebted to Michael Shepanski.
[11] Quine will note that our reactions to our experiences, pain, pleasure, horror, approval ... will vary more than will our acceptance or rejection of observation sentences. That this is so will matter less if we adopt a community-based rather than an individual-based epistemology (see Nelson, 1990).
[12] I am aware that Cartesian scholars will see this reading of the *Meditations* as, at best, attributing to Descartes something that he might have been, but almost certainly wasn't, about.
[13] "Boundaries between disciplines are useful for deans and librarians, but let us not overestimate them – the boundaries. When we abstract from them, we see all of science – physics, biology, economics, mathematics, logic, and the rest – as a single sprawling system loosely connected in some portions but disconnected nowhere" (Quine, 1966, p. 56).
[14] Jules Lapidus, President of the Council of Graduate Schools, likes to remind us that there was a time when we hired faculty to teach our undergraduates and train our graduate students. Now we recruit graduate students to teach our undergraduates and assist faculty with their research.
[15] *On Being a Scientist*, a publication intended for, among others, new graduate students in the sciences, reflects a similar concern about the training of scientists. (Committee on Science, Engineering, and Public Policy, NAS, NAE, IOM, 1995.)

REFERENCES

Borradori, G.: 1994, *The American Philosopher: Conversations with Quine, Davidson, Putnam, Nozick, Danto, Rorty, Cavell, MacIntyre, and Kuhn*. University of Chicago Press, Chicago and London.

Carlson, Bruce M.: 1981, *Patten's Foundations of Embryology*. McGraw-Hill, New York.
Committee on Science, Engineering, and Public Policy: 1995, *On Being a Scientist*. National Academy Press, Washington, DC.
Fausto-Sterling, A.: 1991, 'Society Writes Biology/Biology Constructs Gender', *Daedalus*, 61–76.
Glashow, S.: 1989, Remarks at a Nobel conference, as reported in *The New York Times*, 10/22/89.
Haack, S.: 1993, 'Epistemological Reflections of an Old Feminist', *Reason Papers*, **18**, 31–43.
Harding, Sandra: 1986, *The Science Question in Feminism*. Cornell University Press, Ithaca, NY.
Harding, Sandra: 1993, 'Rethinking Standpoint Epistemology: What is Strong Objectivity?', in L. Alcoff and E. Potter (eds), *Feminist Epistemologies*. Routledge, New York.
Longino, Helen E.: 1990, *Science as Social Knowledge*. Princeton University Press, Princeton, NJ.
Nelson, Lynn Hankinson: 1990, *Who Knows: From Quine to a Feminist Empiricism*. Temple University Press, Philadelphia, PA.
Nelson, L. H. and Nelson, J.: 1995, 'Feminist Values and Cognitive Virtues', in *PSA* Vol. II. PSA, East Lansing, Michigan, MI.
Quine, W. V.: 1960, *Word and Object*. Harvard University Press, Cambridge, MA.
Quine, W. V.: 1963, 'Two Dogmas of Empiricism', in *From a Logical Point of View* (revised edition). Harper & Row, New York (first printing, 1953).
Quine, W. V.: 1966, 'Necessary Truth', in *The Ways of Paradox*. Random House, New York.
Quine, W. V.: 1981a, 'Five Milestones of Empiricism', in *Theories and Things*. Harvard University Press, Cambridge, MA.
Quine, W. V.: 1981b, 'On Austin's Method', in *Theories and Things*. Harvard University Press, Cambridge, MA.
Quine, W. V.: 1981c, 'On the Nature of Moral Values', in *Theories and Things*. Harvard University Press, Cambridge, MA.
Quine, W. V.: 1981d, 'Things and Their Place in Theories', in *Theories and Things*. Harvard University Press, Cambridge, MA.
Quine, W. V.: 1990, *Pursuit of Truth*. Harvard University Press, Cambridge, MA.

SUSAN HAACK

SCIENCE AS SOCIAL? – YES AND NO[1]

Ours is an age in which partial truths are tirelessly transformed into total falsehoods and then acclaimed as revolutionary revelations [Thomas Szasz].[2]

INTRODUCTION

Some feminist philosophers of science claim the insight that science is social.

It is true that the co-operative and competitive engagement of many people, within and across generations, in the enterprise of scientific inquiry, contributes to its success. It is false that social values are inseparable from scientific inquiry; false that the purpose of science is the achievement of social goals; false that knowledge is nothing but the product of negotiation among members of the scientific community; false that knowledge, facts, reality are nothing more than social constructions; false that science should be more democratic; false that the physical sciences are subordinate to the social sciences.

I shall first offer an account of what is epistemologically distinctive about scientific inquiry in which the proposition that science is social, in its true interpretation, plays a significant part. Next, I shall argue that in its other interpretations – the inevitability or desirability of politicized inquiry, social constructivism, "democratic epistemology," and the rest – the proposition that science is social is false. And then, turning to the question, what, if anything, all this has to do with feminism, I shall argue that "science as social" is either a genuine insight, but not a feminist one, or else is no insight at all; and, in conclusion, that feminism has taken a wrong direction in pursuing the project of a feminist epistemology of science – a project that is neither sound epistemology nor sound feminism.

I

[T]he … causes of the triumph of modern science, the considerable numbers of workers and the singleness of heart with which – (we may forget that there are a few selfseekers …; they are so few) – they cast their whole life in the service of science lead, of course, to their unreserved discussion with one another, with each being fully informed about the work of his neighbour, and availing himself of that neighbour's results; and thus in storming the stronghold of truth one mounts upon the shoulders of another who has to ordinary apprehension failed, but has in truth succeeded in virtue of his failure [C. S. Peirce].[3]

A long tradition has taken science to enjoy a peculiar epistemic authority because of its uniquely rational and objective method of inquiry; a long history of failed attempts to articulate what that uniquely rational and objective method is, suggests a

L. H. Nelson and J. Nelson (eds), Feminism, Science, and the Philosophy of Science, 79–93.

need for some re-thinking of the presuppositions of this, as I shall call it, "Old Deferentialist" approach.

Radical critics of the Old Deferentialist picture[4] – the "New Cynics" – conclude that there *are no* objective epistemic standards and that there is *nothing* epistemologically special about science. My view of the matter is much less exciting. There are objective epistemic standards. As I argued in *Evidence and Inquiry*,[5] these standards are not internal to science; they are the standards by which we judge the worth of empirical evidence, and the rigor and thoroughness of empirical inquiry, generally. By those standards, science has succeeded astonishingly well. Science is not epistemologically privileged, but it is epistemologically distinguished.

It is important to distinguish standards for judging the worth of evidence from standards for the conduct of inquiry. The two are run together in much contemporary epistemology, and nearly inextricably confused in most recent philosophy of science; but, though related, they are as different as criteria for judging roses are from instructions for growing them. The former kind of question, though hard enough, is a bit more tractable than the latter. The goal of inquiry is to discover substantial, significant truths;[6] and, since there is a certain tension between the two aspects of the goal – it is a lot easier to get truths if one doesn't mind the truths one gets being trivial – there can be, at best, guidelines, not rules, for the conduct of inquiry. The explication of criteria for appraisal of the worth of evidence is a little more tractable, because focused on only one aspect of the goal, on truth-indicativeness.

The main focus in what follows will be on the harder kind of question, about the conduct of inquiry. What I have to say about this will, however, presuppose a conception of our criteria for appraisal of the worth of evidence according to which the best model of those criteria is not, as much recent epistemology has assumed, a mathematical proof, but a crossword puzzle. Experiential evidence is the analogue of the clues, background information of already-completed entries. How reasonable an entry in a crossword is depends on how well it is supported by the clue and any other already-completed, directly or indirectly intersecting, entries; how reasonable, independently of the entry in question, those other entries are; and how much of the crossword has been completed. An empirical proposition is more or less solid depending on how well it is supported by experiential evidence and background beliefs; how secure the relevant background beliefs are, independently of the proposition in question; and how much of the relevant evidence the evidence includes. How well evidence supports a proposition depends on how much the addition of the proposition in question improves its explanatory integration.[7]

No doubt most scientific theories are, at some stages of their career, no more than tenuously-supported speculations; and no doubt some get accepted, even entrenched, on flimsy evidence. Nevertheless, science has succeeded extraordinarily well, by and large, by our standards of empirical evidence. It has come up with deep, broad and explanatory theories which are well anchored in experiential evidence and which interlock surprisingly with one another. And nothing succeeds like success; as having plausibly filled in long, central entries greatly improves the prospects for completing other parts of the puzzle.

Science has done this, not in virtue of its possession of a uniquely rational method of inquiry, but because of the ways in which it has strengthened, deepened, and extended the method all of us use when we try to figure out some empirical question. There is no such thing as "the scientific method" in the narrow sense in which the phrase purportedly refers to a set of rules which can be guaranteed to produce true, or probably true, or progressively more nearly true, results. No mechanical procedure can avoid the need for discretion, good judgement – as is revealed by the Popperian shift from: make a bold conjecture, test it as severely as possible, and, if counter-evidence is found, abandon it and start again, to: make a bold conjecture, test it as severely as possible, and, if counter-evidence is found, don't give up too easily, but don't hang on to the original conjecture too long. In a broader, vaguer, sense, the sense in which the phrase refers to making conjectures, developing them, testing them, assessing the likelihood that they are true, certainly there is such a thing; but not only scientists, but also historians, detectives, investigative journalists, and the rest of us, use "the method of science" in this sense.

Nevertheless, there *is* something distinctive about inquiry in the sciences – or rather, a lot of things: systematic commitment to criticism and testing; experimental contrivance of every kind; systematic effort to isolate one variable at a time; instruments of observation from the microscope to the questionnaire; all the complex apparatus of statistical evaluation and mathematical modelling; *and* – the engagement, co-operative and competitive, of many persons, within and across generations, in the enterprise of scientific inquiry. The fact that science is, in this sense, social, is an important factor contributing to its epistemological distinction.

There are all kinds of tasks that get done better if several people are involved. But scientific work isn't much like shelling an enormous quantity of peas (which will get done quicker the more people are helping);[8] nor much like carrying a very heavy log (which can be done by several people but not by a single person). It is – of course! – more like doing an enormously complicated crossword puzzle. And the epistemological significance of the social character of science is, correspondingly, quite complex and subtle – no simple case of "many hands make light work." It depends, not purely and simply on the involvement of many people, but on the internal organization of science, as well as its external environment.

Scientific inquiry is forwarded by division of labor. Members of various sub-communities and sub-sub-communities of what philosophers of science sometimes refer to, by a considerable abstraction, as "the" scientific community, work on different problems. Members of each sub-community, in turn, work on different aspects of "their" problem. It is as if different sub-groups, and different persons within them, worked on different parts of a crossword puzzle. The result? – the benefits not only of specialization, but also, provided each individual and each sub-group has access, as needed, to the work of the others, of saving duplication of work in checking the consistency of their entries with other, distant but still obliquely interconnected, areas of the puzzle. Then again, some inquirers are better suited by taste and temperament for deep or broad theoretical speculation, some for precise and patient observation, some for devising complex instrumentation, some for elaborate statistical evaluation, and so forth (rather as if we had whizzes at anagrams, specialists in Shakespearean allusions, devotees of exotic place-names, and so on, working together on part of a crossword).

The social character of science also helps to compensate for individuals' weaknesses and idiosyncrasies. I doubt that criteria of better and worse evidence will yield a linear ordering, and I am sure that no mechanical decision-procedure for theory-choice is to be anticipated. But a community of inquirers will usually, and usefully, include some who are quick to start speculating towards a new theory when the evidence begins to disfavor the old one, and others who are more inclined patiently to try to modify the old. And, though real, imperfect inquirers are seldom, if ever, altogether free of prejudice and partisanship, a community of inquirers will usually, and usefully, include partisans of one approach keen to seek out and expose the weaknesses which partisans of a rival approach are motivated to neglect.

Implicit in all this has been an important distinction between *warrant* and *acceptance*. Warrant is a normative notion; the warrant-status of a proposition is a matter of how good or bad the evidence with respect to that proposition is. Since any warranted empirical proposition will be anchored in part by experience, and since it is individuals who have experience, the warrant-status of a proposition in a community will depend in part on how justified the members' confidence is in each others' reports of observations.[9] Acceptance is a descriptive notion; the acceptance-status of a proposition is a matter of the standing of the claim in the eyes of the scientific community or relevant sub-community: rejected as definitely false; regarded as a possible maybe worthy of further investigation; as a reasonable candidate among several rivals; as probable but not yet acceptable as definitely true; as established unless and until something unexpected turns up; and so on. Ideally, the acceptance-status of a claim will vary concomitantly with its warrant-status. What was suggested rather vaguely above may now be restated a bit more precisely: though individual scientists will likely fall short of the ideal of proportioning their belief to the degree of evidence, in a community of scientists with various prejudices and preconceptions and varying tendencies to over-belief and to under-belief,[10] acceptance and warrant may nevertheless come to be, more or less and by and large, appropriately correlated.

Also implicit above, and also worth making explicit, is the thought that science is not only co-operative but also competitive – in virtue of competition between partisans of rival approaches or theories, and of competition between rival individuals or research teams hoping to be the first to solve this or that problem.[11] And here is as good a place as any to mention how, besides combining co-operation with competition, science combines division of labour with overlap of competencies sufficient to permit justified mutual confidence,[12] and the institutionalized authority of well-warranted results with institutionalized critical scrutiny.

Thus far I have focused on how the social character of science contributes to its success. But both its internal organization and the environment in which scientific work is conducted may be more *or less* conducive to good, honest, thorough, scrupulous inquiry. When one thinks of potential hindrances, the dramatic disasters of Nazi and Soviet science come first to mind.[13] These represent extreme cases of how the politicization of science, by putting scientists under pressure to find evidence favouring a politically desired conclusion, rather than honestly to investigate what hypothesis is best warranted, impedes achievement of the goal of inquiry. Other potential hindrances include: pressure to solve problems which are perceived as socially urgent, rather than freedom to pursue those most susceptible of solution in the present state of knowledge; the necessity to spend large amounts of time and energy on obtaining resources, and to impress whatever body provides the funds, in due course, with one's success; dependence for resources on bodies with an interest in the research coming out this way or that, or in rivals' being denied access to the results; a volume of publications so large as to impede rather than assist communication; and so on. This list of potential hindrances does not particularly encourage complacency about the present condition of science.

The social and human sciences are, in the nature of the case, more susceptible to some of these potential hindrances than the physical sciences, because these are the sciences to which it falls to investigate human biology, psychology, society, and therefore the sciences most concerned with issues on which scientists are most likely to have preconceived ideas, and on which political feelings are most apt to be strong.

II

I must confess that I belong to that class of scallawags who purpose ... to look the truth in the face, whether doing so be conducive to the interests of society or not. Moreover, if I should ever tackle that excessively difficult question, "What is for the true interest of society?" I should feel that I stood in need of a great deal of help from the science of legitimate inference [C. S. Peirce].[14]

Most of the philosophers who have recently insisted on the importance of the social character of science give the claim that science is social one or another (or several) of various interpretations much more radical than the one I have been exploring. I see the social character of science as one of the factors that has contributed to its epistemic distinction; they see it as significantly undermining its epistemic pretensions. For them "science as social" is a key step on the way from the Old Deferentialism to the New Cynicism: to the conclusion that science is not a uniquely rational cognitive enterprise with a special claim to epistemic authority, nor even, as I believe, an imperfect but remarkably successful cognitive enterprise deserving of epistemic respect, but a politics-permeated social institution in urgent need of transformation by an infusion of progressive values.

The fundamental difference between the conception of "science as social" which I think correct, and the radical conceptions which have recently become fashionable, is that whereas mine sees the social character of science as one of the (of course, very fallible and imperfect) factors which help to *keep acceptance appropriately correlated with warrant*, they insist on "science as social" as a way of focusing on *acceptance at the expense of warrant*. In fact, quite a good way to get a grip on these various radical interpretations is to classify them according as they: play down warrant and accentuate acceptance; ignore warrant altogether and acknowledge only acceptance; or attempt to replace the notion of warrant by some socio-political ersatz.

Those who play down warrant and accentuate acceptance insist on the underdetermination of theory by evidence and the inextricability of non-evidential factors in theory-choice. Hence the first of the radical interpretations of "science as social" that I want to consider: that social values are inseparable from scientific inquiry.

What makes this seem plausible appears to be the following thought: evidence never *obliges* us to accept this claim rather than that, and we have to accept *something*, so acceptance is always affected by something besides the evidence,[15] which had better be good, progressive values rather than bad, regressive ones. But we *don't* "have to accept something"; if the evidence is inadequate, why not just acknowledge that we don't know? Not all scientific claims are either accepted as definitely true or rejected as definitely false, nor should they be; evidence may be better or worse, warrant stronger or weaker, and the acceptance status of a claim can, and should, vary accordingly.

Some may feel that I have missed the force of the underdetermination thesis. "The point," they will object, "is not that, in practice, we don't always have enough evidence to decide whether a theory is true, but that, in principle, even all possible evidence is insufficient to decide, that there is always an incompatible, but empirically equivalent, theory." One might reasonably feel that, considering how much weight they are asking it to bear, those who appeal in this context to Quine's thesis of underdetermination-in-principle owe us something more than an appeal to authority – especially as Quine himself has never suggested that his thesis tends to encourage the politicization of science;[16] but set that aside. If the thesis is true, then, for those propositions theoretical enough to fall within its scope, no amount of observational evidence could enable us to tell whether p_1 or empirically-equivalent-but-incompatible p_2 is true. In that case, the most we could learn by inquiry is that either p_1 or p_2. It does not follow, and neither is it true, that we should decide which disjunct to accept by asking which would be politically preferable. Indeed, given that the thesis presumably applies only to the in-principle unobservable, it is not clear that it even makes sense to suppose that political values could have a bearing (is quark-theory or kwark-theory politically more progressive? – the question makes no sense).

But now, perhaps, it will be felt that I have missed the force of the insistence that "we have to accept something." "The point," it may be said, "is that we have to act, and so we have to accept some theory as the basis on which to act." This objection is easily answered by distinguishing between accepting a theory as true (which is the sense relevant to my argument), and deciding, without committing oneself to its truth, to act as if the theory were true (which is the sense in which it is sometimes true that "we have to accept something").[17]

A closely related interpretation of the theme of "science as social" is: that the goal of science is the improvement of society.[18] There is a temptation to respond by pointing to the OED's: "**research** – endeavour to discover facts"; but this would miss the point, which is, I take it, that the goal of science *ought to be* the improvement of society. Even taken as proposing only that science focus its attention on socially urgent problems, this is dubious. Knowledge is interconnected in unpredictable ways, so it is hard to be sure what research will bring social benefits; and focusing scientific effort artificially on problems perceived as socially urgent is apt to mean wasted resources, for the problems which we most want solved are not always those most susceptible to solution in the current state of knowledge. And taken as proposing that those scientific theories should be accepted, the acceptance of which conduces to the interests of society, this is simply a variant formulation of the interpretation of "science as social" just considered, that what scientific theories are accepted should be determined by social values; and is, likewise, untrue.

I turn next to the second kind of radical interpretation of the theme of "science as social," the kind favored by those who, ignoring warrant altogether, acknowledge only acceptance.

The favored phrase is that scientific knowledge is "socially constructed"; but this exploits an ambiguity. In one sense, it is true that scientific knowledge is socially constructed. Science has been the work of many persons, within and across generations; the scientific knowledge we now possess has been achieved, in part, through institutionalized mutual checking and criticism. This doesn't mean that it is communities, rather than individuals, that should be said to have knowledge. On the contrary, it is the knowledge possessed by individuals that is primary. Ordinarily, to say that a group of people knows that *p* is best interpreted simply as a way of saying that each member of the group knows it. There are, however, unusual circumstances in which it might be appropriate to say that a group of people knows that *p*, and not mean, simply, that each of them does. I am thinking of the kind of case where a group of scientists produces a report of results obtained by A's calculations, B's observations, C's instrumentation, etc.[19] *Perhaps*, then, one would appropriately say that the group knows, but not that each member does. A necessary condition of its being appropriate to say this would be that each member of the group be sufficiently justified in believing that each other member is sufficiently justified in believing the propositions that constitute their contribution to the joint result. The observations in section I about experiential evidence, and about mutually overlapping competencies, are apropos.

Often enough, of course, we say, on the basis of the work of some scientific subcommunity, "*we* now know that ..." – even though we who say this could not begin to articulate what the evidence is that warrants scientists' acceptance of the claim in question. By my lights, this is not, strictly speaking, a proper usage.

But these subtleties are not what is at issue with respect to the radical interpretation of the thesis that scientific knowledge is socially constructed. The sense at issue is: scientific knowledge is nothing more than the product of processes of social negotiation.[20] This is doubly false. First: the processes through which scientific knowledge is achieved are not *merely* a matter of social negotiation; they are processes of seeking out, checking, and assessing the weight of *evidence*. Second, not everything that has thus far survived those processes is knowledge; what survives those processes is what *counts as* knowledge, what is *accepted as* knowledge – but not all of it *is*, necessarily, knowledge. Some may, despite surviving those processes, not be warranted; some may turn out to be false.

Some hold not only that knowledge, but also that reality, is socially constructed, thus committing the same kinds of confusion twice over. Scientific theories are devised, articulated, developed, by scientists; theoretical concepts like *electron*, *gene*, *force*, and so forth, are, if you like, their construction. And the entities posited in true scientific theories are real. But it doesn't follow, and neither is it true, that electrons, genes, forces, etc., are constructed by the activity of the scientists who create the theories which posit them.

True, as science proceeds, instrumentation and theory get more and more inextricably intertwined, and one increasingly encounters claims which refer not to natural, but to what one might call laboratory phenomena. But that such phenomena are created in the laboratory does not mean that they are made real by scientists' theorizing.

True, again, social institutions (marriage, say, or banking) and social categories (gender, say, as distinct from sex) are, in a sense, socially constructed; if there weren't human societies, there would be no such things. These are the objects of sociological theories – not so incidentally, the kinds of theory with which social constructivists are most familiar. But, again, they are not made real by scientists' theorizing.

The third kind of radical interpretation of "science as social" engages in a kind of conceptual kidnapping, replacing the concept of warrant by an ersatz of a purely politico-sociological character.[21] This is the strategy of those who urge the merits of "democratic epistemology." Democracy is a political value, and would be apropos if theory-choice in science were a matter of "social negotiation" pure and simple. But it is not; it is a matter of seeking out, checking, and assessing the worth of evidence. Unless one is befogged by the emotional appeal of the word "democratic," it is clear that the idea is ludicrous that the question, say, what theory of sub-atomic particles should be accepted, should be put to a vote. Only those with appropriate expertise are competent to judge the worth of evidence.

True, freedom of thought and speech are important conditions for scientific inquiry to flourish; and it may be that some who favor "democratic epistemology" have confused the concept of democracy with the concept of freedom of thought. If so, the only reply needed is that these are distinct concepts.

True, if we are sociologists trying to understand the institution of polygamy in this society, or of slavery in that, then talking to wives *and* husbands, or to slaves *and* masters, might indeed be desirable as part of our evidence-gathering. But this is just one aspect of the requirement of comprehensiveness of evidence. It has not the slightest tendency to support the idea that democracy could replace warrant – supportiveness, independent security, and comprehensiveness of evidence – as epistemic values.[22]

What, finally, of the thesis that the physical sciences are subordinate to the social sciences? This would be a consequence of the claim that reality is socially constructed; if physical science were, as that claim has it, a kind of myth-making, then, indeed, anthropology would achieve a certain priority over physics. But it is a consequence so grossly implausible as to amount to a *reductio* – albeit a redundant one – of social constructivism.[23] In any event, I shall not linger over the "physics as subordinate to sociology" thesis, since it now appears for what I believe it is: a desperate last-ditch effort to save one or another of the radical interpretations of "science as social" by focusing attention on complications which, if not examined too closely, can give the false impression that in those radical interpretations it is true of the social sciences.

III

When we began theorizing our experience ... we knew our task would be a difficult though exciting one. But I doubt that in our wildest dreams we ever imagined we would have to reinvent both science and theorizing itself to make sense of women's social experience [Sandra Harding].[24]

What has "science as social" to do with feminism? Nothing. It is either a genuine insight but not a feminist one, or else is no insight at all.

Since self-styled feminist philosophers of science have generally not explicitly distinguished the various possible interpretations of "science as social" as I have done, they are apt sometimes to convey the impression that acknowledging "social epistemology," or recognizing that science is the work of a community, is in itself feminist (and even that taking an interest in the logical or the personal dimensions of scientific work betrays hostility to women's interests). But this is completely wrong-headed.

In the modest sense spelled out in section I it is true, and epistemologically significant, that science is social. But there is nothing particularly feminist about this. Peirce, Polanyi, Popper, Quine, come immediately to mind as philosophers neither female nor feminist who have acknowledged, with varying degrees of detail and subtlety, something along those lines.[25]

Some feminist philosophers have apparently been attracted to the idea that science is co-operative, and that scientists must trust each others' work, because they think it suggests the importance of the supposedly feminine virtues of trust and co-operation.[26] But this is triply mistaken. It rests on an old, sentimental stereotype of masculine and feminine qualities: a stereotype which is, incidentally, undermined by the fact that science, which has thus far been conducted largely by men, is, *inter alia*, a co-operative enterprise. It ignores the fact that science is competitive as well as co-operative, and that a realistic account of the social character of science will acknowledge the role of expertise, authority, justified confidence in others' competence, rather than, simply, "trust," and will see, besides co-operation, competition, productive rivalry. And it evades the question, why a theory of scientific knowledge stressing supposedly feminine qualities should be supposed to be a better – truer, more adequate – theory of scientific knowledge.

Most feminist philosophers of science, however, have been attracted to the theme of "science as social" in one and/or another of the radical interpretations discussed in section II. Longino and Nelson are committed to the thesis that social values are inextricable from science, both urging that the underdetermination of theories by data leaves "slack" to be taken up by political considerations.[27] Here the intended connection with feminism is clear enough: "doing science as a feminist," the thought is, requires one to ensure that it is feminist values that inform one's work. But, in the relevant interpretation, the thesis that science is social is false.

Harding is apparently committed to the thesis that "strong objectivity" is achieved by a democratic incorporation of "multiple standpoints," writes approvingly of the social constructivism of Latour *et al.*, and maintains that physics is subordinate to sociology – in all of which interpretations, the thesis that science is social is false. To make matters even more confusing, in Harding the connection with feminism is not achieved by any of these claims, but only by another, which appears, furthermore, to be incompatible with the democratic thrust of the emphasis on multiple standpoints; some standpoints, those of oppressed and disadvantaged classes, women among them, turn out to be, allegedly, epistemologically better. Distinguishing standpoints from perspectives, Harding tells us that what this means is that scientific work is best begun by "thinking from women's lives"; leaving me, I'm afraid, simply baffled by how work on quantum physics, say, *could* be undertaken as she recommends – let alone by why it *should* be.[28]

I think feminism has taken altogether the wrong path here. The problem started when feminist criticisms of sexism in scientific theorizing grew, as the quotation from Harding boasts, into something enormously more ambitious. Some of those criticisms, I think, were (are) correct. In the social and human sciences, theories about women's capacities, or incapacities, *have* sometimes come to be accepted by the relevant scientific sub-community when they were not well-warranted; and the explanation of how this came about *would*, probably, refer to the prejudices and stereotypes common among scientists as well as in the larger society. But this kind of detailed criticism of specific scientific work – which is, I should add, quite difficult, requiring competence in the relevant scientific specialty sufficient to judge the worth of evidence – has now been extrapolated in two exciting-seeming, but illegitimate, directions. The claim began to be heard (significantly, mainly from philosophers, sociologists and literary theorists, not from feminist physicists and chemists) that sexism infects *all* the sciences, the physical sciences included. I think this idea depends on misunderstandings about the role of metaphors in science; but I can't pursue those issues now.[29] It is the second illegitimate extrapolation that concerns me here: the inference that, since what has passed for relevant evidence, known fact, objective truth, and so forth, sometimes turns out to be no such thing, the notions of relevant evidence, known fact, established truth, etc., are revealed to be ideological humbug.

This inference is, of course, fallacious; but it is so ubiquitous that it deserves a name: I call it "the 'passes for' fallacy."[30] Its ubiquity is closely related to the astonishing outbreak of sneer quotes one finds in the literature of feminist philosophy of science ("knowledge," "truth," "reality," "objectivity," etc.), as well as with the pull towards accounts which accentuate acceptance, i.e., what is at a given time taken to be scientific knowledge, over warrant. It is worthy of note that the "passes for" fallacy is encouraged by the idea – which, to repeat, I reject – that epistemic standards are internal to science.

The "passes for" fallacy is ubiquitous; but I want to focus for a while on one, characteristic, instance. It occurs in a paper in the first half of which Ruth Bleier explains why she believes that the claim that there are differences in brain structure and function between the sexes which explain the (as she adds, the presumed) gender-related differences in cognitive ability, is not well-warranted by the evidence. She complains of the "sloppy methods, inconclusive findings and unwarranted interpretations," and the "unacknowledged ideological commitments" on which this supposed knowledge is based. But then, in the second part of the paper, what conclusion does she draw? Not that we need better investigation using rigorous methods, seeking more conclusive findings, based on warranted interpretations and free of ideological commitments – but: that bias is everywhere, that objectivity is impossible, that the "social production of knowledge" is inextricably conditioned by "gender, class, race, and ethnicity, and consequently, a set of values, beliefs, viewpoints, needs, and goals."[31] I shall not pause to protest the egregious assumption that one thinks with one's skin or one's sex organs. The point I want to stress here is that this form of argument, when applied to the concepts of evidence, truth, etc., is not only fallacious; it is also pragmatically self-undermining. For if the conclusion were true, the premiss could only be – as it alleges that the research it criticizes is – an expression of prejudice, as the conclusion takes all "inquiry" to be (here we really need the scare quotes; if you aren't trying to get the truth, you aren't really inquiring).

Furthermore, if the conclusion were true, it would also undermine the possibility of a science informed by feminist values, in which evidential slack was taken up by reference to women's interests. For if there were no genuine inquiry, no objective evidence, we could not know what theories are such that their being accepted would conduce to women's interests, nor what women's interests are.

The position I am articulating here should not be confused with what Harding calls "feminist empiricism," and characterizes as holding that sexism in scientific theorizing is simply bad science, and curable, therefore, by better adherence to the norms of science.[32] I do, indeed, think that sexism in scientific theorizing is often bad science;[33] and that it is curable by seeking out more, or paying closer attention to, evidence. But, first, I do not believe that epistemic standards are internal to science; so I do not think it appropriate to describe the cure simply as "better adherence to the norms of science." And, second, the reason I think sexism in scientific theorizing is often bad science is not that it is contrary to women's interests, but that it is not good, honest, thorough, inquiry; so I do not think it appropriate to describe my epistemological position as "feminist" *anything*.

The point is not that I don't think sexism in scientific theorizing is often bad science; I do. It is not that I don't care about justice for women; I do. It is not that I don't think there are legitimate feminist questions – ethical and political questions – about science; I do.[34] It is, rather, that I see the aspiration to a feminist epistemology of science as pulling towards the politicization of inquiry; which, by my lights, whether in the interests of good political values or bad, is always epistemologically unsound.

And it is no more sound feminism than it is sound epistemology. It would take another paper to spell out in detail why, in my opinion, what is presently conceived as "feminist philosophy of science" is contrary to women's interests; here I can offer only a few sketchy sentences. For generations, talented girls were discouraged from science because of ill-founded ideas about women's (in)abilities. Now there is a danger that talented girls will be discouraged by ill-founded ideas about the masculine or masculinist values with which science is allegedly imbued.[35] And there is a danger of a new kind of sexist science, this time a science supposedly informed by feminist values, which will reinforce the old stereotypes: the sexist science of such works as (heaven help us!) *Women's Ways of Knowing*.[36] Not to mention – but I feel I must – the waste of talent and energy if women interested in the epistemology of science come to feel that they must restrict themselves to approaches certified as "feminist,"[37] or be guilty of complicity with sexism.

Neither sound epistemology nor sound feminism requires us to "reinvent science and theorizing," as Harding's preposterous[38] announcement informed us; on the contrary, both require us to be on our guard against such "total falsehoods ... acclaimed as revolutionary revelations."

University of Miami

NOTES AND REFERENCES

[1] I would like to thank Paul Gross for helpful comments on a draft.

[2] *The Second Sin*, Anchor Books, Garden City, NY, 1974.

[3] *Collected Papers*, eds Hartshorne, C., Weiss, P. and Burks, A., Harvard University Press, Cambridge, MA, 1931–58, 7.51.

[4] In this paper I shall be characterizing as the Old Deferentialism versus the New Cynicism what I formerly called the "Old Romanticism" versus the New Cynicism ('Science "From a Feminist Perspective",' *Philosophy*, 1992, and reprinted in Halfpenny, P. and McMylor, P., *Positivist Sociology and its Critics*, Edward Elgar Press, Aldershot, UK, and Brookfield, VT, 1994; 'Epistemological Reflections of an Old Feminist,' *Reason Papers*, **18**, 1993, and reprinted, modified and abridged, under the title, 'Knowledge and Propaganda: Reflections of an Old Feminist,' in *Partisan Review*, Fall, 1993.) The earlier vocabulary, I now realize, was inappropriate, because, as Leo Marx puts it, "much of today's criticism of science ... may be traced to the ... *romantic* reaction of European intellectuals in the late eighteenth century" ('Reflections on the Neo-Romantic Critique of Science,' in *Limits of Scientific Inquiry*, eds Gerald Holton and Robert S. Morison, Norton, New York, 1978, p. 63; my emphasis).

[5] Susan Haack, *Evidence and Inquiry: Towards Reconstruction in Epistemology*, Blackwell, Oxford, 1993, Chapters 6, 7 and 8. The reader is also referred to my 'Puzzling Out Science,' *Academic*

Questions, **8.2**, Spring 1995, 20–31; to 'Towards a Sober Sociology of Science,' in *The Flight From Reason and Science*, eds Paul R. Gross and Norman Levitt, in *Annals of the New York Academy of Sciences*, **775**, 1996, 259–65, and forthcoming with Johns Hopkins University Press, Baltimore, MD; and to 'The Puzzle of "Scientific Method",' forthcoming in *Revue Internationale de Philosophie*.

[6] "Substantial," here, should not be interpreted as meaning simply, "synthetic." There are, in the sense intended, substantial mathematical truths, and trivial empirical ones. In this I follow Peirce; see *Collected Papers* 4.91: "those who [like myself] maintain that arithmetical truths are logically necessary" are "not *eo ipso* saying that they are verbal in their nature."

[7] I have given here a very brief summary of the much more detailed account to be found in my *Evidence and Inquiry*, Chapter 4. I note that this account acknowledges that there is such a thing as supportive-but-less-than-conclusive evidence, but does not require that there be a formalizable inductive logic.

[8] The analogy is due to Michael Polanyi, from 'The Republic of Science,' in *Knowing and Being*, ed. Marjorie Grene, University of Chicago Press, Chicago, IL, 1969, 49–62.

[9] At the time of writing *Evidence and Inquiry* I pointed out the difficulty of extrapolating the explication there offered of "A is more/less justified in believing that *p*" to the impersonal locution, "*p* is justified." The present paragraph offers an extrapolation to "p is warranted within community C" – I have chosen "warrant" instead of "justified" to mark the important differences between the concepts.

[10] A point explored in more detail in my '"The Ethics of Belief" Reconsidered,' forthcoming in Lewis Hahn, ed., *The Philosophy of Roderick M. Chisholm*, Open Court.

[11] David L. Hull, *Science as a Process*, Chicago University Press, Chicago, IL, and London, 1986, is illuminating on how science balances co-operation and competition.

[12]. See Donald T. Campbell, 'Ethnocentrism of Disciplines and the Fish-Scale model of Omniscience,' in Sherif, Muzafer and Carolyn W., eds, *Interdisciplinary Relationships in the Social Sciences*, Aldine, Chicago, IL, 1969, 328–48, for helpful discussion of the question of overlapping competencies.

[13] On Nazi science, see Alan Beyerchen, *Scientists Under Hitler: Politics in the Third Reich*, Yale University Press, New Haven, CT, 1977, and 'What We Now Know About Nazism and Science,' *Social Research*, **59**, 1992, 616–41. On Soviet Science, see Valery N. Soyfer, *Lysenko and the Tragedy of Soviet Science*, Rutgers University Press, Newark, NJ, 1994.

[14] *Collected Papers*, 8.143.

[15] These steps of this argument are made very explicitly by Mary Hesse in 'How to be a Post-Modernist Without Being a Feminist,' *The Monist*, **77.4**, October 1994, 445–61.

[16] Though the thesis of the underdetermination of theory by data is frequently referred to as "the Duhem-Quine thesis," the attribution to Duhem is a bit misleading; his thesis, that scientific claims are often not testable in isolation but only in conjunction with a bunch of other claims involved in reliance on instruments, is significantly more modest. Even Quine's commitment to the thesis is not unwavering; in 'Empirical Content' (*Theories and Things*, Belknap Press of Harvard University Press, Cambridge, MA, and London, 1981, 24–30) he suggests that what he formerly described as empirically equivalent but incompatible theories would really be verbal variants on one theory (pp. 29–30). This reveals the dependence of the underdetermination thesis on implicitly assumed criteria for the individuation of theories. Since his initial commitment to the underdetermination thesis, furthermore, Quine has shifted away from the unqualified holism of verification which motivated it, towards what he calls "moderate holism," i.e., towards something more like Duhem's position.

[17] This relates to a line of thought with which, like Dewey and Popper, I sympathize: that, since we can seldom be sure what the consequences will be, gradualism is the best policy in social and political change.

[18] This is the position of Karl Pearson, whose *The Grammar of Science*, Adams and Black, London, second edition, 1990, Peirce is criticizing in the passage quoted at the head of this section.

[19] For a vivid example, see 'How Many Scientists Does it Take to Screw in a Quark?,' *Newsweek*, 5.9.1994, 54–5, reporting how "it took 440 physicists from 34 countries ... 17 years" to discover the top quark (p. 54). For a more theoretical discussion of the phenomenon in question, see John Hardwig, 'Epistemic Dependence,' *Journal of Philosophy*, **LXXXII**, 1985, 335–49.

[20] See, for example, Bruno Latour, *Science in Action*, Harvard University Press, Cambridge, MA, 1987; Steven Fuller, *Social Epistemology*, Indiana University Press, Bloomington, IN, 1988.

[21] The strategy is illustrated in a particularly striking way by the terms in which, in *Whose Science? Whose Knowledge?* (Cornell University Press, Ithaca, NY, 1991) Sandra Harding discusses how to "justify" her theory, which unmistakably reveal that she has identified this with the question, how to *sell* her theory to this or that audience.

[22] Some of those who think science would be better if it were more democratic may have in mind, not the epistemological issues on which I have focused, but questions of access to science (or, more likely, they have run the two sorts of question together). Certainly it is desirable that no talented person be excluded from science on the grounds of race, sex, eye-color, or any irrelevant factor; but this thought is rather a meritocratic than a democratic one.

[23] Putting me in mind of C. I. Lewis's shrewd description of the method "which the bigot unconsciously applies": "he simply doesn't believe any evidence which is unfavorable to his bigoted conclusion; and if any such is put forward, he will argue it away by using this same method over again" (*The Ground and Nature of the Right*, Columbia University Press, New York, NY, 1955, p. 32).

[24] *The Science Question in Feminism*, Cornell University Press, Ithaca, NY, 1986, p. 251.

[25] So, in a minimal sort of way, did J. S. Mill, who qualifies as a feminist if any male philosopher does. But this obviously does not establish the required connection.

[26] See, for example, Lorraine Code, *Epistemic Responsibility*, University Press of New England, Hanover and London, 1987 (but note that, though she stresses "trust," and recognizes the co-operative character of science, she is disposed to play down scientific knowledge in favor of the literary). See also my critical notice of this book, *Canadian Journal of Philosophy*, 1990, 91–107.

[27] Helen Longino, 'Can There be a Feminist Science?,' in Ann Garry and Marilyn Pearsall, eds, *Women, Knowledge and Reality*, Allen Hyman, Boston, MA, 1989, and *Science as Social Knowledge*, Princeton University Press, Princeton, NJ, 1990; Lynn Hankinson Nelson, *Who Knows?: From Quine to a Feminist Empiricism*, Temple University Press, Philadelphia, PA, 1990. "Slack" is a term of which Nelson is fond; "doing science as a feminist" is a phrase Longino likes. I note that Longino tends to stress underdetermination *in practice*, whereas Nelson tends to take the Quinean line of underdetermination even *in principle*.

[28] Sandra Harding, *Whose Science? Whose Knowledge?*; 'After the Neutrality Ideal: Science, Politics and "Strong Objectivity",' *Social Research*, **59.3**, 1992, 567–87.

[29] See Paul R. Gross and Norman Levitt, *Higher Superstition*, Johns Hopkins University Press, Baltimore, MD, 1994, pp. 78 ff.; and, on the more general issue of the cognitive role of metaphor, my 'Dry Truth and Real Knowledge: Epistemologies of Metaphor and Metaphors of Epistemology,' in Jaakko Hintikka, ed., *Aspects of Metaphor*, Kluwer, Dordrecht, The Netherlands, 1994, 1–22.

[30] A term I introduced in 'Knowledge and Propaganda: Reflections of an Old Feminist.'

[31] '*Science* and the Construction of Meanings in the Neurosciences,' in Sue V. Rosser, ed., *Feminism Within the Science and Health Care Professions: Overcoming Resistance*, Pergamon Press, Oxford and New York, NY, 1988, 91–104; the quotations below are from pp. 92 and 100.

[32] *The Science Question in Feminism*, Chapter 6. I note that Nelson is not a feminist empiricist in Harding's sense, either; my position is too modest to qualify, hers too radical. Longino's criticism ('Science, Objectivity, and Feminist Values,' *Feminist Studies*, **14.3**, 1988, p. 571), that Harding's concept of "feminist empiricism" seems tendentiously designed so as to be a foil to the feminist standpoint epistemology Harding herself favors, is apropos.

[33] "Often," not "always," because I am using the term "sexist" to refer to claims which are *false* as well as such that their being accepted is contrary to women's interests; and because theories which are false may sometimes be sufficiently well-warranted that one would hesitate to say that their coming to be accepted is bad science. I stress that, in my usage, that a claim is offensive to some women is neither necessary nor sufficient for its counting as sexist.

[34] See section I of 'Science "From a Feminist Perspective",' but note that the "legitimate feminist questions about science" there discussed are all of a social, political, ethical cast – *not* epistemological.

[35] On this point, see Noretta Koertge, 'Are Feminists Alienating Women From the Sciences?,' *Chronicle of Higher Education*, 9.14.94, A80. As she observes, "What young women really need is special encouragement and equal opportunity to learn science, not a feminist rationalization for failure."

[36] Mary Field Belenky, Blythe Mcvicker Clinchy, Nancy Rule Goldberger and Jill Mattuck Tarule, *Women's Ways of Knowing*, Basic Books, New York, NY, 1986 – a remarkable work of sexist pseudo-science. As antidotes, I recommend Carol Tavris, *The Mismeasure of Woman*, Simon and Schuster, New York, NY and London, 1992, especially Chapter 2; and Martha T. Mednick, 'On the Politics of Psychological Constructs: Stop the Bandwagon, I Want to Get Off,' *American Psychologist*, **44**, 1118–23.

[37] Apropos, see Harriet Baber, 'The Market for Feminist Epistemology,' *The Monist*, **77.4**, October 1994, 403–23.

[38] "That is preposterous which puts the last first and the first last.... Valuing knowledge, we *pre-posterize* the idea and say ... everybody shall produce written research in order to live, and it shall be decreed a knowledge explosion" – Jacques Barzun, *The American University*, Harper and Row, New York, NY, Evanston, IL, and London, 1968, p. 221. See also Susan Haack, 'Preposterism and Its Consequences,' *Social Philosophy and Policy*, **13.2**, 1996, 296–315, and in *Scientific Innovation, Philosophy, and Public Policy*, eds Ellen Frankel Paul *et al.*, Cambridge University Press, Cambridge, 1996, 296–315.

LYNN HANKINSON NELSON

EMPIRICISM WITHOUT DOGMAS

1. TOWARDS A MORE NATURALIZED PHILOSOPHY OF SCIENCE

Philosophy is, it seems to me, often in danger of preferring the abstract to the concrete and of concentrating on methodology to the near exclusion of content. That said, methodology is important and what appropriate methodology is or should be in specific subfields of philosophy does not go without saying. In recent years I have addressed various issues in naturalized philosophy of science, a discipline distinguished from traditional epistemology both in its core research questions and its goals.[1] Here I turn to the methodology of naturalized philosophy of science. The methodology I propose is largely constituted by a theory of evidence that construes evidence holistically, and holism generously to include claims and theories informed by social beliefs and values. An implication of holism, construed generously or not, is that it is science communities rather than scientists *qua* individuals that are the appropriate loci of philosophical reconstructions and explanations of scientific practice. A broader implication of holism is that naturalized philosophy of science is a normative and not simply a descriptive enterprise. My case for construing evidence generously to include claims and theories informed by social beliefs and values builds on developments in feminist science scholarship and is illustrated through an extensive case study.

I use these several results to sketch the broad outlines of a social and empiricist naturalized philosophy of science, the core epistemic notion of which is evidential warrant. I conclude with an argument from the perspective of social empiricism to the effect that there are normative questions about the social processes characterizing science that should be pursued in naturalized philosophy of science: precisely those questions that are simultaneously questions about *the bodies of evidence* that support scientific theories and research programs.[2]

I have said that naturalized philosophy of science is not traditional epistemology. The point is worth repeating, for readers who superimpose the goals and core questions of traditional epistemology on the ensuing discussion are likely to miss its point. It is also worth reminding readers that the argument that the philosophy of science should be naturalized originates in empiricist philosophy of science, and specifically in arguments advanced by W. V. Quine for the view that the philosophy of science is continuous with science and should be pursued as such. The argument here advanced is that further naturalization – specifically, the abandonment of the traditional commitments to epistemological individualism and to a hard and fast boundary between good science and social beliefs and values – is called for.

My discussion presupposes several broad commitments. These include a realism that is naturalistic, i.e., a realism supported by the evidence supplied by science but

L. H. Nelson and J. Nelson (eds), Feminism, Science, and the Philosophy of Science, 95–119.

not transcending it. On this view, the evidence for the objects of science and of common sense is that theories that include them allow us to explain and predict much of what happens. On the same grounds, i.e., the evidence supplied by science, I take relativism to be implausible. As an empiricist, I assume that experience is the final arbiter of claims and theories, and successful explanation and prediction the end points of science. Empiricism is, of course, a theory of evidence and the epistemic notion central to all of the arguments I advance is evidential warrant – construed naturalistically.

The shifts in naturalized methodology I am advocating, from an individual-centered to a social empiricism and away from a firm boundary between good science and social beliefs and values, are partly a function of my views about the demands appropriately made of the methods employed in the philosophy of science. These views reflect converging arguments in naturalized and feminist philosophies of science to the effect that methodological principles in the philosophy of science should be adopted, revised, or abandoned on the basis of their evidential warrant. It is with these arguments that I begin.

The Import of Philosophical Methodology

The methodological principles employed in the philosophy of science carry substantial normative import, on two levels. Most obviously, such principles constitute norms for practitioners of the philosophy of science, prescribing and, by implication, circumscribing, the questions and considerations legitimately pursued and emphasized. These prescriptions and circumscriptions may be explicit – as is, for example, the "arationality principle" advocated by Larry Laudan in *Progress and Its Problems*, which maintains that explanatory principles incorporating social processes, beliefs, and values are to be utilized in theorizing about science only in cases in which the beliefs to be explained cannot be explained "in terms of their rational merits" (Laudan, 1977, 202). More often, such prescriptions and circumscriptions are implicit, as are those imposed by longstanding methodological commitments to epistemological individualism and to a boundary between good science and social beliefs and values. The norms prescribed by these commitments are no less consequential for being implicit, and have functioned as effectively as Laudan's principle to confine the values, beliefs, and social processes figuring in the explanatory principles employed in the philosophy of science to a relatively small set deemed to be those "constitutive of" science.[3]

By way of such prescriptions and circumscriptions, methodological principles substantively shape philosophical explanations of scientific practice and the criteria that function in philosophical evaluations of that practice. That is to say, philosophical methodologies have serious implications for our understandings of science. Feminist scientists and science scholars have identified a wide range of cases in which social processes, internal and external to science, and social beliefs and/or values have been at work in cases taken to constitute good science.[4] Given traditional methodological commitments, we must either revise our assumption that the cases in question do constitute good science, deny the findings of the scientists, historians, and philosophers who claim a substantive role for such factors, or boldly assert that even if such factors were at work in these cases, they are not *relevant* to our reconstructions and explanations of these cases, or to our understanding of why these are cases of good science.

The first two alternatives will not work for a wide range of cases, cases where values and/or social factors traditional principles did not allow for clearly are at work and the cases clearly are instances of good science. So it is the third alternative we must investigate. What support might be advanced for this position? One might appeal to *a priori* definitions of rationality according to which non-constitutive values and most social processes are compromising of or unnecessary to rationality.[5] But appeals to *a priori* definitions are suspect if not wholly beside the point in naturalized philosophy of science.

A more promising defense would invoke the claim that philosophers of science can successfully explain cases of good science *without* incorporating any role for such factors. This would constitute an empirical justification of a philosophical methodology, a justification that does warrant consideration. But it is a central thesis of my larger discussion, and a consistent theme in feminist philosophy of science, that the empirical results are otherwise.[6] I will argue that "explanations" and "reconstructions" of good science that do not recognize more kinds of social factors to be constitutive of science than traditional methodologies allowed for are empirically inadequate, and that there are many cases of good science that cannot be adequately explained without including a substantive role for social beliefs and values. Methodological principles that prescribe the exclusion of such factors generate incomplete explanations and, by ceding those cases that cannot be reconstructed using narrow construals of rationality to the sociology of knowledge, lead us to ignore the presence of good theorizing and sound research in cases of unsuccessful science. That is, such principles leave little room for distinguishing unsuccessful science from bad science.

It may now seem that the claim that methodological principles should be assessed on the basis of their empirical warrant is non-controversial. After all, a methodology that led to accounts of science obviously incommensurate with its history or current practice would be regarded by philosophers of science as suspect for that very reason. Witness the criticism that methodological commitments in the sociology of knowledge so minimize the role of the world in constraining scientific theorizing that science's predictive success comes to look like magic.

Still, it is likely that many will not agree that the warrant for philosophical methodologies is, in the end, empirical. Laudan's arationality principle, for example, is advanced in a work that advocates the meta-methodological principle that philosophers "identify and impose preferred pre-analytic intuitions about scientific rationality" (Laudan, 1977, 160). I doubt there are any such intuitions, but whatever the source of Laudan's principle, I take the vision of the philosophy of science as a discrete enterprise, a vision with roots in logical positivism, to have limited the attention paid to the empirical warrant or lack thereof for the methodologies adopted in the discipline.

I have noted that feminist philosophy of science constitutes one exception to this general tendency. Naturalized philosophy of science, to which we next turn, constitutes another such exception.[7]

Methodology Naturalized

The naturalistic tradition in the philosophy of science traces its roots to W. V. Quine's arguments that the philosophy of science is continuous with its subject matter and should be pursued as such.[8] I contend that the continuity Quine claimed and advocated is twofold.[9] First, that the philosophy of science should draw on relevant scientific research and theories (e.g., research in empirical psychology and/or social psychology) in constructing its theories of science, and abandon the pretense of providing an "extra-scientific" explanation of or justification for the sciences, and that its own methods and theories are to be judged by the standards, e.g., explanatory power and empirical adequacy, used to evaluate research and theories in the sciences proper. I claim, more controversially, that Quine's proposal implies a second continuity: that like the sciences, the philosophy of science is normative, using criteria such as empirical success to judge whether an episode of theory adoption in science is progressive, a decision to pursue a particular line of research warranted, and to judge whether the processes that characterize science are those likely to produce the best theories and research programs. Many view the acceptance of the first continuity as ruling out any normative function for the philosophy of science.[10] But naturalized philosophy of science as I have described it, takes the goal of both science and the philosophy of science to be the construction of theories that organize, explain, and predict experience, with efforts and results to be evaluated on the basis of how effectively they *do* these things, a claim for which my discussion as a whole will provide support.

Of relevance to the present topic, the criteria to be used in assessing philosophical methodologies, three criteria have evolved in naturalized philosophy of science and figure in recent discussions and debates within the discipline: (i) commensurability with the actual history and contemporary practice of science; (ii) grounding in scientific research and theories that carry implications for a theory of science (e.g., in empirical psychology, social psychology, cognitive science, evolutionary biology, and/or sociology); and (iii) consistency of methodological principles, i.e., a consistent approach to the role of social factors and values in episodes deemed progressive and less than progressive, and in periods of consensus and dissent.[11]

Note that each of these criteria would assess philosophical methods on the basis of their empirical warrant. The first builds on the assumption that the degree to which a philosophical methodology is naturalistic is a function of the degree to which it does not force or facilitate "excessively rationalized" reconstructions of scientific practice; the second prescribes that philosophical methodology be grounded in relevant empirical research. Those of us advocating consistency of methodological principles maintain that the recognition of social factors and values only in non-progressive episodes represents an un-naturalistic imposition of philosophical proclivities for viewing rationality in individualistic terms and values as inherently compromising of it (Solomon, 1994), and, for the reasons earlier outlined, leads to distortions in philosophical reconstructions and explanations of scientific practice.

I do not claim that these criteria constitute an algorithm or formula for assessing the methods employed in the philosophy of science. Indeed, the criterion of consistency, which I advocate, remains controversial. What I claim is that these criteria reflect a commitment to assessing philosophical methods on the basis of their empirical warrant. I also do not claim that the philosophy of science can avoid rational reconstruction altogether. Indeed, for reasons to emerge more fully in the discussion ahead, I do not believe we should try to avoid all reconstruction. What is at issue is *excessive* rationalization.

2. TWO METHODOLOGICAL PRINCIPLES

I have elsewhere advocated two methodological principles as elements of feminist naturalized philosophy of science.[12] I described the latter as a research program that starts from the naturalistic thesis that the philosophy of science is continuous with science, is subject to the criteria outlined above and contributes to their refinement, and takes the practices of feminist scientists and results in feminist science scholarship, resources largely untapped within the discipline, as among its primary resources. The importation of such resources, I argued, is commensurate with the naturalist thesis that the philosophy of science is continuous with science and with the view of both as evolving enterprises.

The methodological principles I advocated together constitute a social empiricism and I here argue for their adoption more broadly. I do so on the grounds that they meet the criteria for naturalized methodology outlined in the preceding section and constitute a further naturalization of the philosophy of science.[13]

A Social Empiricism

A widely recognized implication of four decades of research in the philosophy of science is that observation sentences, hypotheses, sentences of theories, and sentences expressing methodological or broad metaphysical commitments do not, in Quine's words, "face the tribunal of experience" individually, but do so as part of a larger body of current theory. This larger body is, for Quine and Duhem, in principle all of science;[14] for those who recognize a role for "auxiliary" statements and theories, some significant part of science; for Kuhn and sociologists of science, the disciplinary matrices which define a normal science tradition; and for semantic theorists, background assumptions which mediate the relationship between data and models.

A widely recognized implication of two decades of feminist science scholarship is that social beliefs and values concerning gender have informed various of the sciences: shaping the research questions identified and pursued, the design and interpretation of experiments, broad and specific ontological and methodological commitments, and theories generated. Significant disagreement remains, however, concerning the implications of such episodes, including whether they can or should be construed as "bad science" and/or as idiosyncratic, or reflect something of broader significance about science.

My first methodological principle is that

> The evidence supporting a specific theory, hypothesis, or research program is constituted by observation, itself largely structured as current theories would have it, and other theories that together constitute a current theory of nature, inclusive of those informed by social beliefs and values.

In other words, there are two kinds of evidence for individual theories, research projects, methodologies, and claims. One of these is, as traditional empiricism would have it, experience parsed in terms of observation sentences – though not, as traditional empiricism would have it, "pre-theoretic" experience or observation. The second is a body of accepted methods, standards, and theory – though not, as other versions of holism would have it, exclusive of those informed by social beliefs and values.[15] On this view, evidential warrant is constituted by both empirical success and integration within a body of accepted research, standards, and theories.

This account of evidence constitutes an alternative to models of evidential relations that construe evidence narrowly (e.g., as exhausted by "data" or observation sentences), relegating other theories and broad methodological and metaphysical commitments recognized to figure in the adjudication of an individual theory or hypothesis to a realm of "background" assumptions or "auxiliary" theories. It also constitutes an alternative to theories of evidence that presume a sharp distinction between observation sentences and theoretical sentences, as it maintains that observation and sentences reporting it are structured by bodies of current theory and standards. And it constitutes an alternative to theories of evidence that demand a hard and fast boundary between good science and social values and beliefs. It is significant because it provides an account of how, without these several distinctions, theories and research programs are constrained by their observational consequences.

I advanced two lines of argument for this account of evidence.[16] One builds on Quinean holism: the thesis that individual sentences, including observation sentences, have empirical content only as part of a body of theory.[17] This view of empirical content is the core of Quine's argument that a substantial body of theory is the unit of empirical significance: that is, that it is such bodies that yield observational conditionals and that figure in the adjudication of any particular hypothesis or theory. What I add to Quine's argument is that holism is best understood as taking the *evidence* for a specific claim or theory to include *both* the observational consequences of that claim or theory (together with the larger theory or theories within which it is embedded) *and* the relationship of the claim or theory in question to other accepted theories, methods, and standards.[18] I use a case study in the next section to demonstrate that modest bodies of theory function in practice to adjudicate individual hypotheses and theories, that there is more disunity in science than Quine's initial arguments for holism suggested, and that modest holism can be used to assess the evidential warrant of competing research programs and theories.[19]

I built on feminist science scholarship to maintain that the kinds of claim or theory that can constitute evidence include those informed by social beliefs and values – not just in episodes in which scientific practice fails to adhere to standards taken to be constitutive of good science, but when, with the obvious exception of "value-neutrality", such standards are met. The case study is representative of others on which feminist scientists have focused in demonstrating that social beliefs and values can do more than motivate research questions, the application of results, or the choice between theories in cases in which that choice is not dictated by available evidence. It demonstrates that such beliefs and values can be *integrated* in research questions, experiments, methodologies, and ontologies that meet general and discipline-specific standards for good science.

I have claimed that modest inclusive holism sustains the empiricist norms of explanatory power and empirical success, a claim the case study will demonstrate. But one question this account of evidence immediately raises is how broad a body of theory and results needs to be considered in the assessment of a specific research program or theory. My methodological principle leaves this question unanswered, speaking of "a current theory of nature". This reflects the full scope of the holistic claim and that demarcation will be case-specific. I use the case study to identify some of what should guide us in such demarcations, building in part on a second methodological principle:

> The appropriate loci of philosophical analyses of science are science communities, with the standards, theories, and practices of such communities the appropriate loci of philosophical explanations and evaluations of scientific practice.

This principle constitutes an alternative to methodologies that take scientists *qua* individuals as their loci. I contend that the epistemological primacy of science communities is an implication of a holistic account of evidence.

The theories and standards that this account of evidence recognizes as constituting part of the evidence for specific hypotheses and theories are not the "property" or achievement of the scientist who uses them, but the products of collective efforts to explain and predict experience, themselves undertaken within, and structured and constrained by, traditions of inquiry. Even construals of evidence that limit it to "data" or observation sentences cannot plausibly construe the "auxiliary" or "background" theories and standards that mediate the relationship of these to particular theories or models in individualistic terms.[20]

I claim more controversially that holism dictates that observation, the second component of evidence, is best parsed as social, and this constitutes my most significant break with traditional empiricist accounts of evidence. One consequence of a holistic account of empirical content – but, we should add, this is a consequence of the more general developments that led to the demise of logical positivism – is that there is no determinate list of sensory stimulations from which a particular claim is derived or derivable (Quine, 1960 and 1969).[21] The observations an individual or a community will countenance as evidence for a theory are themselves largely structured by a larger body of accepted theory. On this view, the experiential evidence we garner for our theories consists of observations of *features of the world* – features posited by the bodies of theory established and maintained by communities, experienced as communal standards of observation and the world as experienced ordain, and justified on the basis of their contribution to our ability to make sense of the world.

It is the social nature of both kinds of evidence that forms the core of my argument that philosophical reconstructions, explanations, and assessments of science should take science communities, rather than scientists *qua* individuals, as their primary loci.

The nature and implications of the shift in focus I recommend are perhaps most clear when contrasted with alternatives. One such alternative, advocated by Quine, would have us construe observation as the "triggerings of exteroceptors", evidence as observation so defined, and the subject matter of naturalized philosophy of science as the relationship between the triggerings of exteroceptors and "stimulatory situations" (Quine, 1990 and 1993). On this view, empirical psychology will serve as the primary resource for naturalized philosophy of science and individualism, of a fashion, will be preserved. A second alternative would have us take research and theories in cognitive science – some version of decision theory, for example – as primary resources for explanations of specific episodes of theory adoption and pursuit and/or for the construction of a general theory of science. This approach might also be understood to preserve individualism.[22] Cognitive processes and properties – decision-making procedures, for example, and "cognitive salience" – are typically attributed to individual scientists *qua* individuals.

The view I am advocating incorporates the naturalized account of the relation between "stimulatory situations" and the triggerings of exteroceptors as a part, and I assume that results in cognitive science will also constitute part of a general theory of science. But there is more for naturalized philosophy of science to do. We must explain theory generation and adoption, and the explanations we generate must have room for the role of bodies of accepted theory, science education, experiments and instruments, research traditions, peer review and funding mechanisms, and other features of the world. Such explanations of course presuppose some account of the triggerings of exteroceptors in response to stimulatory situations and some account of decision processes. But most of their content will be at the level of both the esoteric posits of science and the less esoteric posits of ordinary life. There is no reason to think that we could *start from* the triggerings of exteroceptors or decision processes and build to explanations of these posits and our theory building involving them.[23] There *is* reason to think that the kinds of explanation afforded by empirical psychology and cognitive science will only emerge apace with these higher level explanations. It is at this higher level, the level at which we engage in explanations and evaluations of episodes of *theory formation, adoption, and change*, I am arguing, that the appropriate loci of our analyses are science communities, for the reasons earlier cited and illustrated by the case study in the next section.[24]

The shift in focus to communities will mean that sciences and research programs that can provide insight into the social processes through and by which knowledge is generated, social psychology for example, will serve as our primary resources. The shift also brings a concomitant change in our normative questions, from those built on the assumption that the discernment and evaluation of evidential relations is in principle an 'individual' undertaking, to questions about whether the social processes that currently characterize scientific practice – those involving the recruitment or education of scientists, for example, peer-review mechanisms, and so on – are the processes that *should* be at work – are likely to produce the *best* theories and research programs – and, if not, what changes should be made.

As modest holism would have it, the parameters of the communities serving as the loci of philosophical analyses will be case specific, a function of shared or partially overlapping research questions and traditions, standards, theories, and methodologies. In some cases, the relevant community will be a narrowly defined scientific discipline; in others, a research community or tradition traversing more than one discipline or science; in still others, the effect one tradition or science has on another (that, for example, which an influx of physicists in the 1930s had on developments in American biology); and at times, two or more communities whose theories conflict. In the case study, I follow Richard Burian and Miriam Solomon in using institutional factors (the existence of disciplines, journals, departments, and professional associations), the core research questions of disciplines, research collaborations, published results, citations, and conferences to identify relevant communities and bodies of evidence (Burian, 1993 and Solomon, 1994).

Beyond the emphasis on the role of shared bodies of theory and standards, at times inclusive of those informed by social beliefs and values, I have left the designation of "social factors" imprecise. The open-endedness with which I treat the category reflects the view that *naturalized* philosophy of science must allow the details of individual episodes to dictate which, if any, of the many kinds of social factor cited in recent science scholarship were of import, in what ways, and to what degree, rather than adopt methodological principles that prejudge this question.

In their broad outlines, the changes I recommend in philosophical methodology – the shift in focus from individuals to communities and the abandonment of the commitment to a hard and fast boundary separating good science and social beliefs and values – are commensurate with other recent work in naturalized and feminist philosophy of science to develop a social empiricism (e.g., Longino, 1990 and Solomon, 1994). As I see things, a holistic and inclusive theory of evidence is the most *substantive* basis for social empiricism and a normative philosophy of science. I use the case study to which we now turn to demonstrate that when evidence is construed holistically, science communities are not the "closed systems" that radical social constructivists and sociologists of knowledge suggest, evidence is communal *and* substantive, and research programs and theories – including those in which social beliefs and values are integrated – can be assessed on the basis of their evidential warrant.

3. CASE STUDY

Organizer hypotheses emerged in endocrinology and empirical psychology, particularly in research concerned with sexual differentiation of the rat brain, and were subsequently extended into investigations of other species, including some primates. Most attribute what Robert Goy and Bruce McEwen describe as an "organizing effect" to androgens that mitigates what they call "the intrinsic tendency [of the fetus] … to develop according to a female pattern of body structure and behavior".[25] One effect that has been attributed to testosterone is right-hemisphere lateralization, and some organizer hypotheses posit testosterone as the cause of right-hemisphere dominance in human males and the latter as the biological foundation for sex-differentiated visuo-spatial and mathematical abilities. The formulation we consider is representative of this group and was advanced by Norman Geschwind and Peter Behan in 1982.[26]

In a study reporting correlations among left-handedness, immune-system disorders, and learning abilities, Geschwind and Behan proposed the effects of testosterone *in utero* as an explanation of "the biological foundations of laterality" – by which they meant right-hemisphere lateralization in human males (Geschwind and Behan, 1982, 5099). The evidence they cited for this hypothesis came from several sources. Their own study and others found left-handedness, immune-system disorders, and learning abilities to be more common in men and boys. Four additional studies reported asymmetries in human fetal brains. One, Chi *et al.* (1977), reported that two convolutions of the right hemisphere in areas linked to language develop several weeks earlier than corresponding convolutions of the left. Another, Ounsted and Taylor (1972), used results indicating that convolutions in areas of the brain associated with language were more common in boys in the first year of life as evidence of a sex difference caused by testosterone *in utero*, and related both to sex-differentiated lateralization.[27] A fifth study cited by Geschwind and Behan, Diamond *et al.* (1981), had reported that two areas of the cortex of male rat brains are 3 percent thicker on the right side than the left (an asymmetry not found in female rats) and had proposed that the thickness was caused by androgens and related to right-hemisphere lateralization. Diamond *et al.*'s evidence for this hypothesis included their ability to reverse the lack of asymmetry in female rat brains by removing the ovaries at birth; other studies reporting sex differences in asymmetries in weight, structure, and size of the hemispheres in rats and some primates; and studies in empirical psychology linking areas of the right cerebral cortex in male rodents to spatial integration and preference.

Building on these results and hypotheses, Geschwind and Behan proposed that the differential rate of development in human brains reported by Chi *et al.*, and others is caused by testosterone slowing the development of the left hemisphere, and further proposed that this slower development was the foundation of right-hemisphere lateralization. They did not link such lateralization to mathematical ability in the study;[28] but two years later they appealed to a study in empirical psychology that reported "a marked excess of males" among mathematically gifted children as further evidence that testosterone causes right-hemisphere lateralization.[29]

In the 1980s, feminist biologists criticized this and other hypotheses linking right-hemisphere lateralization and mathematical abilities to testosterone. Of this specific formulation, they pointed out that no causal mechanism was identified for the relationship posited between testosterone and the slower development of the left hemisphere, that Diamond *et al.*'s research had established a thickness in small areas of male rodent brains but not a relationship between that thickness and lateralization, and that the hypothesis of sex-differentiated lateralization to which Diamond *et al.*, Ounsted and Taylor, and Geschwind and Behan appealed remained controversial in empirical psychology. Finally, they pointed out that the study by Chi *et al.* cited by Geschwind and Behan had in fact reported differential development in the hemispheres in both male and female brains, and also stated that investigators could find no sex differences in 507 human brains of 10–44 weeks' gestation.[30]

In arguments challenging the evidence for organizer hypotheses more generally, feminist biologists criticized the emphasis on the organizing effects of androgens and challenged various hypotheses concerning their effects, pointing to continuous conversions of some forms of sex hormones to others as presenting difficulties for both. They also challenged the linear explanatory model organizer hypotheses presumed, and in particular the extrapolation of the model to humans, citing experimental results indicating complex and often non-linear interactions between cells, and between cells and the maternal and external environments, during every stage of fetal development. Finally, they pointed out that hypotheses positing sex-differentiated lateralization themselves rely on controversial hypotheses concerning sex differences in cognitive abilities. Many questioned the rationale for looking for a biological foundation for the sex differences alleged given that a substantial body of research documents significant differences in relevant socialization, that differences among males and among females appear to be more significant than the differences between these groups, and that studies claiming to establish such differences typically assume that gender is a sufficient variable and use criteria for cognitive abilities that are themselves controversial. On the basis of these several lines of critique, some feminist biologists concluded that organizer hypotheses of the sort Geschwind and Behan proposed lacked evidential warrant and that their driving force was largely political.[31]

Considered in isolation, Geschwind and Behan's hypothesis might appear to lack evidential warrant. Given the nature of the topic, the gender stereotypes informing it, the staggering number of explanations proposed in the history of science for alleged sex differences in cognitive abilities, and the misstatement or misrepresentation of Chi *et al.*'s results, its driving force can be seen, as some feminist biologists have seen it, as largely political. But in my initial analyses of this case, I argued that when evidence is taken to include *both* empirical results *and* integration within a body of accepted research, standards, and theories, Geschwind and Behan's hypothesis is revealed to have enjoyed substantial evidential warrant at the time it was proposed, notwithstanding Chi *et al.*'s results. This is not to say, I argued, that social beliefs and values did not constitute part of its evidential warrant; construing evidence holistically reveals that these had a role both more subtle and pervasive than an analysis focusing on this hypothesis in isolation would reveal (Nelson, 1990 and 1995).

Results, hypotheses, and core research questions in three research traditions provided evidential support for this hypothesis. The core questions, standards, and theories of neuroendocrinology, the discipline within which Geschwind and Behan were working, concern relationships among hormones, neural events, and behavior, and there is a long-standing commitment to sexual dimorphism and to designating hormones, and mechanisms and lines of fetal development, as "male" and "female". *Given* these commitments, males and females appear to provide a "natural" base line for investigating the relationships with which the discipline is concerned, and sex differences have functioned as support for its core hypotheses positing such relationships. That is to say, Geschwind and Behan's investigation and hypothesis fell well within the parameters of neuroendocrinology. Further, the prevalence of some cognitive and other disabilities among boys and men revealed by their research provided additional warrant for the investigation of the effects of hormones on neuro-organization and function.

In addition, research based on a male/female dichotomy, and extending this dichotomy to hormones, lines of fetal development, neuro-organization, and behavior, had yielded results in closely allied disciplines and research collaborations. Research in reproductive endocrinology two decades earlier had led investigators to posit a "male" rodent brain as resulting from the influence of androgens and a "female brain" as resulting from their absence.[32] And, as indicated by the studies which Diamond *et al.* and Geschwind and Behan cited, investigations in endocrinology and empirical psychology (most involving rats) had linked androgens to morphological sex differences in the brain, to "cognitive capacities" (e.g., maze-negotiating abilities), and to behavior (e.g., "aggression" in laboratory animals). Studies in psychology claiming to establish clear sex differences in cognitive abilities in humans, and in particular differences in visuo-spatial and mathematical abilities, were also numerous in the 1970s and 1980s. And research into cerebral dominance involving collaborations across these disciplines and neuroanatomy, research in which Geschwind had been involved, had found structural and functional asymmetries in the human brain. By 1979, three years before Geschwind and Behan's proposal, two landmark studies in reproductive endocrinology could claim that sexual differentiation of brain morphology and function was established and call for further investigations into its mechanisms.[33]

When evidence is construed holistically, Geschwind and Behan's hypothesis is revealed to represent the synthesis of core research questions, current hypotheses, and experimental results in three research traditions, aspects of which reflected cultural assumptions about sex differences (e.g., that males have superior spatial and mathematical abilities, that there is a biological foundation for these sex differences, etc.). In short, their hypothesis enjoyed substantial evidential warrant.

A Dissenting Community

That said, an analysis of the evidence appealed to by feminist biologists critical of this hypothesis also reveals a substantial role for unquestioned assumptions about gender and a fair dose of androcentrism in the research providing evidential warrant for Geschwind and Behan's hypothesis. Recall that feminist biologists criticized the emphasis organizer hypotheses place on the effects of so-called "male" hormones. Androcentrism characterized research into fetal development more generally.[34] Research concerned with sexual differentiation of the brain has been characterized by an emphasis on the effects of male hormones, and much of the research ostensibly concerned with human fetal development has been devoted to the sex-determining function of the Y chromosome and H-Y antigen in the organization of the primitive gonad into the testes, and to the subsequent effects of testosterone on the developing brain and somatic sex organs. Until relatively recently, when as Anne Fausto-Sterling describes it, "a positive role for estrogen began to creep into parts of the literature" in fetal development, what research there has been into the role of estrogen has been largely devoted to the question of why the developing male embryo is not feminized by them. Well into the late 1980s, relatively little was known about the directions or mechanisms of female fetal development (Fausto-Sterling, 1987). Even with the creeping in of interest that Fausto-Sterling notes, a survey of articles and textbooks indicates that the conflation of male fetal development with human fetal development has been widespread.[35]

In the research background earlier summarized, gender dimorphism, inclusive of social connotations, was also commonly imposed on other species as well as objects that are not sexed. The association of males with activity and of females with passivity has been pervasive in research into fetal development in endocrinology and empirical psychology. These associations have shaped descriptions of laboratory animal behavior and facilitated extrapolations of that behavior to human behavior. One reason why so little was known about the mechanisms or directions of female fetal development was that it had been conceptualized as occurring passively in "the absence of instructions" from sex hormones designated as male (Fausto-Sterling, 1987) – a methodological commitment John Money calls "the Adam principle" and describes as the principle "that you have to do something to get a male" (quoted in Fausto-Sterling, 1987).

The commitment to sexual dimorphism and to extending it to biological entities and processes is also pervasive. Androgens and estrogens, for example, are consistently designated as respectively "male" and "female", even though males and females produce both (it is the amounts that differ); the circulating form of progesterone, which is metabolized to testosterone, the major androgen, is also metabolized to estradiol (the major estrogen); and among the three families of sex hormones, there are continuous conversions of some forms to others. The difficulties in isolating the effects of these hormones, and the role of the assumption of sexual dimorphism and presuppositions about gender in the interpretation of research results involving them, are attested to by the recent reversal of claims that an "organizing effect" on fetal rat brains that researchers had attributed to androgens and linked to rodent behavior they described as "masculine", is now attributed to estrogen converted from testosterone by brain cells (Fausto-Sterling, 1987).

Feminist biologists have provided detailed analyses that demonstrate that conclusions cannot be drawn on the basis of research apparently establishing the organizing and activating effects of androgens given that a similar amount of research has not been devoted to the effects of estrogen, or to the mechanisms and directions of female fetal development. They have also demonstrated that androcentric descriptions of behavior, and the imposition of gender on biological entities and process, facilitated unwarranted extrapolations of explanatory models positing linear causal relationships from hormones, to neural events, to cognitive capacities and behavior from rodents to humans.[36]

In my initial analysis of this case I offered the following assessment. First, notwithstanding the problems feminist scientists have identified, the organizer hypothesis is *not* plausibly written off as "bad science". There was substantial evidence for it, constituted by research traditions and experimental results in several disciplines, and widely accepted assumptions, within the relevant sciences and the broader social community, about gender differences and their source in biology. In 1981, the year before Geschwind and Behan proposed their formulation of the organizer hypothesis, a number of the articles in a special issue of *Science* devoted to sex differences took sexual differentiation of the brain and its source in the organizing effects of androgens as *established* for many species and as *well supported* for humans.[37]

Second, viewed in the light of the research questions, results, and theories brought to bear by feminist scientists, that hypothesis is revealed to be considerably less promising than its proponents took it to be, and aspects of the research background providing its evidential warrant are problematic in the ways just outlined. It might not be reasonable to expect endocrinologists in the 1970s and early 1980s to know and consider sociological studies suggesting alternative explanations for the sex differences alleged, to know of the critiques offered by psychologists of research into sex differences in cognitive abilities and lateralization, or even to know of the several levels of critiques offered by feminist colleagues in biology. But relevant studies and critiques were sufficiently publicized by the late 1980s to make it reasonable to expect those pursuing biological explanations for sex differences to show why such explanations constituted or were likely to constitute better explanations than sociological explanations, to attend to the unwarranted imposition of gender, the critiques of the commitment to sexual dimorphism and of its imposition on biological entities and processes, the lack of research into female fetal development, and so on. And while holism does not take the borrowing of hypotheses criticized by feminist biologists to be itself problematic, I argued that those key to the rationale of organizer hypotheses – hypotheses positing sex differences in lateralization and in cognitive abilities – would need to be confirmed, and the substantial problems in the research to date involving them would need to be resolved, to warrant future organizer hypotheses.

Finally, I argued that this case, along with others focused on by feminist scientists, carried implications for the role of social beliefs and values in science. The role of androcentrism, gender dimorphism, and of the impositions of gender and its social connotations on other species and biological entities and processes just outlined, *and* the identification of these factors by feminist scientists, demonstrate that social beliefs and values can be *integrated* in research questions, experiments, methodologies, and ontologies that meet general and discipline-specific standards. Such cases indicate that neither the role of social and political factors of various kinds, nor that of values, can serve as a litmus test for good science or be ignored in philosophical theorizing about science.

In the interim, Geschwind and Behan's hypothesis has lost support among biologists, with that loss of support apparently related to the failure to identify the causal mechanisms it posited. Because of this, and given the misstatement of the results claimed by Chi *et al.*, some scientists have objected to my assessment that the hypothesis did enjoy evidential warrant. I think the proper response to this objection is along the lines of "hindsight is 20/20". Darwin had no mechanism to explain inheritance, but we don't find the provisional acceptance of natural selection pending the discovery of that mechanism unwarranted. Had a mechanism been found in the present case, it is likely that, despite Chi *et al.*'s results, Geschwind and Behan would have been credited with synthesizing what was in fact a broad spectrum of results and hypotheses, and with providing a biological explanation for sex differences in cognitive abilities – an explanation in which there continues to be considerable interest in biology and psychology.

Two other objections were common. Given that this hypothesis is no longer viewed as viable, some argued, it is rather like the hypotheses that intelligence is a function of brain size and that both are differentiated by class, race, and sex – of historical interest perhaps, but not a case to take as providing insights into science. In a more generalized version of this objection, scientists and philosophers committed to a hard and fast boundary between good science and values often maintain that this case and others feminists cite as demonstrating relationships between social values and beliefs, and research questions, methods, and theories in science, are idiosyncratic, cases of bad science, or both. One formulation of such criticisms maintains that the problems feminist scientists identified would be recognized by anyone "practicing good science".[38]

Responses to some of these objections are straightforward. Neither this case nor craniometry is idiosyncratic; three decades of science scholarship have revealed numerous cases, both historical and current, in which social values and beliefs are integrated in research questions, the design of experiments and interpretation of results, and theories. As much to the point, to paraphrase Stephen Jay Gould's argument that there are lessons to be learned from craniometry, organizer hypotheses have been taken seriously by a substantial number of scientists, with research assuming or proposing them published in science journals, the focus of special issues of *Science*, and so on. They have been taken seriously, as Gould demonstrated was the case with hypotheses in craniometry, because they represent a logical extension of broader bodies of accepted research questions, standards, and theories – much of which remains in place in the present case. If we take the integration of social beliefs and values as a litmus test, the amount of research and results we would need to designate as "bad science" is far more than one or even several hypotheses and extends well beyond research specifically devoted to establishing or explaining differences in women's and men's cognitive abilities.

Finally, the case we have considered is not one in which scientists without social values or preconceptions about gender were, *on that basis*, able to recognize the role of such beliefs and values in their colleague's research. An analysis of the evidence feminist scientists appealed to reveals the conscious integration of theories and results in their scientific disciplines with theories and results produced by an evolving community of feminist scientists and science scholars, *and* that it was such integration that led to the identification of the problems in this research and to the identification of alternative models.

The critiques feminist biologists offered of androcentrism in research into fetal development, for example, cited and built on critiques offered by feminists in anthropology, sociology, psychology, and primatology documenting the role of androcentric assumptions in shaping observations, research emphases, and theories in their respective disciplines. Feminist biologists used this larger body of critique to frame their analyses of the nature of androcentrism in biology and as evidence that the rationale for organizer hypotheses – the alleged "universality" of sex-differences in behavior and cognitive abilities across cultures and species – is itself a result of research in other sciences and disciplines shaped by androcentric assumptions, questions, and methods.[39]

Similarly, the critiques leveled by feminist biologists of the imposition of gender and its social connotations on hormones, brains, behavior, and fetal development frequently cited and built on critiques offered by feminists in primatology, anthropology, archaeology, and sociology detailing the imposition of gender dimorphism and social connotations of gender on other species and cultures. These critiques also evolved apace with critiques of the imposition of gender on entities in developmental and molecular biology (where, for example, the nucleus and cytoplasm were long designated as male and female respectively – and, accordingly, as active and passive, and dominant and subordinate).[40]

The critiques leveled against the explanatory model assumed by organizer hypotheses evolved apace with and frequently cited a broader body of criticism by feminist scientists that challenged the adequacy of models positing linear, hierarchical relationships: those focused on the emphasis on "dominance hierarchies" in primatology, animal sociology, anthropology, sociobiology, and sociology, and those criticizing uni-directional models of gene action presupposed in sociobiology and "master molecule" theories of cellular protein synthesis as over-simplifying complex biological processes.[41] Finally, a central focus in each line of critique is the role of androcentrism, of the imposition of gender and gender dimorphism, and of over-simplified models of biological processes in lending credence to biological determinist theories of gender.

We cannot, in other words, disassociate the critiques feminist biologists levelled against the organizer hypothesis from *feminism* any more than we can disassociate them from the *science* on which these scientists rely.

The broader implications of this case are the following. First, explanations and reconstructions that focused on individual scientists and/or that excluded the social beliefs and values integrated in the research within which organizer hypotheses developed would be *empirically inadequate*. And explanations and reconstructions of the identification of these factors that excluded the role of an evolving body of feminist science scholarship analyzing relationships between gender and science, and undertaken by and publicized among feminist scientists and science scholars, would be similarly inadequate.

Second, a substantial benefit of holism is that it provides the basis for the kinds of assessment just concluded. In this case, holism allowed us to distinguish between the "local" body of theory and standards that constituted evidence for the organizer hypothesis, and the broader body of theory and standards brought to bear by feminist scientists in light of which aspects of the former are problematic. Such analyses and discriminations are necessary if we are going to be able to distinguish successful from unsuccessful science, and both from bad science (i.e., science characterized by the flagrant violation of norms, the ignoring of relevant and available results, and the like). What is good but unsuccessful science can be seen to be so, because it can be seen to fit with local theories and standards but to fail when the scope of holism is enlarged.

Finally, while there is no algorithm in the foregoing for demarcating evidence and communities, there are guidelines. I used the core research questions of disciplines and research traditions, research collaborations, published results, citations, and conferences, to identify relevant communities and bodies of theory, and the latter to assess evidential warrant. I also used the two methodological principles I advocate, and turn now to the implications of the analysis just completed for the normative function of naturalized philosophy of science.

4. METHODOLOGY NATURALIZED AND NORMATIVE

I began this discussion with the claim that naturalized philosophy of science is not traditional epistemology. It is also not one of the sciences. It is a discipline that takes the sciences as its subject matter, hence the distinction, but is also continuous with them as it can claim no "higher" norms for assessing theories and practices, whether of common sense, philosophy, or the sciences, than those that evolve apace with our efforts as laypersons, philosophers, and scientists to explain and predict experience. For empiricists, these norms at least include empirical success and explanatory power. Both the distinction and the continuity argue for the normative function of naturalized philosophy of science, one I will parse in terms of the methodological shifts I advocate.

As a discipline, naturalized philosophy of science attempts to explain and evaluate the efforts made within the sciences to construct empirically adequate theories. Like the sciences, its methods and research focuses are those that prove (or that we assume will prove) most effective in achieving its goals. Accordingly, we find philosophers of science, but usually not practicing scientists, concerned with providing general accounts of evidence and of cognitive agency. Scientists could, of course, engage in such research just as philosophers of science could pursue degrees in science. But for the most part, the questions and research philosophers of science pursue, like those scientists pursue, presuppose a tradition of inquiry and the core research questions, categories, unresolved issues, and overlapping bodies of theory that constitute their discipline.

My assessment of the evidential warrant for the organizer hypothesis and of aspects of the research background within which it emerged are obviously different from those of the scientists directly involved because I considered theories, research, and questions of a broader reach. And while these included those brought to bear by feminist scientists, I did not always concur with the conclusions reached by these scientists. Both differences are traceable to the methodological principles I utilized, and these in turn to developments within my own discipline. Scientists claim and attribute results in a more individualistic manner than I assumed in my analysis and recommend for naturalized philosophy of science. Relatedly, many construe evidence as exhausted by empirical success and remain committed to the view that research and theories in which values are integrated cannot constitute good science. I have recommended that philosophers of science part company with scientists on each of these issues.

If we begin from the view that the goal of science is to generate theories that organize, explain, and predict our experience, and presumably the best possible theories, and that the goal of the philosophy of science is to produce theories that explain and evaluate these efforts, then the shifts in methodology I recommend suggest two sets of deeply related questions. We should ask whether the social processes that characterize science – those involving the recruitment and education of scientists, for example, peer review mechanisms, the ways in which research is reported and disseminated, and differences in prestige that separate specialties and sciences – do or do not contribute to that goal. For example, do current mechanisms for peer review insure that the most promising research is published and/or funded, or are they such that very good research, if it challenges entrenched models or powerful networks, will not receive funding? That science has achieved empirical success, and substantial empirical success, is not debatable – but this does not entail that the theories and research programs it has generated are the best that might be generated.

As we turn to individual cases, our questions about the social processes that characterize science will take on more specific content. The case just considered suggests that we should ask if the processes that characterize science education and recruitment, or current standards governing scientific practice, are likely to foster awareness among scientists of the social values and beliefs they bring to science, or of those integrated in the research questions and theories of their disciplines. Does science education as currently organized preclude study in the humanities and, if so, if it is reasonable to expect scientists to be experts on social policy or to be in a position to assess the impact of their highly-specialized endeavors on the broader social community, and vice versa? How do prestige hierarchies among the sciences and specialization impact on the directions of research? Do they contribute to cases such as that just considered in which unquestioned assumptions about gender deeply inform research in biology, and in which a substantial body of research in the social sciences relevant to the existence and explanation of sex difference is not considered by those positing biological explanations of these – or, to take a more extreme case, to the confidence among some geneticists engaged in socio-biology that they can *do* sociology and anthropology, and recommend social policy?

Let us be very clear here. From the point of view of the holistic account of evidence I have advocated, *the questions just listed about social processes are simultaneously questions about bodies of evidence*. To see this, consider the question of how it was that feminist scientists in a range of fields and disciplines recognized the role of gender and values that their colleagues did not, or why the designation of different strains of *E. coli* as "male" and "female" in a 1986 edition of a widely-used text in molecular biology was dropped in the second edition, or why the following passage, which biologist Scott Gilbert points out was read by most embryologists educated through the 1970s, has recently been dropped from a widely-used embryology textbook:

> In all systems that we have considered, maleness means mastery; the Y-chromosome over the X, the medulla over the cortex, androgen over estrogen. So physiologically speaking, there is no justification for believing in the equality of the sexes.

These are *successes*.[42] *How* did they happen? I have argued that answers to that question will invoke social factors (changes in broader social views about gender, the presence of scientists who identified themselves as feminists, and so on) and *simultaneously invoke bodies of evidence*: the bodies of accepted results and theory, internal and external to science, within which scientists work and, in the case here considered and others, bodies of theory maintained by two or more research communities which overlap in many ways and differ in others.

If we locate our work in naturalized philosophy of science, our work to identify and answer questions about the social processes and bodies of evidence that characterize science will involve empirical research, rather than armchair reasoning. And the answers bench scientists give to these and other questions we identify will need to be part of what we consider. But when we disagree, such disagreements should be pursued. And if, as philosophers of science, we conclude that our answers are more empirically warranted than those scientists offer, we should advocate changes in scientific understandings and practices.

I am arguing, of course, that given that the goal of both the philosophy of science and science is the production of theories that explain and predict our experience, philosophical analyses of science will be by their very nature normative. This is an obvious consequence of construing evidence holistically – but it is no less obviously a consequence of empiricism.

ACKNOWLEDGEMENTS

This discussion extends and revises Nelson, 1995. I am grateful to Karen Barad, Michael Flowers, Ronald Giere, Elizabeth Potter, Joseph Rouse, Naomi Scheman, Pauline Sargeant, Nancy Tuana, Amanda Vizedom, and an anonymous reviewer for constructive criticisms of earlier drafts. Jack Nelson read many versions and his advice and criticism have been invaluable. Earlier versions were presented to the Nordiskt Nätverk för Feministisk Epistemologi och Feministisk Vetenskapsteorito, and the Institutionen för Filosfi och Vetenskapsteorito, at Umeå Universitet, in October 1994, at the University of Pennsylvania in September 1995, at Swarthmore College in November of 1995, at the Centre for Philosophy of Science at the University of Minnesota in March 1996, and to the departments of philosophy at Illinois State University and Illinois Wesleyan in March 1996. I am grateful to members of these audiences for challenging and insightful criticisms. Correspondence with Jim Maffie also improved sections of this discussion.

NOTES

[1] Following Richard Burian, I take disciplines to be "organized and institutionalized bodies of research focussed on a core set of questions" (Burian, 1993: 387–388). I outline the core questions of naturalized philosophy of science below.

[2] To the best of my knowledge, Solomon (1994) is the first to use the phrase 'social empiricism' but not the first to argue that empiricism can and should be separated from individualism. Helen E. Longino's *Science as Social Knowledge* (1990) and my *Who Knows* (1990) each advanced a version of social empiricism, though different from one another and from Solomon's. I take holism to constitute the most substantive basis for social empiricism and to demonstrate that the claim "knowledge is social" need be neither trivial nor vacuous.

[3] The denotations of "social factors" and "social processes" vary widely in recent literature, encompassing peer review and funding mechanisms, science education, the "internal politics" of disciplines and sciences, prestige hierarchies among the sciences, divisions in cognitive labor and authority within research communities and between scientists and the lay public, and features of the broader social community within which science communities are embedded (e.g., gender and other social relations). (See, e.g., the works by Burian, Downes, Longino, Solomon, and Stump listed in the references.) As much to the point, there are deep disagreements as to whether the factors so designated are epistemologically significant, i.e., have a bearing on the content of scientific knowledge. The open-endedness with which I treat the category reflects the view that *naturalized* philosophy of science must allow the details of individual episodes to determine which, if any, such factors were of import, in what ways, and to what degree.

[4] These are by no means the only such cases. See, for example, recent analyses of the plate tectonic evolution (Solomon, 1992 and 1994) and of the developments in biology leading to the discovery of the double helix (Burian, 1985 and 1993; Keller, 1983; Olby, 1974; Sapp, 1983 and 1987).

[5] Appeals to *a-priori* notions are neither new nor novel in philosophy, but what is somewhat surprising are claims that an *a-priori* conception of rationality can be used to explain cases of good science even when factors the conception takes to be antithetical to rationality are admitted to be at work in those cases.

[6] See, e.g., the articles in the special issue of *Synthese* devoted to Feminism and Science, *Synthese*, **104**(3), September 1995, and Wylie, 1995.

[7] There are other exceptions of course, perhaps most notably Thomas Kuhn's arguments in *The Structure of Scientific Revolutions*. But I take Kuhn to have shared more with positivism than is often acknowledged (Nelson, 1990), and differences between his positions and those I advocate will emerge.

[8] Quine, 1960 and 1969.

[9] The next several paragraphs closely parallel Nelson, 1995.

[10] See Maffie (1990) for an overview of these debates.

[11] The first criterion is advocated in Giere, 1988; Nelson, 1990; Solomon, 1994; and Stump, 1992. The second is advocated in Giere, 1988; Kornblith, 1985; Maffie, 1991; Nelson, 1990; Quine, 1969; Solomon, 1994; and Stump, 1992. All three are advocated in Nelson, 1990 and 1995; and Solomon, 1994. See also Downes, 1993; Fuller, 1988; and Stump, 1992.

[12] Nelson, 1995.

[13] The formulations of these principles next presented have been revised since Nelson, 1995.

[14] As I note below, Quine has long maintained that a "relative" or modest holism suffices (Quine, 1960 and 1981c) and has recently described "a tempering of the extreme holism" advocated in 'Two Dogmas of Empiricism' (Quine, in Borradori, 1994: 36).

[15] My use of 'theory' to denote methodological commitments and standards reflects the view that these, like parts of theories narrowly construed, are sentences (Nelson, 1990 and 1995).

[16] Nelson, 1995.

[17] Quine, 1963 and 1981b,c. In a line of reasoning similar to that I will take below, Quine (1960) maintains that while the stimulus conditions of an occasion sentence (i.e., an observation sentence) may be specifiable, the specification will be by reference to the behavior of a *community* with the empirical content determined in part *by the theories* that community maintains (Quine, 1960: 35–45). Note that while a notion of "observation sentences" has been, in Quine's words, "dredged out", the sentences so designated are theoretical and the empirical content an observation sentence *holds for an individual* is partly determined by the theories and standards generated by a community.

In apparent contrast, Quine, 1981b and 1990 define an observation sentence in terms of *individuals*, and specifically as a sentence that "*taken as an undivided whole* commands [an individual's] assent consistently or dissent consistently when the same global sensory stimulation is present" (Quine, 1981b: 26, emphasis added; cf. Quine, 1990: 1–8). But Quine also maintains in these later discussions that *observation sentences share terms with theory* – indeed, it is on the basis of such connections, he notes, that observation is "relevant to scientific theory" – and that when viewed *analytically*, observation sentences are "theory-laden" (Quine, 1990: 7). It is when viewed holophrastically (i.e., taken as a whole and "as conditioned to stimulatory situations") that Quine maintains that such sentences are "theory-free" (*ibid.*, 7). Hence, the shift *in focus* in these later definitions – from communities to individuals – is really a shift from viewing sentences analytically to holophrastically, and does not change the fact that the empirical content an observation sentence holds *for an individual* – the stimulatory situations in light of which she or he will assent to or dissent from it – is determined in part by a body of accepted theory.

[18] I retain evidence as a technical notion. In contrast, as I discuss below, Quine describes observation, and "evidence, if that was observation", as the "temporally ordered set of all those of [a subject's] exteroceptors that triggered on [an] occasion", and maintains that, so defined, both evidence and observation "drop out" as technical notions. "We can make do instead", he maintains, "with the notion of observation sentences" (Quine, 1990: 2).

[19] See also Nelson, 1990 and 1995; Klee, 1992; Nelson and Nelson, 1995; Quine, 1960, 1981c, 1990; and Quine, in Borradori, 1994.

[20] This is demonstrated in, for example, Solomon, 1994.

[21] See n. 17.

[22] This approach might be so understood, but Giere (1988) and Solomon (1994) demonstrate that it need not. Both draw heavily on cognitive science and also insist on the constitutive role of social factors.

[23] That is, once we acknowledge (as Quine himself insists) that we can expect no sameness of physical state (i.e., not the triggering of the "same" exteroceptors) among similarly placed observers or for the same observer on individual occasions in response to some "fixed range of stimulations", that the *stimuli* that shapes our positing of objects, events, and causal relationships *includes* the sentences constituting the methods, core questions, and theories of established traditions of inquiry, and that decision-making and other cognitive processes take place within and are constrained by such traditions, it is clear that the emphasis on individuals for naturalized accounts of theory generation and adoption is misplaced.

[24] I have argued that research and analyses undertaken by feminist science scholars further the case for our needing such higher level explanations and the case for taking communities as their primary loci (Nelson, 1990 and 1995). In a line of reasoning I will pursue in the case study, feminist scientists and scholars have documented ways in which social arrangements and processes, including divisions in cognitive labour and authority, have affected both the directions and the content of scientific knowledge (e.g. Addelson, 1983; Keller, 1985; Potter, 1993; Tuana, this volume and works cited in n. 40).

[25] Goy and McEwen, 1980.

[26] In summarizing this research, its theoretical background, and feminist critiques of both, I draw most heavily on a special issue of *Science* (Vol. **211**, 1981) focusing on sex differences (in particular, Ehrhardt and Meyer-Bahlburg, 1981; MacLusky and Naftolin, 1981; and Wilson *et al.*, 1981); and Bleier, 1984 and 1988; Chi *et al.*, 1977; Diamond *et al.*, 1981; Fausto-Sterling, 1985, 1987, and 1993; Geschwind and Behan, 1982 and 1984; Geschwind and Galaburda, 1984; Gorski *et al.*, 1978; Gorski, 1979; Harris and Levine, 1965; and Hubbard *et al.*, 1979.

[27] In addition to Chi *et al.* and Ounsted and Taylor, Geschwind and Behan cited LeMay and Culebras, 1972; Wada *et al.*, 1975; and Galaburda and Kemper, 1979.

[28] But, in describing this study a year later in *Science*, Geschwind suggested that the effects of testosterone *in utero* can produce "superior right-hemisphere talents such as … mathematical talent" (Kolata, 1983: 1312).

[29] Geschwind and Behan, 1984: 221; the study they cited was Benbow and Stanley, 1983.

[30] Representative critiques are found in Bleier, 1984 and 1988; Fausto-Sterling, 1985; Hubbard *et al.*, 1979; overviews of aspects of the debate concerning organizer hypotheses are provided in Longino, 1990 and Nelson, 1990.

[31] See the works cited in n. 30.

[32] Harris and Levine, 1965; cf. Fausto-Sterling, 1987. Harris and Levine also investigated the role of estrogens, but this represented an exception rather than the rule as I outline below. Overviews are provided in Bleier, 1984 and 1988; and Fausto-Sterling, 1985, 1987, and 1993.

[33] The first of these studies announced that "the concept of the sexual differentiation of brain function is now well established" (Gorski *et al.*, 1978: 334); the second called for a search for "a clear morphological signature of sexual differentiation in the brain" (Gorski, 1979: 114).

[34] I rely here primarily on Bleier, 1984 and 1988; Fausto-Sterling, 1985, 1987, and 1993; Hubbard *et al.*, 1979; and the special issue of *Science* (Vol. **211**, 1981) focusing on sex differences (in particular, Ehrhardt and Meyer-Bahlburg, 1981; MacLusky and Naftolin, 1981; and Wilson *et al.*, 1981); Geschwind and Galaburda, 1984; Gorski *et al.*, 1978; Gorski, 1979; and Harris and Levine, 1965.

[35] In MacLusky and Naftolin's 1981 overview of the current state of research into the effects *in utero* of sex hormones on the central nervous system, for example, the subject is presented in these general terms, but the authors move on in the next sentence and subsequent discussion to the effects of androgens secreted by the testes on the development of the central nervous system.

[36] See the works cited in n. 34.

[37] Thus, for example, the introductory section of one article states, "While the role of social learning is much greater in human behavior than in subhuman mammals [for which the existence of sex differences in neuro-organization is no longer a question], there is sufficient evidence to suggest that biological factors influence psychosexual differentiation in human beings" (Ehrhardt and Meyer-Bahlburg, 1981: 1312).

[38] This was the response of the anonymous reviewer of this volume to the case as outlined in an earlier version of this essay.

[39] Representative analyses include Bleier, 1984 and Fausto-Sterling, 1985.
[40] See, for example, the analyses undertaken in Bleier, 1984; Fausto-Sterling, 1985; Hubbard *et al.*, 1979; Gilbert *et al.*, 1988, and the citations in these works, for evidence of the relationships I am outlining. See, also, the Program Abstracts for the "Women, Gender, and Science Question" Conference, May 1995, University of Minnesota, where the relationships among these developing lines of critique are reflected in conference papers and citations.
[41] See the works cited in n. 34, Keller, 1985, and Nelson, 1990 for evidence of the relationships here claimed.
[42] The molecular biology text is Darnell *et al.*, 1986 and 1990, and the change is reported in Spanier, 1987; the quoted passage is from *Embryonic and Fetal Development*, edited by C. R. Austin and R. V. Short (London: Cambridge University Press, 1972) and appears in an article by R. V. Short (Scott Gilbert, personal correspondence).

REFERENCES

Addelson, K. P.: 1983, 'The Man of Professional Wisdom', in S. Harding and M. Hintikka (eds), *Discovering Reality*. D. Reidel, Dordrecht.

Biology and Gender Study Group: 1988, 'The Importance of Feminist Critique for Contemporary Cell Biology', in N. Tuana (ed.), *Feminism and Science*. Indiana University Press, Bloomington, 61–76.

Benbow, C. and J. Stanley: 1983, 'Sex Differences in Mathematical Reasoning Ability: More Facts', *Science*, **222**, 1029–1031.

Bleier, Ruth: 1984, *Science and Gender: A Critique of Biology and Its Theories on Women*. Pergamon Press, New York.

Bleier, Ruth (ed.): 1988, *Feminist Approaches to Science*. Pergamon Press, New York.

Borradori, G.: 1994, *The American Philosopher: Conversations with Quine, Davidson, Putnam, Nozick, Danto, Rorty, Cavell, MacIntyre, and Kuhn*. University of Chicago Press, Chicago and London.

Burian, R. M.: 1985, 'The "Internal Politics" of Biology and the Justification of Biological Theories', in A. Donagan *et al.* (eds), *Human Nature and Natural Knowledge*. D. Reidel, Dordrecht.

Burian, R. M.: 1993, 'Technique, Task Definition, and the Transition from Genetics to Molecular Genetics', *Journal of the History of Biology*, **26(3)**, 387–407.

Campbell, R.: 1994, 'The Virtues of Feminist Empiricism', *Hypatia*, **9(1)**, 90–115.

Chi, J. G., E. C. Dooling, and F. H. Giles: 1977, 'Gyral Development of the Human Brain', *Annals of Neurology*, **1**, 86–93.

Darnell, J., H. Lodish, and D. Baltimore: 1986, *Molecular Cell Biology*. Scientific American Books, New York (2nd edition, 1990).

Diamond, M. C., G. A. Dowling, and R. E. Johnson: 1981, 'Morphologic Cerebral Cortical Asymmetry in Male and Female Rats', *Experimental Neurology*, **71**, 261–268.

Downes, S. M.: 1993, 'Socializing Naturalized Philosophy of Science', *Philosophy of Science*, **60**, 452–468.

Ehrhardt, A. E. and H. F. L. Meyer-Bahlburg: 1981, 'Effects of Prenatal Sex Hormones on Gender-Related Behavior', *Science*, **211**, 1312–1318.

Fausto-Sterling, Anne: 1985, *Myths of Gender: Biological Theories about Women and Men*. Basic Books, New York.

Fausto-Sterling, A.: 1987, 'Society Writes Biology/Biology Constructs Gender', *Daedalus*, **116**, 61–76.

Fausto-Sterling, A.: 1993, 'The Five Sexes: Why Male and Female Are Not Enough', *The Sciences*, **33(2)**, 20–24.

Fuller, Steve: 1988, *Social Epistemology*. Indiana University Press, Bloomington, Indianapolis.

Galaburda, A. M. and T. M. Kemper: 1979, *Annals of Neurology*, **6**, 94–100.

Geschwind, N. and P. O. Behan: 1982, 'Left-handedness: Association with Immune Disease, Migraine, and Developmental Learning Disorder', *Proceedings of the National Academy of Sciences*, **79**, 5097–5100.

Geschwind, N. and P. O. Behan: 1984, 'Laterality, Hormones, and Immunity', in N. Geschwind and A. M. Galaburda (eds), *Cerebral Dominance: The Biological Foundations*. Harvard University Press, Cambridge, MA and London.

Geschwind, N. and A. M. Galaburda (eds): 1984, *Cerebral Dominance: The Biological Foundations*. Harvard University Press, Cambridge, MA and London.

Giere, Ronald N.: 1988, *Explaining Science: A Cognitive Approach*. University of Chicago Press, Chicago and London.

Gibert S. *et al*: 1988, 'The Importance of Feminist Critique for Cell Biology', in N. Tuana (ed), *Feminism and Science*, Indiana University Press, Bloomington.

Gorski, R.: 1979, 'The Neuroendocrinology of Reproduction: An Overview', *Biology of Reproduction*, **20**, 111–127.

Gorski, R., J. H. Gordon, J. E. Shryne, and A. M. Southam: 1978, 'Evidence for a Morphological Sex Difference Within the Medial Preoptic Area of the Rat Brain', *Brain Research*, **148**, 333–346.

Goy R. and B. S. McEwen (eds): 1980, *Sexual Differentiation of the Brain*. MIT Press, Cambridge.

Harding, S.: 1983, 'Why Has the Sex/Gender System Become Visible Only Now?', in S. Harding and M. Hintikka (eds), *Discovering Reality*. D. Reidel, Dordrecht.

Harris, G. W. and S. Levine: 1965, 'Sexual Differentiation of the Brain and its Experimental Control', *Journal of Physiology*, **181**, 379–400.

Hubbard, R., M. S. Henifin, and B. Fried (eds): 1979, *Women Look at Biology Looking at Women: A Collection of Feminist Critiques*. G. K. Hall & Co., Boston.

Keller, E. F.: 1983. *A Feeling for the Organism*. Freeman, San Francisco.

Keller, E. F.: 1985, *Reflections on Gender and Science*. Yale University Press, New Haven.

Kinsbourne, M.: 1980, 'If Sex Differences in Brain Lateralization Exist, They Have Yet to be Discovered', *Behavioral and Brain Sciences*, **3**, 221–242.

Klee, R.: 1992, 'In Defense of the Quine–Duhem Thesis', *Philosophy of Science*, **59**, 487–491.

Kolata, G.: 1983, 'Math Genius May Have Hormonal Basis', *Science*, **222**, 1312.

Kornblith, Hilary (ed.): 1985, *Naturalizing Epistemology*. MIT Press, Cambridge.

Kuhn, Thomas, S.: 1970, *The Structure of Scientific Revolutions*. University of Chicago Press, Chicago.

Laudan, Larry: 1977, *Progress and Its Problems*. University of California Press, Berkeley.

LeMay, M. and A. Culebras: 1972, *New England Journal of Medicine*, **286**, 168–170 (cited in Geschwind and Behan, 1982, without title).

Longino, Helen, E.: 1990, *Science as Social Knowledge*. Princeton University Press, Princeton.

MacLusky, N. J. and F. Naftolin: 1981, 'Sexual Differentiation of the Central Nervous System', *Science*, **211**, 1294–1311.

Maffie, J.: 1990, 'Naturalism and the Normativity of Epistemology', *Philosophical Studies*, **56**, 333–349.

Maffie, J.: 1991, 'What is Social about Social Epistemics?', *Social Epistemology*, **5**, 101–110.

Maffie, J.: 1995, 'Towards an Anthropology of Epistemology', *Philosophical Forum*, **XXXVI(3)**, 218–241.

Money, J. *Love and Lovesickness*. Johns Hopkins University Press, Baltimore, p. 5.

Nelson, Lynn Hankinson: 1990, *Who Knows: From Quine to a Feminist Empiricism*. Temple University Press, Philadelphia.

Nelson, L. H.: 1993, 'Epistemological Communities', in L. Alcoff and E. Potter (eds), *Feminist Epistemologies*. Routledge, New York and London, pp. 121–160.

Nelson, L. H.: 1995, 'A Feminist Naturalized Philosophy of Science', *Synthese*, **104(3)**.

Nelson, L. H. and J. Nelson: 1995, 'Feminist Values and Cognitive Virtues', *PSA II*, Philosophy of Science Association, East Lansing, MI.

Olby, R.: 1974, *The Path to the Double Helix*. University of Washington Press, Seattle.

Ounsted, C. M. and D. C. Taylor: 1972, *Gender Differences: Their Ontogeny and Significance*. Churchill-Livingstone, Edinburgh.

Potter, E.: 1993, 'Gender and Epistemic Negotiation', in L. Alcoff and E. Potter (eds), *Feminist Epistemologies*. Routledge, New York and London.

Quine, W. V.: 1960, *Word and Object*. Harvard University Press, Cambridge, MA.

Quine, W. V.: 1963, 'Two Dogmas of Empiricism', in *From a Logical Point of View* (revised edition). Harper & Row, New York (first printing, 1953), pp. 20–46.

Quine, W. V.: 1969, 'Epistemology Naturalized', in *Ontological Relativity and Other Essays*. Columbia University Press, New York.

Quine, W. V.: 1981a, *Theories and Things*. Harvard University Press, Cambridge, MA.

Quine, W. V.: 1981b, 'Empirical Content', in *Theories and Things*. Harvard University Press, Cambridge, MA.

Quine, W. V.: 1981c, 'Five Milestones of Empiricism', in *Theories and Things*. Harvard University Press, Cambridge, MA.

Quine, W. V.: 1981d, 'On the Nature of Moral Values', in *Theories and Things*. Harvard University Press, Cambridge, MA.

Quine, W. V.: 1981e, 'On the Very Idea of a Third Dogma', in *Theories and Things*. Harvard University Press, Cambridge, MA.

Quine, W. V.: 1990, *Pursuit of Truth*. Harvard University Press, Cambridge, MA.

Quine, W. V.: 1993, 'Three Indeterminancies', in R. Barrett and R. Gibson (eds), *Perspectives on Quine*. Blackwell, Oxford, UK, pp. 1–16.

Sapp, J.: 1983, 'The Struggle for Authority in the Field of Heredity: New Perspectives on the Rise of Genetics', *Journal of the History of Biology*, **16(3)**, 311–342.

Sapp, J.: 1987, *Beyond the Gene: Cytoplasmic Inheritance and the Struggle for Authority in Genetics*. Oxford University Press, Oxford.

Solomon, M.: 1992, 'Scientific Rationality and Human Reasoning', *Philosophy of Science*, **59(3)**, 439–455.

Solomon, M.: 1994, 'Social Empiricism', *Nous*, **XXVIII**, 325–343.

Solomon, M. Unpublished manuscript, 'A More Social Epistemology'.

Spanier, B.: 1987, 'Gender and Ideology in Science: A Study of Molecular Biology', *NWSA*, **3(2)**, 167–198.

Stump, D.: 1992, 'Naturalized Philosophy of Science with a Plurality of Methods', *Philosophy of Science*, **59**, 456–460.

Synthese 104(3): Special Issue on Feminism and Science (September 1995).

Thagard, Paul: 1992, *Conceptual Revolutions*. Princeton University Press, Princeton.

Wada, J. A., A. Clarke, and A. Hamm: 1975, *Archives of Neurology*, **32**, 239–246 (cited in Geschwind and Behan, 1982 without title).

Wilson, J. D., F. W. George, and J. E. Griffin: 1981, 'The Hormonal Control of Sexual Development', *Science*, **211**, 1278–1284.

Wylie, A.: 1995, 'Doing Philosophy as a Feminist', *Philosophical Topics*, special issue on "Feminist Perspectives on Language, Knowledge, and Reality", edited by Sally Haslanger.

ELIZABETH POTTER

UNDERDETERMINATION UNDETERRED

INTRODUCTION

The underdetermination thesis, every hypothesis/theory is underdetermined by the data, has been used by non-feminist and feminist science scholars in case studies designed to show that factors other than evidentiary ones sometimes enter the justificatory work of scientists. Recent attacks on the underdetermination thesis (Laudan, 1990; Laudan and Leplin, 1991) have been taken to show that it is no longer viable to use the thesis in this way (Laudan, 1990; Fuller, 1990; Pinnick, 1994). In this essay I will argue that the most recent attack upon one version of the underdetermination thesis fails. Not only does the attack by Laudan and Leplin (1991) fail to touch the fundamental version of the thesis, the attack fails to defeat the version Laudan and Leplin consider.

Factors other than evidentiary ones range from the pragmatic virtues through professional interests to gender politics, for although many philosophers agree that pragmatic considerations enter the "context of justification" of scientific hypotheses, others further along the rationalist spectrum do not. Arch-rationalist attacks on science scholars for questioning attempts to find a logic of science are not new; nor do attackers always make the political stakes clear in the way that Imre Lakatos did (1970, p. 93) when he accused Thomas Kuhn (1962) of giving support to "contemporary religious maniacs" and "student revolutionaries." The philosophical stakes, however, have been as clear as they are venerable: find an account of scientific rationality narrow enough to rule out the possibility that rational methods in science include pragmatic and social factors. Moderate rationalists wish to include pragmatic factors, yet rule out social ones. But both groups attack those of us who suspect that social factors sometimes enter the "context of justification" on the grounds that we make scientific work appear to be irrational and so undermine the cognitive authority of science (Laudan, 1984, p. 72). This dispute among rationalist and anti-rationalist philosophers is a deep one over the scope of rationality. I suggest that many rationalists mistake the possibility of a philosophical account of scientific rationality for scientific rationality itself, and this mistake has led them to confuse our rejection of their narrow definitions of scientific rationality with a rejection of scientific rationality itself. I have argued elsewhere that when we show how social or political considerations, including considerations of gender, sometimes constrain good scientific decisions, we do not thereby show that good science is bad science (Potter, 1995a,b). And because we need not confuse philosophers' narrow accounts of scientific rationality with scientific rationality itself, the failure of the rationalist project has no consequences for the cognitive authority of science. The rationality and cognitive authority undoubtedly possessed by science do not

L. H. Nelson and J. Nelson (eds), Feminism, Science, and the Philosophy of Science, 121–138.

depend upon the rationalist project; therefore, feminist and other science scholars who have given up the rationalist project do not make science irrational.

SECTION ONE: FEMINIST AND NON-FEMINIST USES OF UNDERDETERMINATION

Andrew Pickering's work on high-energy physics nicely exemplifies a non-feminist use of the underdetermination thesis. Pickering describes the production of the current "standard model" in high-energy physics and argues against the traditional view that experimental facts, themselves unproblematically read off from experiments, "dictate which theories are to be accepted and which rejected." Instead he argues:

> even if one were to accept that experiment produces unequivocal fact, it would remain the case that choice of a theory is underdetermined by any finite set of data. It is always possible to invent an unlimited set of theories, each one capable of explaining a given set of facts. Of course, many of these theories may seem implausible, but to speak of plausibility is to point to a role for scientific *judgment*: the relative plausibility of competing theories cannot be seen as residing in data which are equally well explained by all of them. Such judgments are intrinsic to theory choice, and clearly entail something more than a straightforward comparison of predictions with data (Pickering, 1984, pp. 5–6).

Pickering's strategy is to point out crucial decisions in the history of how the new physics replaced the old; against the view of both philosophical rationalists and of the scientists themselves that the experimental data obliged the scientists to opt for a particular theoretical model, Pickering points out that the physicists could have interpreted the data according to some other theory. Given that the choice of theory is underdetermined by the data, Pickering's analysis is that scientists choose in accordance with their professional interests. Experimentalists are trained in the use of certain types of apparatus, techniques, etc., while theoreticians learn to use certain analogies, mathematical approaches, and so on. The training provides resources to be exploited as the opportunity arises; for example, experimentalists will try to use the combination of resources provided by their own research tradition when new theoretical models suggest new work to be done, and theoreticians will act in a similar way when there are new experimental results to be explained. Pickering dubs this approach to explaining the choices of scientists "opportunism in context."

Opportunism in context, like other interest theories, shows us scientific decisions which are relative to the interests of the scientists who make those decisions; since different scientists have been trained in different research traditions, they will have different interests and, at crucial junctures, will likely make different decisions about how to interpret experimental results and, given the underdetermination of theory by data, about which theories to accept or reject.[1]

I have also used the underdetermination thesis (although my understanding of it differs from Pickering's) to help make clear how gender politics influenced the production of Boyle's Law of Gases (Potter, 1989, 1993). That law was produced as part of the new paradigm unfolding in seventeenth-century science, the mechanical philosophy. When Robert Boyle chose between the old hylozooist system of natural knowledge and the new mechanical philosophy (not yet a well worked-out scientific system), he took into consideration not only observational and experimental data, but also the social meaning of each paradigm. For in mid-seventeenth-century England, hylozooism supported the religious and social aspirations of radical protestant sectarians (as well as the old Aristotelian natural philosophy upon which conservative Catholic and Anglican religious and social philosophies rested). Boyle shared with the radicals and others the widely held conviction that the natural order on one hand and the moral and social order on the other are mutually reflective. Sectarian political views reflected and were supported by a natural philosophy that grew out of certain theological heresies involving hylozooism, the principle that all matter is alive. Instead of the hierarchical society logically supported by the doctrines of the Church of England, the sectarians rejected the state church and preferred a more democratic society, logically supported by their belief that God is in all his saints so they do not need priests of the established church to mediate between them and God.

Sectarian emphasis upon the individual soul and the Spirit within had important implications for sectarian women. The seventeenth century saw the development of the ideal woman as a bourgeoise who was to marry and to stay at home minding the house; while married, she was to own no property. She had no voice in the Church or State. But sharing at least spiritual equality, all members of sectarian congregations, including women, debated, voted, prophesied, preached and even travelled abroad to spread their message. Moreover, sectarian women were among those London women who, throughout the 1640s and into the 1650s, petitioned and demonstrated at Parliament complaining of the "decay of trade" and the high price of food due to the war.

Boyle, and others who later became mechanists, were concerned not only about the democratic social aspirations of the radicals, but also about the outrageous activities of women at Parliament and in their churches. It was during this period that Boyle wrote what we might call his "essays on women" in which he exposes a relentless concern that women occupy the domestic space being created for them by bourgeois liberal ideology.

Boyle was well equipped to negotiate against the radicals' class and gender struggles by defeating the natural philosophy used to support them. I suggest that suppressing these struggles provided *one* of the constraints upon the decision by Boyle and other mechanists to adopt the hypothesis that matter is inert and to argue against hypotheses supporting hylozooism.

Boyle confirmed an early version of the law that bears his name on the basis of experiments performed to defeat a hylozooist hypothesis while he was engaged in a dispute with Franciscus Linus over the proper explanation of a number of experiments involving suction phenomena, the Torricelli tube (predecessor of the barometer) and the air pump. (Many of these experiments are discussed in Shapin and Schaffer, 1985.) As a mechanist, Boyle held that matter is passive, brute or dead; that change is ultimately due to an external, nonmaterial entity, God; and that all the properties of matter can be explained, as he put it, by the "shape, size, motion, and other primary affections of the smallest parts of matter" (Boyle, 1772, p. 37). And using these principles, he sought to explain the experimental and observational data. Linus, on the other hand, claimed to explain the same data by appealing to a number of Aristotelian hypotheses and in particular to the hypothesis that a funiculus (a cord-like vapor) is given off by the mercury in the Torricelli tube as the air is pumped or sucked from it.

In the mid-seventeenth century the experimental and observational phenomena in question did not determine between their competing explanations. Typically Boyle said of the experiments he performed, not that the results refuted Linus' funicular hypothesis, but only that the results of his experiments "agree with and confirm" and "agree rarely-well" with his mechanistic hypothesis. Moreover, Boyle recognized that his dispute with Linus could not be settled by one experiment or by twenty; he did not claim that Linus' theory is false, but that it is unnecessary, precarious, rests on an unproven (in the strong sense of having eliminated alternate explanations) assumption and that it is unintelligible – by which he meant that it is not explained mechanically and that it has problems (which it does). Finally, Boyle said that he and Linus differ "so generally" because each has "reasoned closely" from "his own principles" which are "*contrary suppositions*." Laudan finds in such remarks evidence for Boyle's holding an early version of the underdetermination thesis: "Boyle does not believe that theories will arise ready-made from the data, or that the data will uniquely determine any single theory" (1966, p. 87). Both hypotheses covered the available data fairly well, and the funicular hypothesis enjoyed the advantage of belonging to the broader Aristotelian paradigm which, for all its anomalies, had more explanatory power than the nascent mechanical philosophy. Nevertheless, Boyle argued strongly for his mechanistic hypotheses and against Linus.

Linus' Aristotelian science offered strong support to a religious and political outlook to which Boyle objected, and in arguing against the science that supported the religion and politics, Boyle helped undercut them. Nor is it unreasonable, when the data do not uniquely select one theory and the competing accounts have comparable empirical adequacy, to select the theory that *coheres* with one's world view. As we mentioned above, people of the seventeenth century understood the natural and social orders to reflect one another; therefore, one could reasonably choose as an explanation of controversial natural phenomena that account carrying the most congenial social meaning.

SECTION TWO: THE POST-KUHNIAN TURN FROM NARROW DEFINITIONS OF SCIENTIFIC RATIONALITY

Antagonists on both sides of the current dispute over underdetermination agree only that, as the Quine–Duhem thesis shows, the isolated hypothesis is not the unit of empirical test because we cannot derive predictions from an isolated hypothesis; instead, we derive predictions from the hypothesis plus a set of auxiliary hypotheses. The scope of the set is controversial, however; it might include all of science (Duhem, 1981, Ch. 6), but it is probably a limited subset of the set of all scientific beliefs (Quine, 1975, pp. 314–315). Pragmatists have argued that when a proposed hypothesis is confronted by recalcitrant observations, we can hold to it *if we are willing to make the necessary changes* in the set of auxiliary hypotheses used to make the predictions. And here the fundamental sort of underdetermination arises: both choices, the choice to reject the hypothesis under consideration, leaving the set of auxiliary hypotheses unchanged, and the choice to hold to it and redistribute truth values throughout the set of auxiliary hypotheses, will produce empirically adequate systems. The choice between the two systems is underdetermined by the data.

What makes it difficult for philosophers to accept Pickering's narrative strategy, i.e., pointing out that physicists have such a choice following each experiment, is that as a science develops, the number of auxiliary hypotheses it is unwilling to revise increases. Fewer and fewer choices depend solely upon one set of data. Instead, the scientist's choice to accept or reject a hypothesis is constrained by the experimental results *plus* established auxiliary hypotheses which scientists are less and less willing to jettison. Quine and Ullian refer to this constraint as "conservatism" (1970, p. 66f.). Furthermore, they note that when scientists do decide to keep a proposed hypothesis even though the experimental results it predicts fail to obtain, they must make changes in the set of auxiliary hypotheses used to derive the hypothesis and they do so according to a maxim of minimum mutilation: rescind the fewest possible hypotheses or the ones least crucial to the overall theory while restoring consistency to the overall theory (Quine, 1992, pp. 14–15). Philosophers part company at this point over whether the decision to accept or reject a hypothesis in the face of a disconfirming observation is constrained by evidential considerations alone, by evidential and pragmatic considerations, or by evidential, pragmatic and/or non-technical (e.g., professional, social, political, etc.) considerations. The first extended discussion of this question, and the one that turned twentieth-century philosophy of science on its head, was Kuhn's. In 1962 he began to examine the conditions under which scientists do actually hold to a hypothesis "come what may" (e.g., its predictions fail repeatedly) (Kuhn, 1962). His conclusion was, roughly speaking, that they do so (a) when their theoretical system is faced by serious anomalies, i.e. observations that resist continuous attempts to reconcile them with the current system and (b) when there exist at least the rudiments of an alternative system which might adequately account for them. We should note that this is not a situation in which both theoretical systems are empirically equivalent, for Kuhn points out that the choice is made well before the new system has been developed. Moreover, (a) and (b) are not sufficient to cause scientists to make such a choice. Thus, if the current theoretical system, current methods and current standards all constrain the rejection of the proposed hypothesis and maintenance of the current system, the philosophically and sociologically interesting question is not why scientists reject recalcitrant observations and maintain the current system intact. The interesting question is why they decide to make a change in the system.

Inasmuch as scientific rationality was supposed by philosophers to be captured by one or another logic of science, and Kuhn's arguments made it clear that these logics were incapable of accounting for major scientific changes that include revolutions in methods and standards, Kuhn's view was attacked for making science appear "irrational" and for making scientific decisions appear to be the results of "mob psychology." But he persuaded many that "[i]ndividual scientists embrace a new paradigm for all sorts of reasons and usually for several at once." And he suggested that to find reasons for the choice, we might have to look at science education, professional loyalties, aesthetic sensibilities and at desiderata lying "outside the apparent sphere of science entirely," including for example, religious, personal and national loyalties (Kuhn, 1970, pp. 152ff.). Of course, the choice to keep a troublesome hypothesis and to change the truth-value or the meaning of one or more auxiliary hypotheses does not always precipitate a major change throughout the system; such a change can be relatively minor. Moreover, sociologists of science have found that cases in which such a change is proposed by some scientists and contested by others are likely to yield examples of professional interests or other "non-scientific" desiderata for the choice (Shapin, 1982).

Laudan has been foremost among philosophers of science who for the last thirty years have rejected such efforts by sociologists on the grounds that they reduce scientific change to "a whimsical change of style or taste" and depict scientific decision-making as "fundamentally capricious" (1984, p. 72). Laudan has pushed ahead with the rationalist project begun by logical empiricism: to come up with an account of scientific rationality narrow enough to exclude desiderata lying "outside the apparent sphere of science entirely." But persuaded of the Duhem–Quine thesis that no hypothesis faces the tribunal of experience in isolation, he joined Lakatos and others in moving from the hypothesis as the unit of scientific rationality to a larger unit, the research program or research tradition, a large unit including theories, methods, standards and goals (Laudan, 1977, 1984). Laudan's first attempt to capture the rationality behind scientific theory choice distinguished two sorts of problems that occur in research traditions, empirical problems arising when current theories are unable to explain certain empirical findings, and various kinds of conceptual problems, notably consistency within a theory or among theories (Laudan, 1977). One research tradition is more rational than another if it solves more of the important empirical problems and has fewer conceptual problems. Of significance for those of us who want to understand how science is actually done, Laudan agreed with Lakatos that internal history of science is sufficient to explain successful research traditions; such an internal history does not deny that "social causes" might have brought a scientist or scientists to believe a successful hypothesis. But we have no need for such an explanation if we can reconstruct the case to fit our account of scientific rationality. Case studies of rationally reconstructed successful science are perfectly adequate even though they omit the part played by social considerations in the actual decisions scientists made. However, we do need such an explanation, to be provided by an external history or a sociology of science, for unsuccessful research traditions, i.e. false or irrational beliefs. To explain these failures, we might well look to personal or social factors that influenced the judgement of scientists. Laudan thus marks out an "arationality" principle for explaining the history of scientific beliefs and sets up a neat division of labor: philosophers explain good science; sociologists and others explain poor science.

Laudan's arationality principle establishes an asymmetrical approach to explaining scientific beliefs and is diametrically opposed to David Bloor's "impartiality" and "symmetry" principles for explaining those beliefs. According to Bloor, such explanations "would be impartial with respect to truth and falsity, rationality or irrationality, success or failure. Both sides of these dichotomies will require explanation." Moreover, explanation of scientific belief "would be symmetrical in its style of explanation. The same types of cause would explain, say, true and false beliefs" (Bloor, 1976, pp. 4–5). The intuition underlying the symmetry principle is that the truth and/or rationality of a scientific theory is not its own explanation; rather, how a theory came to be regarded as true or as rational needs explaining in the same terms by which we explain how a theory came to be regarded as false or irrational. Laudan, on the other hand, maintained that "when a thinker does what it is rational to do, we need inquire no further into the causes of his action. ..." (1977, p. 188) (see also Laudan, 1981 and Bloor, 1981).

In 1984 Laudan did not include the underdetermination thesis in his attack against the possibility that successful scientific work could be explained by social, political or other "irrational" factors. Instead, he charged that because Bloor, Hesse and Quine ignore the distinction between theory choice and theory preference, they are led to argue that, in Laudan's words, "since theory choice is underdetermined by methodological rules, it follows that no rational preference is possible among rival theories, which entails, in turn, that every theory is as well supported as any other, and that every party to a scientific debate is thus as rational as every other" (Laudan, 1984, p. 30). (The difference between the argument Laudan attributes to these thinkers and the argument they, along with Kuhn, actually make is a difference of framework or scope. Kuhn does make such an argument about *rival paradigms*; and, as we shall see below, Hesse and Quine would agree that *global* theory choice is underdetermined by methodological rules.) Nevertheless, to block the slide from underdetermination (here left unquestioned) to what he perceived as pernicious relativism, Laudan distinguished (a) strict underdetermination, whereby two or more hypotheses are (perpetually) evidentially equivalent in that no conceivable evidence could ever discriminate between them and (b) underdetermination by current rules and existing evidence: although we can *conceive* of evidence that would differentially support one hypothesis, we do not now have it, in which case he suggested that "the participants will simply have to agree to disagree – pending the accumulation of further evidence" (Laudan, 1984, p. 28). (This is a normative, rationalist suggestion, not a description, since scientists rarely wait for further evidence; instead, they accept and work with one or the other hypothesis. Why they do so is precisely what is at issue between rationalists and anti-rationalists, and is open to scrutiny by sociologists who may find pragmatic or social criteria at work.) In both these cases, he says, our *belief* in one theory is not rationally dictated by the rules or by the evidence. But in all other cases, i.e. those not actually underdetermined, even though we can conceive of evidentially equivalent rivals, we are merely choosing among actual rivals, and it is permissible and rational to *prefer* (though not to believe) the hypothesis that has a greater degree of empirical support.

SECTION THREE: THE LAUDAN–LEPLIN ATTACK ON DEDUCTIVE MODELS OF SCIENTIFIC REASONING

Recently Laudan and Leplin (1991), have attacked the underdetermination thesis itself by rejecting the view that there are empirically equivalent rivals to every successful theory. Their attack rests on the following set of assumptions about the underdetermination thesis and the pragmatist argument for it: the underdetermination thesis states that

(1) every theory is underdetermined by the empirical evidence which is entailed by
(2) every theory has empirically equivalent rival(s)
(3) the empirical evidence for a theory T is contained in the set of empirical consequences of T
(4) the set of empirical consequences of T is the subset of its logical consequences expressible in observation sentences
(5) empirical equivalence is having the same set of empirical consequences, i.e. having the same subset of logical consequences expressible in observation sentences.

With these assumptions in mind, Laudan and Leplin (hereinafter L&L) offer counterexamples to prove that (3) is wrong and by arguing that, if (3) is wrong, the concept of empirical equivalence is not applicable, so (2) is false and the pragmatists' argument for (1) fails. Their analysis of how so many philosophers could go so wrong is that those philosophers have confused semantics with epistemology because they modeled the epistemology of science semantically. Hence, the underdetermination thesis is an artifact of semantic models. As rationalists, L&L maintain that "evidentially probative grounds" can be found for distinguishing between empirically equivalent rivals (1991, p. 450). We should note that L&L do not deny that there might be cases in which two or more incompatible theories are empirically equivalent (this might be Laudan's reading of the dispute between Boyle and Linus); rather, they deny that every theory has such empirically equivalent rivals. They seek to cast doubt on the applicability of the concept "empirical equivalence" by arguing that, even if there were empirically equivalent theories, that equivalence is defeasible hence underdetermination never obtains; it never obtains because, they assume, underdetermination obtains only when empirically equivalent rivals remain empirically equivalent forever. L&L demonstrate the defeasibility of empirical equivalence by pointing out that, sooner or later, change will occur both in the range of observable phenomena for which the theories account and in the set of auxiliary hypotheses necessary for deriving the observable consequences of each of the empirically equivalent theories. (We will see below the significance of L&L's

admission here that auxiliary hypotheses are necessary for deriving the observable consequences of a theory.)

The response to L&L's defeasibility attack has been to point out that, even supposing it true that proponents of the underdetermination thesis had overlooked the fact that the range of the observable can change and that any set of auxiliaries is likely to be augmented or diminished, we can construe empirical equivalence as a relation between indexed theories and go on as before (Kukla, 1993, pp. 1–2; see also Bergström, 1993, pp. 340 and 340n). Responding in turn, L&L (1993) offer no further argument against empirical equivalence and underdetermination construed in this way; instead, they simply maintain that

> ... if the observational consequence classes of theories ever diverge, then those theories are *never* empirically equivalent. EE [the thesis that every theory has observationally equivalent rivals] is clearly intended, by proponents and detractors alike, as an atemporal thesis.... It denies for any theory the possibility of (ever) observationally discriminating it from some rival theory (1993, p. 8).

For the purposes of feminist and other science scholars, the interesting cases are those in which there are or could be empirically equivalent rival theories, however short-lived the equivalence. For us, the interesting question is on what grounds scientists decide in favor of one or another rival. It is noteworthy that Laudan (1984) agreed, and that his 1991 argument represents a change in his understanding of underdetermination. In 1984 Laudan recognized a problem of "strict," i.e. perpetual, underdetermination for some hypotheses, though not for all, but he also admitted temporary underdetermination: "Instances of such (temporary) underdetermination arise fairly often, and they are, of course, the most interesting ones for historians and sociologists of science to examine" (1984, p. 28, parentheses in original). By 1991, temporary underdetermination arising from temporary empirical equivalence has disappeared and an argument against "strict" underdetermination and perpetual empirical equivalence is put forward instead. In the following discussion, empirical equivalence and underdetermination will be understood as time-indexed.

To support and illustrate the notion of empirical equivalence, philosophers have presented both algorithms for producing empirically equivalent theories and actual examples from science. L&L attempt to undercut this support by attacking both the philosophical algorithms and the scientific examples. Here, I will simply agree with them that what we want are "genuine" rival theories, and I agree with Lars Bergström that philosophers cannot construct good examples because we would have to act as scientists and produce good theories in empirical science (Bergström, 1993, p. 341). "Fakes" include logically or conceptually equivalent theories or theory formulations; theories whose possibility rests on the Lowenheim–Skolem theorem; instrumentalized theories (in which an existing theory is stripped of some or all of its unobservable structure) and theories constructed by Goodmanesque maneuvers (Kukla, 1993; Laudan and Leplin, 1993; and Hoefer and Rosenberg, 1994). This leaves us with L&L's attack on the claim that there exist scientific examples of empirically equivalent theories. They take up a favorite of philosophers, Newtonian mechanics. And following van Fraassen, they set out two versions of Newton's laws of gravity and motion (*TN*): *TN* + *R* adds the hypothesis that the center of mass of the expanding universe is at rest in absolute space, and *TN* + *V* adds the hypothesis that the center of mass of the universe has constant absolute velocity *V*. Against van Fraassen's claim that these two theories are empirically equivalent, L&L imagine an extension of *TN* + *V* to nonmechanical phenomena: absolute motion gives rise to a new, detectable particle; thus, detection of the particle will confirm *TN* + *V*. Moreover, because the positive absolute velocity of the universe represents energy available for creation of these particles, *V* can be measured by counting the particles. Hoefer and Rosenberg (hereinafter H&R) object to this argument by pointing out that when philosophers like van Fraassen use this example, they are assuming that *TN* is true and so its empirical consequences are true, particularly the claim that differences in absolute velocity have no observable effects. L&L's argument for the defeasibility of *TN* rests on *TN*'s being false; in their imagined extension of *TN* + *V* (and of *TN* + *R*), it turns out to be false that differences in absolute velocity have no observable effects. H&R note that

> [i]n general whether two theories are equivalent may depend on external factors, such as what other theories and facts are assumed as auxiliaries for the derivation of observable consequences. But it does not depend on whether the theories are true; to say that two theories are empirically equivalent is to say that they predict the same observable phenomena, regardless of whether the predictions are correct.

And H&R conclude, as I think correctly, that *TN* + *R* and *TN* + *V* constitute a historically reasonable example of empirical equivalence despite their known falsehood[2] (H&R, 1994, p. 599).

We begin to see the depth of the disagreements between L&L and proponents of the underdetermination thesis when we examine the notion of empirical equivalence itself. Quine has defined the empirical content of a testable sentence or set of sentences for a speaker as "the set of all the synthetic observation categoricals that it implies" (synthetic observation categoricals take the form, "when a willow grows at the water's edge, it leans over the water"). Empirically equivalent sets of sentences, then, imply the same synthetic observational categoricals (Quine, 1992, p. 17). And again, he says that when two theories are empirically equivalent, "whatever observation would be counted for or against the one theory counts equally for or against the other" (Quine, 1992, p. 96). As counterexamples, L&L provide cases in which the evidence supporting a hypothesis is not found among the logical consequences of that hypothesis and cases in which some of the logical consequences of a hypothesis provide no evidential support for it. The cases designed to show that not all evidence for a hypothesis is a logical consequence of it are schematized by L&L as follows:

> Theoretical hypotheses H_1 and H_2 are empirically equivalent but conceptually distinct. H_1, but not H_2, is derivable from a more general theory T, which also entails another hypothesis H. An empirical consequence e of H is obtained. e supports H and thereby T. Thus, e provides indirect evidential warrant for H_1, of which it is not a consequence, without affecting the credentials of H_2 (L&L, 1991, p. 464).

For example, the theory of continental drift (T) tells us that

every region of the earth's surface has occupied both latitudes and longitudes significantly different from those it now occupies. It is thereby committed to two general hypotheses:
H_1: There has been significant climatic variation throughout the earth, the current climate of all regions differing from their climates in former times.
H_2: The current alignment with the earth's magnetic pole of the magnetism of iron-bearing rock in any given region of the earth differs significantly from the alignment of the region's magnetic rocks from earlier periods (L&L, 1991, pp. 461–462).

Studies of remnant magnetism yielded evidence for H_2; this evidence is not a consequence of H_1, but by confirming T, which entails H_1, it also provides support for H_1. This case would fit the schema produced by L&L if H_1 had an empirically equivalent rival H, for then the indirect evidence provided by remnant magnetism would favor H_1 over H. The hypotheses would no longer be empirically equivalent and underdetermined by the evidence.

In their attack on Quine's deductive model, L&L ignore the distinctions among a hypothesis, an isolated theory, a large group of theories – what Quine sometimes refers to as a "critical semantic mass" or a "pretty big" conjunction of sentences (1992, p. 17), and a global theory or system of the world. L&L's discussion and conclusions mention only theories, but their counterexamples turn on hypotheses. They are correct in pointing out that (a) not all the logical consequences of a hypothesis or isolated theory provide evidence for it and that (b) sometimes the evidence supporting a hypothesis or isolated theory is not a logical consequence of it. Quine himself agrees with (a) for, as we have seen, he limits the logical consequences that provide evidence for a hypothesis or theory to the subset consisting of "all the synthetic observation categoricals that it implies". But if Quine denies (b), he is wrong and if he agrees with (b), then the deductive model is, as L&L remark, unperspicacious (1991, p. 461), but it is unperspicacious only for picturing the relations between hypotheses or isolated theories and the evidence supporting them. The model does, however, picture the relations between global theories or systems of the world and evidence. A global theory or system of the world can be understood to be the complete set of scientific theories that we have now. (Alternative understandings will be discussed below.) And the set of all theories together does imply the set of observation categoricals that provide support for it. On this understanding of "theory", i.e., "global theory", the counterexamples offered by L&L fail to defeat the claim that every theory has empirically equivalent rival(s) and the claim that the choice between such theories is underdetermined by the evidence.

This understanding of "systems of the world" and this defense of empirical equivalence and underdetermination should not be confused with a similar yet importantly different defense by H&R. H&R argue that L&L have shrunk the scope of the putatively empirically equivalent theories and have offered arguments only against the underdetermination of "partial" theories as opposed to "global" theories. But H&R agree with L&L that theories are only empirically equivalent if they are so perpetually; therefore, they agree that partial theories are not (perpetually) empirically equivalent. However, H&R argue, underdetermination is a thesis about "empirically adequate total science; it is a thesis about what Quine calls 'systems of the world' – theories that comprehensively account for all observations – past, present and future" (H&R, 1994, p. 594). We can see immediately that the counterexamples offered by L&L do not cut against this understanding of empirically equivalent theories, for on this understanding, H_1 and H_2 both belong to one, global, empirically adequate theory of the world and would not be candidates for equivalence or underdetermination in the first place[3] (H&R, 1994, p. 600).

SECTION FOUR: THE IRRELEVANCE OF EMPIRICAL EQUIVALENCE TO UNDERDETERMINATION

As we have seen, L&L attack the notion of empirical equivalence because they believe that it has provided the most important argument for underdetermination. But we shall see that, although the notion of empirical equivalence provides a general argument for the underdetermination of global theories, the notion of empirical equivalence is irrelevant to the underdetermination of hypotheses, isolated theories and "pretty big" theories (i.e., conjunctions of theories large enough to make predictions with which to test a hypothesis or isolated theory). Despite the examples put forward by L&L in which they say that two hypotheses are empirically equivalent, empirical equivalence cannot be attributed to hypotheses or isolated theories. No isolated hypothesis or theory can make the prediction(s) that will test it. If the prediction fails, the conjunction of the hypothesis (or isolated theory) plus all the necessary auxiliary theories is falsified and, if the prediction pans out, it is the same conjunction that is supported. It follows that empirical equivalence (on any definition that refers to having the same evidence – whether as logical consequences, as providing the same probability or as counting equally for or against each rival) is properly attributable only to the conjunction of the hypothesis or isolated theory plus all the auxiliary theories necessary to make the predictions that provide support for it or disconfirm it.

This means that rival hypotheses and rival isolated theories are not empirically equivalent, but it does not mean that rival hypotheses and rival isolated theories are not underdetermined by the evidence. Just as scientists can choose to accept or reject a hypothesis or isolated theory regardless of how its predictions turn out, they can choose a hypothesis or its rival *if they are willing to make the necessary changes in the rest of their system*. If we did characterize two such rivals as "empirically equivalent," it would be in the very different sense that each one leads to a different overall system that is *empirically adequate*. But if, when we accept one of the rivals, we make many changes in our overall system, the two rivals cannot be empirically equivalent because they do not predict the same data, are not supported by the same data, do not cover the same data. This is because our theories tell us what the data are (there is no sharp distinction between theory and observation or data) and if we have incompatible/very different theories, we will have incompatible/very different data. Nevertheless, the rivals are underdetermined by the data in the sense that the original data plus the backlog of theory used to predict the data do not determine which choice we should make and either choice leads to an empirically adequate overall system of theories.

Pretty big theories are like global theories in that they are big enough to include all the auxiliary theories necessary to make predictions for testing a hypothesis or isolated theory. This means that two such theories could be empirically equivalent – a possibility that L&L admit when they say, "We maintain not that there are no cases of empirical equivalence, but that the claim that there are is defeasible" (1991, p. 459). And if they were empirically equivalent, they would be underdetermined. But whether such a theory has an empirically equivalent rival or not, it is underdetermined in the same fundamental way that every hypothesis and every isolated theory is underdetermined.

SECTION FIVE: A NON-DEDUCTIVE MODEL OF SCIENTIFIC REASONING

Not all semantic models are susceptible to the criticism L&L level against Quine's deductive model. Hesse's (1974) Network Model is a non-deductive way of picturing the relations among beliefs, specifically, as she presents it, the complex relations among scientific laws. Although she does not say so explicitly, it is not a model of one scientist's system of beliefs – not even the "ideal" scientist's – because scientific knowledge does not reside in any one scientist's head (Potter, 1993; Nelson, 1993; Longino, 1990). Scientific knowledge is public and, in this sense, objective. Whether a particular belief should be included in the public, legitimate body of knowledge is arguably the most important aim of scientific discussion and debate. Thus, as the body of scientific knowledge changes, the model changes. The model need not be understood as static; rather, it can be understood to be constantly changing as scientific knowledge shifts and changes. Finally, unlike Quine's model, Hesse's model is non-deductive; she says,

> [t]he logical system that immediately suggests itself for the explication of inference in a network of theory and observation is the theory of probability. This is an obvious generalization of the deductive model of science, which, though inadequate as it stands, may certainly be seen as a first approximation to adequacy as a logic of science. In the deductive model, theory and observation are seen in terms of a hierarchy ordered by deductive inferences from theory to observation, but the deductive model as such gives no account of logical inference from observation to theory. In a probabilistic inductive model ... the hierarchy is replaced by a system in which all statements are reciprocally related by conditional probability, of which deductive entailment is the limiting case. Probabilistic inference can therefore be seen as a generalization of deduction, permitting inductive and analogical as well as deductive forms of reasoning in the theoretical network (1974, p. 5).

Thus, the relations among scientific beliefs, including deductive, inductive and analogical relations, can all be captured by the relations among statements in the model. Hesse argued in 1974 for a particular theory of probability with which to analyse the Network Model (and for a Bayesian personalist interpretation of probability), but nothing in the model dictates which theory must be used.

We see that Hesse can easily model the examples L&L produce to show that evidence may support a hypothesis of which it is not a consequence; introducing their first example, L&L remark:

> We begin by noting that instances of a generalization may evidentially support one another, although they are not consequences of one another. Previous sightings of black crows support the hypothesis that the next crow to be sighted will be black, although that hypothesis implies nothing about other crows. Supposing this evidential connection to be uncontroversial, we ask why, then, in the case of universal statements it should be supposed that evidential support is limited to logical consequences (L&L, 1991, p. 461).

Let us see how Hesse's model accommodates this case. Here our beliefs include:
(1) $x \ldots xn$ are crows
(2) $x \ldots xn$ are black
(3) $x \ldots xn$ are crows and $x \ldots xn$ are black
(4) $xn + 1$ resembles $x \ldots xn$ in some non-color features
(5) $xn + 1$ will be black.
The relations among these beliefs can easily be modeled by Hesse as: (3) rests on the conjunction of (1) and (2); and given belief (3), belief (4) has probability P.

L&L tell us that all their examples of the second sort follow the same schema (cf. p. 130 above) which presents no problem for Hesse's model. Any evidence e supports hypothesis H and theory T probabilistically and so makes hypothesis H_1 more probable. That is, e has a *direct* evidential relationship to both H and H_1. It follows that H_1 and H_2 were not empirically equivalent in the first place, if empirical equivalence is defined as having the same evidential support.

SECTION SIX: THE INDEPENDENCE OF UNDERDETERMINATION AND SEMANTICS

L&L argue that the underdetermination thesis depends on the reduction of epistemology to semantics (1991, p. 467), and they imply that all semantic models are subject to the confusion of epistemology with semantics.[4] On the basis of their claim that semantic models cannot capture epistemological insights, L&L conclude that most twentieth-century epistemology and philosophy of science has been confused in the same way and has been "arbitrarily and unreasonably constrained." It has deviated from the proper track it was on in the nineteenth century when theories were understood to be explanations of all relevant phenomena, not logical antecedents of a hodgepodge of relevant and irrelevant logical consequences (L&L, 1991, pp. 466–472). As it stands, the allegation that pragmatists have confused semantics with epistemology is simply false, for it is the allegation that philosophers who use semantic models have confused the model with the thing being modeled. Although semantic models may not be the best way to understand scientific work, the semantic models of Hesse, Quine and other pragmatists do not confuse epistemology and semantics.

For example, on Hesse's Network Model of the body of scientific knowledge, hypothesis H has probability P given evidence e and given the background assumptions, H_1–H_n. Now certainly, that fact does not entail that scientists individually or collectively believe H because it has *P* given e and H_1–H_n. The actual reasons for which each scientist holds a belief might or might not accord with the standards of the relevant epistemic community. Thus, the model does not describe actual relations among beliefs held by scientists as individuals or even as groups. Rather, we should understand it as describing the relations scientists claim obtain among beliefs when they argue in public to the relevant community. The model pictures relations among beliefs argued for on the basis of shared standards of proof. The model does assume that the beliefs to be proven, the beliefs used to prove them and the relations among them can be stated. If these assumptions are correct, then the beliefs and the relations among them constituting public or shared scientific knowledge can be semantically modeled.

Contra L&L, the thesis that every theory is underdetermined by the evidence does not arise from a semantic model, despite Quine and others modeling the thesis semantically. To state that when they run an experiment testing a prediction, scientists assume a very large number of laws, principles, hypotheses, facts, and so on, both about nature and about their experimental equipment, is not to make a semantic point, but a point about scientific reasoning. Notoriously, scientists do not follow the laws of deductive logic (or of probability) in their reasoning, but logic and probability theory can help us to make their reasoning specific. Thus, the point that when a prediction fails to obtain, scientists must decide whether it was somehow wrong or whether some other assumption (probably about their equipment) was faulty is not a logical point, even though we can use logic to make the point. When we test a hypothesis (or an isolated theory or a pretty big theory) and get a particular experimental result, our options include (A) accepting the results and the hypothesis if we decide that the results fit within the parameters of what we expected to see given the hypothesis and given our understanding both of our area of science and of our equipment; or (B) rejecting the results and the hypothesis because we decide that the results fail to fit our prediction. But more importantly, (C) if we want to keep our hypothesis and accept experimental results that fall outside the parameters of our prediction, we can do so if we are willing to change enough of our science so that the results fit our expectations. Or (D), if we want to keep our hypothesis and reject the results that it predicted, we can do so if we are willing to change enough of our science so that we do not predict those results. The changes we make in our science in cases (C) and (D) will be more or less drastic. But the decisions between (B) and (C) and between (B) and (D) are underdetermined by the results and even by the rest of our science.[5] These are both instances of the underdetermination thesis which states that every hypothesis and every theory is underdetermined by the empirical evidence, and the argument leading to it is a standard way to reach the thesis that, contra L&L, does not depend at all upon the concept of empirical equivalence or upon semantics. Therefore, Laudan (1990) and L&L (1991) have failed to show that the underdetermination thesis is false. Of course, this little argument does not settle the question *whether there is a gap between theory and evidence*, for this is one of the most pressing problems in twentieth century philosophy of science. It has been discussed most often in semantic terms but it is neither a semantic issue nor an artifact of semantics. The claim that it is amounts to rationalist wishful thinking.

CONCLUSION

It may not be possible to offer an argument for the underdetermination thesis that will be acceptable to rationalists, for it is part of a picture of science in which not all good or successful scientific decisions are made solely on the basis of experimental or other observational data or on the basis of the data plus technical background assumptions, i.e. those lying *inside* "the apparent sphere of science." But case studies have shown us that the picture fits many scientific decisions. And unless the term "rational" is limited to decisions made solely on the basis of data or data plus technical background assumptions, it can also be applied to decisions based on a combination of the data plus background assumptions plus other considerations including pragmatic ones, professional interests, gender considerations, etc. Reluctance to admit that good decisions can be constrained by, for example, gender considerations arises in part from the belief that any considerations of gender constitutes bias and since biased science is bad science, decisions constrained by gender politics are *ipso facto* bad ones. But in what sense are they bad decisions? Such decisions do not necessarily produce false or unsuccessful hypotheses as we see in the case of Boyle's decision to accept the inertia of matter. Hypotheses whose choice is constrained in part by gender politics can still be, as I have argued elsewhere, empirically adequate, fruitful, general, and so on (Potter, 1995a,b, 1989; cf. also Longino, 1983). The underdetermination thesis remains a useful tool for supporting the feminist hunch that scientists make good choices and rational decisions that are nonetheless influenced by gender politics.

Mills College,
5000 MacArthur Blvd,
Oakland CA 94613,
USA

ACKNOWLEDGEMENTS

This essay has benefited immeasurably from the criticisms of Lynn Hankinson Nelson and Jack Nelson.

NOTES

[1] Pickering's understanding and use of underdetermination here is not unproblematic. For example, he does not make it clear that choice of a competing theory would have repercussions throughout a theoretical system. If the repercussions are major, we will end up with a very different overall theoretical system and, since theories tell us what count as data, very likely what count as data for the new system

will be very different from data for the old system. In this case, *the same data* are not "equally well explained by" the competing theories. (Nevertheless, the competing theories, nestled within their respective theoretical systems, may each be empirically adequate, i.e. cover all the relevant data.) A further problem with Pickering's analysis will be discussed below.

I apologize to Pickering for butchering his carefully constructed book in the foregoing summary. He can take cold comfort from the way I butcher my own in the following summary.

[2] Bergström doubts that there are any "very good" examples of empirical equivalence, but that is because he prefers to limit empirical equivalence to those theories having not only the same observational consequences but also the same balance of pragmatic virtues such as simplicity (1993, pp. 334ff.).

[3] Bergström, on the other hand, argues that if the underdetermination thesis were about theories that account for all observable events, it would be uninteresting since we are unlikely ever to come across such a theory (1993, p. 333). Instead, he notes, the underdetermination thesis is primarily concerned with what Quine calls "our system of the world" which presumably "is not complete in the sense that it can account for all observable events" (Bergström, 1993, p. 332n). Bergström proposes instead that a global theory or system of the world be understood as one that "contains or could contain the totality of someone's beliefs about the world" (1993, p. 332). The difference between the H&R and the Bergström interpretation of global theories or systems of the world is that, on the H&R interpretation, a system of the world could only be captured by the fictional end of inquiry, whereas for Bergström, our system of the world is the system of hypotheses we have now which does not account for all the observations possible at the fictional end of inquiry when we have a totally empirically adequate global theory, but instead accounts for all the observations possible for us according to our system of the world (presumably fewer than would be possible at the fictional end of inquiry). On this understanding of a system of the world, the counterexamples presented by L&L can also be seen to be examples of partial theories contained within our system of the world.

[4] There are two uses of "semantic model" at work in L&L; the first is for models such as those of Quine or Hesse that picture the relations among scientific beliefs as subject to the laws of deductive logic or of probability theory and the second is for van Fraassen's semantic theory of theories (L&L, 1991, p. 451n.).

[5] H&R argue, and attribute the argument to Quine, that "as the portion of science that we are inclined to hold constant and unrevised increases ... [t]hough it may be permissible in principle to preserve a favored hypothesis 'come what may', it would eventually be irrational to do so" (H&R, 1994, p. 593). It is not at all clear that this is Quine's view, but in any case, it leads to the arch-rationalist view that major scientific revolutions, however successful, are irrational.

REFERENCES

Bergström, Lars: 1993, 'Quine, Underdetermination, and Skepticism', *Journal of Philosophy*, **90:7**, 331–358.

Bloor, David: 1976, *Knowledge and Social Imagery*. Routledge and Kegan Paul, London, Henley and Boston.

Bloor, David: 1981, 'The Strengths of the Strong Program', *Philosophy of the Social Sciences*, **11**, 173–198.

Boyle, Robert: 1772, 'Of the Usefulness of Natural Philosophy', in Thomas Birch (ed.) *The Works of the Honourable Robert Boyle*, II. Facsimile edition published by Georg Olms Verlagsbuchhandlung, Hildesheim, 1965.

Duhem, Pierre: 1981, *Aim and Structure of Physical Theory*, trans. Philip P. Wiener. Atheneum, New York.

Fuller, Steve: 1990, 'They Shoot Dead Horses Don't They?: Philosophical Fear and Sociological Loathing in St. Louis', *Social Studies of Science*, **20**, 664–681.

Hesse, Mary: 1974, *The Structure of Scientific Inference*. University of California Press, Berkeley and Los Angeles.

Hoefer, Carl and Alexander Rosenberg: 1994, 'Empirical Equivalence, Underdetermination, and Systems of the World', *Philosophy of Science*, **61**, 592–607.

Kuhn, Thomas S.: 1962 and 1970, *The Structure of Scientific Revolutions*. University of Chicago Press, Chicago.

Kukla, André: 1993, 'Laudan, Leplin, Empirical Equivalence and Underdetermination', *Analysis*, **53:1**, 1–7.

Lakatos, Imre: 1970, 'Falsification and the Methodology of Scientific Research Programs', in I. Lakatos and A. Musgrave (eds), *Criticism and the Growth of Knowledge*. Cambridge University Press, Cambridge, UK.

Laudan, Laurens: 1966, 'The Clock Metaphor and Probablism: The Impact of Descartes on English Methodological Thought, 1650–65', *Annals of Science*, **22:2**.

Laudan, Larry: 1977, *Progress and Its Problems*. University of California Press, Berkeley, Los Angeles, London.

Laudan, Larry: 1981, 'The Pseudo-Science of Science?', *Philosophy of the Social Sciences*, **11**, 173–198.

Laudan, Larry: 1984, *Science and Values*. University of California Press, Berkeley, Los Angeles, London.

Laudan, Larry: 1990, 'Demystifying the Underdetermination Thesis', in C. Wade Savage (ed.), *Minnesota Studies in the Philosophy of Science*, **14**. University of Minnesota Press, Minneapolis.

Laudan, Larry and Jarrett Leplin: 1991, 'Empirical Equivalence and Underdetermination', *Journal of Philosophy*, **87:9**, 449–472.

Laudan, Larry and Jarrett Leplin: 1993, 'Determination Undeterred: Reply to Kukla', *Analysis*, **53:1**, 8–16.

Longino, Helen: 1983, 'Beyond Bad Science', *Science, Technology and Human Values*, **8:1**, 7–17.

Longino, Helen: 1990, *Science As Social Knowledge*. Princeton University Press, Princeton.

Nelson, Lynn: 1993, 'Epistemological Communities', in L. Alcoff and E. Potter (eds.), *Feminist Epistemologies*. Routledge, New York and London.

Pickering, Andrew: 1984, *Constructing Quarks*. University of Chicago Press, Chicago.

Pinnick, Cassandra L.: 1994, 'Feminist Epistemology: Implications for Philosophy of Science', *Philosophy of Science*, **61**, 646–657.

Potter, Elizabeth: 1989, 'Modeling the Gender Politics in Science', in N. Tuana (ed.), *Feminism and Science*. Indiana University Press, Bloomington and Indianapolis.

Potter, Elizabeth: 1993, 'Gender and Epistemic Negotiation', in L. Alcoff and E. Potter (eds), *Feminist Epistemologies*. Routledge, New York.

Potter, Elizabeth: 1995a, 'Good Science and Good Philosophy of Science', *Synthese*, **104:3**.

Potter, Elizabeth: 1995b, 'Methodological Norms in Traditional and Feminist Philosophy of Science', *PSA 1994*, 2.

Quine, W. V.: 1975, 'On Empirically Equivalent Systems of the World', *Erkenntnis*, **9**, 313–328.

Quine, W. V.. 1992, *The Pursuit of Truth*. Harvard University Press, Cambridge, MA.

Quine, W. V. and J. S. Ullian: 1970, *The Web of Belief*. Random House, New York.

Shapin, Steven: 1982, 'History of Science and Its Sociological Reconstruction', *History of Science*, **20**.

Shapin, Steven and Simon Schaffer: 1985, *Leviathan and the Air-Pump*. Princeton University Press, Princeton.

ILKKA NIINILUOTO

THE RELATIVISM QUESTION IN FEMINIST EPISTEMOLOGY

INTRODUCTION

Relativism has been an important issue in epistemology ever since the dispute between Plato and Protagoras. Relativism, after struggling with counterarguments by various kinds of objectivists and realists, survives today in several different variants. The lively discussions within feminist epistemology give new interesting material and insights for this long-standing debate. As feminism *prima facie* seems to be committed to the idea that knowledge is in some way gender-specific, it is no wonder that current forms of relativism have attracted some advocates of feminist epistemology. However, besides facing the danger of incoherence, the most radical versions of relativism undermine the whole project of criticizing "male bias" in science and thereby weaken the emancipatory potential of the feminist movement. Somewhat surprisingly, what is usually considered as the most prominent and best developed alternative to relativism, *viz.* the feminist standpoint epistemology, can be seen as a transformation of the Protagorean *homo mensura* doctrine into the questionable thesis that "woman is the measure of all things". Feminist empiricism has criticized the standpoint theory, but has so far left the relativism–realism issue unsettled. In this paper I conclude that it would be worthwhile to try to develop the philosophy and methodology of Women's Studies on the basis of critical fallibilist scientific realism. Another suggestion is that politics for women could be served by feminist applied research.

1. VARIETIES OF RELATIVISM

Relativism is a bundle of different doctrines: in the general scheme "*X* is relative to *Y*" we may choose *X* and *Y* in several alternative ways (cf. Niiniluoto, 1991b). Depending on the choice of *X*, relativism may be cognitive or moral in the broad sense (cf. Meiland and Krausz, 1982). In *cognitive* relativism, *X* is a category of *ontology* (object, space, fact, world, reality), *semantics* (reference, truth, meaning), *epistemology* (perception, belief, justification, knowledge), or *methodology* (theory, inference, rationality, progress). In *moral* relativism, *X* is a cultural or social category (custom, values, ethics, law, politics, religion).

On the other hand, *Y* might be taken to be an individual person; that is usually called *subjectivism* or *Protagoreanism* (Margolis, 1991), while F. C. S. Schiller called it "humanism". *Y* might be a group of persons, defined by some common characteristics – such as nationality, gender (*gender relativism*), social class (*class relativism*) or a whole species (*specieism*). In Nietzsche's *perspectivism*, *Y* is a perspective or a point of view. Further, in *cultural relativism*, *Y* is a form of life or a

L. H. Nelson and J. Nelson (eds), Feminism, Science, and the Philosophy of Science, 139–157.

culture (cf. Hollis and Lukes, 1982); in *conceptual* or *framework* relativism, *Y* is a language, a conceptual framework, a theory, or a paradigm (cf. Bernstein, 1983); and in *social* relativism, *Y* is defined by social and political structures, interests or values (cf. Brown, 1984).

By these choices for *X* and *Y* we may easily generate hundreds of variants of relativism. But, depending on philosophical background assumptions, there are also systematic interconnections between different relativist theses. For example, if a point of view is defined by a class position or by the possession of a conceptual system, perspectivism becomes identical with class relativism or framework relativism, respectively. If it is assumed that women have a language characteristically different from the male language (cf. Spender, 1982; Harding and Hintikka, 1983), then linguistic relativism entails gender relativism.

It is also clear that some variants of relativism exclude each other. For example, Protagoreanism, which claims that each individual person has his or her subjective beliefs or values, is incompatible with class and gender relativisms.

Some variants of relativism may seem to be almost trivially true. For example, we now recognize that human customs, values, belief systems, religions, and laws have in fact been different in many past and present cultures. However, it is important to distinguish such factual diversity or *relativity* from *relativism*. As a philosophical doctrine, the thesis that "*X* is relative to *Y*" claims that *Y* is the necessary or ultimate *medium* for the existence of *X* (for example, there cannot be visual perceptions without a viewpoint or location of the perceiver), or *Y* is the best, only, or ultimate *standard* or *measure* for *X* (for example, values as human constructions have to be decided by persons or cultures). It was precisely in this sense that Protagoras presented his *homo mensura* doctrine: "man is the measure of all things". Putnam's (1990) internal realism (objects and facts are relative to languages and theories) is an example of a relativist doctrine in the same sense, even though Putnam at the same time rejects the cultural relativity of truth.

Moreover, it is important to separate relativism from such negative doctrines as *scepticism* and *nihilism* which deny the existence of both absolute and relative standards (Meiland and Krausz, 1982; Margolis, 1991). For example, Paul Feyerabend's (1987) position is closer to Pyrrhonian scepticism than relativism. Thus, subjectivist, perspectival, cultural, and social forms of cognitive and moral relativism do not deny the existence of knowledge and values, but rather claim that their truth or rightness is always relativized to some individual, point of view, culture, or social interest. Hence, relativist doctrines are positive, non-trivial, controversial, and interesting. In principle, each form of such views has to be studied and evaluated separately.

We may distinguish *local* and *global* forms of relativism. The former restricts its claim to a specific category *X*, while the latter generalizes this claim to all *X*. For example, global subjectivism asserts that everything is relative to individual persons. A local gender relativism might be restricted to morality only.

It is also important to distinguish between radical and modest forms of relativism. Suppose that *X* is relative to *Y* in the sense that *Y* is the measure of *X*. Then *radical* forms of relativism assert that all of the different instances of *Y* are equally good (or equally bad). For example, a radical framework relativism about truth would claim both that truth is always relative to a conceptual system *L* and that all alternative conceptual systems *L* are equally good. Thus, there are concepts of relative truth "true-in-*L*" for each *L*, but we have no way of having rational preferences among the different frameworks *L*. Similarly, radical moral relativism asserts that all value systems are equally well justified. *Modest* relativism instead allows for the relativity to *Y*'s, but accepts the possibility of having at least some principles of comparing the instances of *Y* with respect to their adequacy. For example, a modest framework relativist thinks that our cognitive categories are relativized to conceptual systems, but still there are rational criteria of at least preferring some languages to others in some situations (cf. Niiniluoto, 1991a). Similarly, a modest moral relativist claims that our values are always relativized to some systems of moral ideas, but the comparison of such systems allows us to speak of moral progress (Niiniluoto, 1991b).

2. RELATIVISM VS. REALISM

The traditional argument against global forms of relativism, due to Plato, is their *incoherence* (see Siegel, 1987). Variants of this incoherence argument can be directed against many current forms of relativism (see Harris, 1992; Niiniluoto, 1991a).

Suppose, for example, that a feminist relativist makes the claim

(GR) All beliefs are relative to gender.

If the relativist now admits that GR itself is true in an absolute, gender-independent sense, the game is lost, since GR is thereby falsified. It may be possible to save the coherence of gender relativism by accepting that its basic thesis GR is valid only for the relativist (cf. Preston, 1992), i.e., true-for-women, but then this is a "self-vitiating" doctrine that can never be presented in a convincing way to men. The same argument goes through, if "woman" and "man" are interchanged. Hence, global gender relativism is either incoherent or irrelevant in communication across genders.

In my view, modest moral relativism is a plausible view: statements of the form "*a* is good" and "*a* is right" are incomplete, unless complemented by reference to some axiological or normative system, e.g., Christian religion or the moral code accepted in Finland (Niiniluoto, 1991b). This social constructivism with respect to morality, i.e., values and norms are human-made social artifacts, is compatible with realism with respect to statements of the form "In a moral system *S*, *a* is good".

The issues concerning cognitive relativism depend on controversial questions about reality, truth, justification, and the scientific method.

The so-called *Strong Programme* of the sociology of knowledge, developed with vigour by the Edinburgh school (Bloor, Barnes, Shapin), has conducted a series of historical case studies, intended as an inductive proof of the thesis that the beliefs of scientists have to be explained causally by appealing to the same type of social factors (cf. the discussion in Brown, 1984). For example, social class turns out to be a relevant factor in the explanation of beliefs about phrenology in nineteenth-century Scotland. At least in this case, beliefs have been to some extent "relative" to social factors. But this is not yet enough to establish a *causal* connection between class status and beliefs. It is also still far away from the general claim that social factors *always* influence beliefs in the scientific community. Moreover, even this factual claim would not suffice to prove the *normative* thesis of cognitive social relativism that social position and interests are standards for scientific beliefs.

The relativist's appeal to Quine's underdetermination thesis is not convincing, either (see Laudan, 1990). An anti-relativist may point out that many scientific controversies in fact have been settled by rational epistemological and methodological arguments. In cases where strictly logical and empirical criteria do not decide the choice between theories, there may be more general cognitive standards for the tentative acceptance of scientific hypotheses. If there still remain gaps in the available evidence and arguments, it is more rational to suspend judgement between the alternatives than to fill these gaps with politics.

The "Empirical Programme of Relativism", defended by Harry Collins (1981), makes the further – by now quite famous – claim that in natural science nature makes no contribution to the content of the beliefs of the scientists.

If intended as a factual empirical assertion about science, Collins's claim is in direct conflict with one of the most pregnant characteristics of scientific methods. As formulated by Charles Peirce, the creation of scientific knowledge is based upon causal interaction between the scientist and the part of reality under investigation. This interaction makes the transmission of information possible, since it forces the scientist to modify his or her beliefs under the external constraint. The testing of scientific hypotheses by observation and experiment is designed so as to satisfy precisely this condition. In this sense the method of science is based on the principles of *scientific realism*.

Critical scientific realism has to admit, with Peirce's *fallibilism*, that all factual beliefs in science are fallible, liable to error – uncertain, probable, or truthlike (see Niiniluoto, 1984). It is also crucially important that, as Popper has insisted, the results of individual scientists are not yet accepted until they have survived critical examination and discussion by other scholars. As Peirce and later advocates of social epistemology have emphasized, the *scientific community* is the true subject of scientific knowledge.

3. THE TRUTH ABOUT RELATIVISM

Joseph Margolis (1991) has recently promised "the truth about relativism". He admits that strong forms of cognitive relativism with relativized truth-values are self-defeating, but argues for a "robust" relativism with many "truth-like values" instead of bivalence. He grounds this view by an ontological picture of the world as a flux without permanent invariances, but does not otherwise develop it in detail.

To investigate the possible forms of cognitive relativism, let us recall that, according to the classical definition, *knowledge* is justified true belief. If this definition is applied to scientific knowledge, we have to understand that the subject *S* of knowledge is the scientific community. Then it becomes inappropriate to speak about the "beliefs" of the members of *S*: rather, the scientists propose hypotheses, pursue research programs to test them, and eventually prefer or "accept" those that are better than their rivals. But in most cases they do not believe that even these accepted hypotheses are strictly true. Today we know that, e.g., Newton's theory is only approximately correct in its everyday applications and quite far from the truth in some other applications. According to critical scientific realism, to say that "*S* knows that *p*" implies at best that *p* is *truthlike*, i.e., close to the truth in relevant respects (see Niiniluoto, 1984, 1991a).

The *justification* of scientific hypotheses is clearly relative to the standards and methods used in the scientific community. As these standards have changed in the historical development of science, relativism of justification has to be accepted. When we say that agent *S* was justified in believing that *p*, or that it was rational for *S* to accept *p*, we thus have to take *S* to be a historically and socially situated subject and judge *S*'s cognitive states relative to the standards and methods of the relevant community. But this is a *modest* form of relativism only, since the rational and critical evaluation of these standards is possible in epistemology and philosophy of science. Scientists of the earlier generations can be assessed also by our current standards. But the present methods actually used in the scientific community do not give the final word about this matter (cf. Boyd, 1984).

Some philosophers take the modest methodological relativism of justification to support stronger forms of semantic and ontological relativism. The move from the former to the latter is often made without any argument at all, and sometimes by linking the concepts of truth and reality to the concepts of justification, acceptance, assertion, or belief. This is the strategy of contemporary pragmatism and social constructivism.

This inference can be avoided by defining *truth* as the *correspondence* between a statement and reality. This means that, in analysing the meaning of "*S* knows that *p*", the truth or truthlikeness of "*p*" is taken to depend on the relation between "*p*" and the part of the world under investigation, not on the relation between *S* and "*p*". This relation is mediated by an interpretation of the language, which is a human construction. But as soon as the interpretation is fixed, the truth and falsity of statements is decided by reality. This account of truth can be formulated by following the model-theoretic version of Tarski's definition (cf. Niiniluoto, 1994b). Then claims of the form "it is true that *p*", where "*p*" is a statement in an interpreted language, make sense and have truth-values independently of any relativization to belief systems, theories, or social interests.

No one has so far formulated a satisfactory theory for a relative truth-claim "*p* is true-for-*S*", where *S* is a person or some other cognitive agent, unless this simply means that "*S* believes that *p*" (cf. Meiland and Krausz, 1982). Similarly, relative truth for subjective value statements could be trivially reduced to acceptance:

"Strawberries taste good" is true-for-me in the sense that I like the taste of strawberries. Statements with indexical expressions seem to provide a special case where the relative truth predicate makes sense ("I am a Finnish male" is true-for-Ilkka-Niiniluoto and false-for-Barbara-Streisand), but in the logician's sense they correspond to open formulas (which lack truth values) rather than genuine sentences. As statements like "Ilkka Niiniluoto is a Finn" are objective non-relative truths, this case does not give any support to relativism concerning truth. In particular, this means that truth is not gender-relative.

It is important – but more controversial – to add that this view of truth applies not only to nature (Popper's World 1) but also to the contents of human mind (World 2) and to cultural and social reality (World 3). Even though my objectivity in the sense of impartiality may become a serious problem, I can make assertions about my own mental states ("I feel miserable today") and about social or institutional states of affairs that I (among others) have helped to generate ("The Christina Institute was founded at the University of Helsinki in 1991") – and these statements are true or false in the correspondence sense. The same holds of statement about other minds and social facts. Hence, scientific realism can be defended not only in natural science, but in psychology (cf. Niiniluoto, 1994b), history and the social sciences (cf. Niiniluoto, 1985) as well.

4. WOMEN'S STUDIES AND FEMINISM

Feminism, in the original meaning of the term, has usually been conceived as the ideology of a political movement which aims at improving the status of women in society. As with other emancipatory programs, there are facts about the historical development and the current state of women's status, values concerning the ideal situation, and group consciousness of the discrepancy between the current status and the ideal. In spite of their commitment to the values of democracy and equality, the Western societies can be claimed to be still "androcentric", dominated by male power and interests in most domains of life. (It is another matter that the position of women is even worse in many other societies.)

One of these "androcentric" domains is education, academic learning, and scientific research. As a part of the ideological superstructure, science also influences other sectors of society. Therefore, especially since the 1960s, the science question in feminism has become increasingly important (cf. Harding, 1986). It includes at least two important problem areas.

First, one should try to develop a new field of research which studies questions relevant to the struggle of women for better positions. It should study in novel ways women in society and history, give new knowledge about their situation, and help them to understand their lives and conditions better. This is the self-understanding of *Women's Studies*, as it was introduced in the late 1960s. Secondly, it can be argued that science as it is practiced and applied today, as well as the scientific community within academic institutions, is still a stronghold of "sexist" attitudes and structures and thereby serves "regressive" social tendencies. Therefore, the whole of science, including its methods and institutions, should be radically changed by a critical *feminist science*.

The Scandinavian countries, which used to be proud of their democratic welfare states, illustrate this situation. The young men of Finland first went to study at the University of Paris in the fourteenth century. The first university in Finland was founded in 1640, and it accepted its first female students with special permission in the late 1870s, and with equal rights in 1901. The women of Finland were the first in Europe to gain suffrage, in 1906. In the 1920s females came to constitute more than 50 percent of the students in the humanities. In the 1950s they constituted more than 50 percent of all university students, and today they constitute 74 percent of the students in the humanities and 62 percent of all university students. In 1994, 80 percent of the masters degrees awarded in the arts were awarded to women, as were 43 percent of the doctoral degrees. While the percent of doctorates being awarded to women has been increasing, it has yet to have a major impact on the academic marketplace: in the humanities 46 percent of all teaching assistants are women and 14 percent of full professors are women. Across all disciplines, 31 percent of teaching assistants are women as are 9 percent of full professors.

The first institutes of Women's Studies were established in Finland in the late 1980s. They run multi-disciplinary study programs in topics relevant to the position of women in society. When the Christina Institute at the University of Helsinki recently appointed the first Professor of Women's Studies, the range of the research areas of the applicants is illustrative of the field today: the status of women in ancient Rome, women as creators and carriers of folklore within the Finno-Ugric tribes, substance use and addiction among women, lesbian subcultures in modern cities, education of school girls, feminist mothers, gender and modern dwelling, female painters in art history, gender issues and existentialism, gender and political theory.

Moderate and radical understanding of the nature of Women's Studies were present right from the beginning (see Bowles and Klein, 1983). According to the moderate view, Women's Studies is an addition to the standard curriculum and research agenda. Done mostly (but not necessarily) by women, coming from different disciplines with their theories and methods (such as history, sociology, philosophy, psychology, literature, jurisprudence, theology), it teaches and studies problems that are ignored by the normal "malestream" science. The radical view sees Women's Studies as a new autonomous discipline which adopts the feminist perspective and aims at the revolutionary transformation of all academic "male-defined" knowledge. In particular, feminist philosophy wants to correct the "male-bias" in philosophy, not only by re-evaluating the (to us often surprisingly strange) views that the Ancients had on women, but also by re-examining its basic problems and key concepts (see Griffiths and Whitford, 1988).

The task of *feminist epistemology* can now be understood in two different senses. First, Women's Studies is a new academic program, perhaps even a new discipline, and its special nature, aims, and methods should be systematically studied. In brief, feminist epistemology in this narrow sense is *the philosophy and methodology of Women's Studies*. If Women's Studies is understood in the moderate sense, this is perhaps not very exciting, since the methods are then mainly borrowed from the existing academic disciplines. But the radical interpretation suggests a second potential task for feminist epistemology: it should try to show that there is a special kind of *woman's knowledge*, obtained by typically feminine methods or means. This second task could follow the line of naturalized epistemology, by studying the actual processes by means of which women's knowledge is formed. But it can also assume a normative stand and claim that this women's knowledge serves as a model for all science, not only for Women's Studies.

The third line in contemporary feminism, inspired by French post-structuralism, challenges the dominant male discourse without assuming any autonomous and unitary notions of feminine gender identity (see Irigaray, 1985; Fraser and Nicholson, 1990).

It is now evident that relativism is bound to be an important theoretical problem for feminist epistemology, since the whole project involves at least a lure of *gender relativism*. Being a woman is supposed to make a relevant difference in the theory of knowledge. On the other hand, radical gender relativism seems to make the emancipatory enterprise of feminism pointless: the talk about male-*bias* does not make sense any more on the radical understanding, since that would presuppose some objective standard of truth and falsity. (For a criticism of Spender's radical gender relativism about beliefs in this respect, see Heinämaa, 1994.) If the results of "androcentric" science are true-for-males, then it does not help very much to claim that they are false by the feminine standards. And if the feminist presents her own claims (e.g., the facts about the position of women in the academic institutions) as merely true-from-the-female-viewpoint or as causal effects of her cultural and social position, not as assertions which correspond to reality, who would take them seriously?

5. IS THERE A FEMININE METHOD?

The first partisans of women's liberation argued that the exclusion of women from academic posts is unwarranted and unjust, since women are *equally good* for teaching and research positions as men. There is nothing that prevents women, when they have the right education, from using the methods of science as well as men. Indeed, whatever differences there may be between the two sexes, they are irrelevant for the evaluation of a person's academic potentials.

This view was further supported by the argument that the dominant male philosophers, often unintentionally, have tended to associate characteristics considered masculine (such as rationality and objectivity in a special sense) to their allegedly universal picture of "man", "human being", and "reason", and thereby justified the exclusion of the more emotional and subjective women. But when these stereotypical *gender differences* were emphasized, it was a short step to the "gynocentric" conclusion that after all they are crucially important. Thus, it was claimed that the current methods of science are "sexist", and should not be assumed to be a valid model of research.

Feminist epistemology took the task of proving that there is a feminine language, mind, and thinking different from the masculine. Biological and physiological explanations of these differences were generally rejected. Anthropologists suggested that this is a mentality difference, due to the traditional roles of men and women in social occupations and domestic life. Nancy Chodorow, in her psychoanalytic object-relations theory, argued that male and female personalities are constructed in different ways, because sons and daughters have different relations to their mother. This explains, Chodorow asserts, why men are attached to the subject–object distinction and to the urge of dominating and controlling external reality. Carolyn Merchant (1983) and Evelyn Fox Keller (1985) pointed out that precisely these features are typical of the experimental method of modern natural science, as formulated by Francis Bacon in the early seventeenth century. The same charge that men are guilty of destroying nature has later inspired the ecofeminist movement.

Merchant's historical analysis is very interesting and important, since modern science indeed gave up the ancient idea of nature as a maternal or female organism. But, for the same reason, I think she takes too literally the crude sexual metaphors that Bacon used in his description of the experimental method: modern philosophers did not take any more seriously the idea that nature really is female. The Baron of Verulam was a man of his time, and in advertising his method of learning he chose his rhetorics for his male audience. Men have later used more refined and elegant ways of talking about their passionate "love of truth" or their wish to "uncover" the "hidden secrets" of nature. Instead of domination, this relation is often one of admiration and worship. Even Bacon knew that "nature to be commanded must be obeyed".

It is easy and fascinating to play with the idea that the appeal of scientific methods could have a link to the patterns of human sexuality. If the manipulation and cross-examination of nature in experiments is seen as masculine, then similarly the hermeneutic method of *Verstehen* or emphatic understanding (derived from "to stand under") has been regarded as typically feminine. This method employs emotions, instead of excluding them, and aims at the unity of subject and object. It has also been suggested that, instead of propositional knowledge, female knowledge is typically intuitive and tacit.

The suggestion that there are *masculine and feminine scientific methods* in science is in fact one of the possible forms of gender relativism in science. Thereby this kind of feminist program has also become a target of the criticism that the feminine research methods relying on "feminist consciousness" (cf., for example, Stanley and Wise, 1983) are too "soft" and "personal" to be scientific. It is no wonder that many feminists have wished to reconsider or denounce such views. In her Introduction to the influential book *Feminism and Methodology* (1987), Sandra Harding argues "against the idea of a distinctive feminist method of research" (p. 1).

6. WOMAN IS THE MEASURE OF ALL THINGS

Relativism is usually understood in its radical form in feminist writings. Dale Spender's (1982) slogan against objectivity is "Not right or wrong, but equal".

Anne Seller (1988) defines relativism first as "the view that every woman's experience is valid, not false, illusory or mistaken, and that all ways of making sense of the world are equally valid", but changes this incoherent subjectivist position to the view that "the truth of a claim is relative to the group within which that claim is made". She thinks that a "politically adequate" democratic epistemology, where the researchers are willing to learn from other members of the group, "to listen, to care for and to support each other, and to express ourselves honestly" (p. 179), makes the distinction between realism and relativism irrelevant.

This is not convincing: any actually existing group of persons may be mistaken in its shared group beliefs. As every fallibilist has to admit, history of science gives us enough evidence for this. When Peirce, Habermas, and Putnam have tried to characterize truth and reality in terms of accepted beliefs, they have referred to an *ideal*, indefinitely large community which approaches to the truth only asymptotically in its ultimate opinion. And, as Peirce himself observed, even this ideal scientific community is destined to reach truth not necessarily, but rather only with probability one (cf. Niiniluoto, 1984, Ch. 3).

Sandra Harding's *feminist standpoint theory* accepts fallibilism, but asserts that "women's experiences, informed by feminist theory, provide a potential grounding for more complete and less distorted knowledge claims than do men's" (Harding, 1987, p. 184). "Women's and men's characteristic social experiences provide different but not equal grounds for reliable knowledge claims", she adds, so that "we all – men as well as women – should prefer women's experiences to men's" (*ibid.*, p. 10).

Harding says that "standpoint theorists are not defending any form of relativism", since women's and men's experiences are not claimed to be equal grounds for knowledge claims (*ibid.*, p. 186). This means that in her vocabulary relativism is always assumed to have the radical form (see also Harding, 1993, p. 61). In my terms, however, Harding's basic position here is modest gender relativism, since she insists on the difference between two types of experiences, but also expresses her preference ordering of them. At the same time, Harding's bold thesis that women's experiences should be preferred both by men and women makes her a sort of objectivist, since it asserts the existence of a privileged framework for doing science. This is not only concerned with social research, since "a critical and self-reflective social science should be the model of all science" (Harding, 1986, p. 44). To paraphrase Protagoras, as far as there is scientific knowledge about society and nature, woman is the measure of all things.

My wife has a T-shirt with the text: "In order to be successful in our society, women have to be three times more intelligent than men. Fortunately that is very easy." This is not Harding's thesis: she is not urging that women are more intelligent than men, or equipped by superior mental powers. Rather her argument is based on the *social position* of women as "oppressed" and "marginalized": in fighting against their oppressors the women achieve "a truer (or less false) image of social reality".

As the women's position in society is not a historical constant, Harding's position is not objectivist in Bernstein's (1983) sense, which requires the existence of a "permanent, ahistorical matrix or framework" for doing sciences. But this seems to mean also that women would lose their epistemic advantage, if the feminist would take power in the society (Pinnick, 1994).

Nancy Hartsock (1983) derives the feminist standpoint view directly as a transformation of the *Marxist* view that the proletariat in the capitalist society is in an epistemically privileged position in relation to the bourgeois. Similarly, Marxist ethics combined moral class relativism with an objectivist twist: the morality of the "most progressive class" is the right one, since it is in the direction of the presumed objective laws of history (see Redlow *et al.*, 1971).

The Marxists had a grand narrative to support the privileged historical position of the working class. Similar stories about the feminine class are hardly more plausible, even if they can be used as rhetorical devices within political consciousness raising. This is not the only problem in the relation of Marxism and feminism.

A radical version of Western Marxism argued that the "logic of the capital" deforms or distorts all human thinking, and therefore there is a difference between "bourgeois physics" and "socialist physics" (cf. discussions in Sandkühler, 1975). Another version asserted that the economic value forms are constitutive of the forms of thinking in natural and social science (cf. Pietilä, 1981). This emphasis on social factors reminds one of the Soviet doctrine in Stalin's era: that social practice may change the laws of genetics was a background of the Lysenko case (cf. Roll-Hansen, 1989). But here the academic Marxist school was even more orthodox than the official Marxist-Leninists in the Soviet Union: Stalin himself wrote in 1950 an article in support of the view that the laws of logic and grammar are, like the laws of geometry, independent of social and economic structures; this argument liberated the study of logic in the Soviet Union (see Klaus, 1958, p. 13).

Pierre Duhem's polemics against "German science" and the accusations of Einstein's "Jewish physics" in the Third Reich are notorious examples of the ways in which science has been relativized to nationalist and racist purposes. To many of us they sound outdated, something that should be eliminated from science – even though the strong relativist programmes in the sociology of science have started to urge that such a relativism to social factors is inevitable. Is the standpoint thesis of the superiority of "feminist research" to bad "androcentric science" any better than the idea of "socialist physics"?

Why should we think that the marginalized people at the bottom of social hierarchies are epistemologically superior? As Pinnick (1994) notes, hardly any empirical evidence is given for this factual claim. Harding's answer seems to be based on the assumption that social science studies "problems" of some individuals or groups: "a problem is always a problem *for* someone or other" (Harding, 1987, p. 6). Therefore, these different social groups have an important role in the context of discovery, i.e., in the generation of questions that are asked in science. Further, when hypotheses about the life of these groups are investigated, it is certainly important to test them against the experience of the members of these groups.

Arguments along these lines are sufficient to show that some research problems may be generated by *problems for women*, and that *hypotheses about women* (e.g., how women experience motherhood) should be tested by studying women's experiences. This is what was originally in the agenda of moderate Women's Studies. But they do not yet show why women are the best researchers in the study of women's problems. Any argument to this effect would most likely be transformable to support the corresponding claim that men are the best researchers in Men's Studies. Indeed, Harding admits that "men sympathetic to feminism" might do some useful "phallic critique" (Harding, 1987, p. 11).

Harding's thesis that all science should preferably be based on *women's experiences* remains unwarranted. All marginalized people are not women, and not all members of the "oppressor group" are male. And all research problems are not generated from people's everyday experiences. As Thomas Kuhn and the so-called erotetic logicians have shown, research problems are also generated by earlier theories and paradigms. (Why is grass green? is a respectable research problem, even though probably no one experiences it as personally problematic.)

Another argument suggests that men fail to recognize social problems, since they still rule and largely create the societal structures. But, again, certainly there are also marginalized men. Moreover, this argument in fact turns upside down the traditional conception of maker's knowledge (see Hintikka, 1974): the more men participate in the making of society, the more (not less) knowledge they should have about it.

Harding (1993) has recently clarified her position by saying that "marginalized lives provide the scientific problems and the research agendas – not the solutions – for standpoint theories". But, if this is her view, then it is difficult to see what remains of her original bold thesis that women's experiences should always be preferred as "grounds of knowledge claims". If we really wish to "maximize objectivity" in social research, it no doubt is desirable to have persons with different social background among the community of investigators (i.e., the membership in the scientific community should be open to all who have the will and talent for higher academic education). This diversity may enhance the critical discussion and questioning within the scientific community. It may be beneficial to have members with "antiauthoritarian" and "emancipatory" values (cf. Harding, 1986, p. 27). But, by this principle, a community which includes *both men and women* is better than any genderwise homogeneous group. And if "strong" or self-reflective objectivity is not achieved, because social values held by the whole scientific community will not be identified (Harding, 1993, p. 57), then the research community of Women's Studies should include non-feminists, too.

Maybe the only way to support the claim that a purely feminine community of investigators is preferable to one with mixed genders would be a transformation of the Marxist doctrine that the "logic of the capital" distorts human thinking. Thus,

Du Bois (1983) believes that the force of patriarchy deforms all thinking: "We are observer and observed, subject and object, knower and known. When we take away the lenses of androcentrism and patriarchy, what we have left is our own eyes, ourselves, and each other." In other words, when male domination is eliminated, the way is reopened to a Cartesian transparent subject and foundationalist naive empiricism. This kind of return to feminine objectivism is rejected by Harding.

Harding's (1993) aim is to show that it is possible to have genuine knowledge that is "fully socially situated". This makes her position resemble what the Marxist epistemologists used to call the "double determination view" of scientific knowledge. The knowledge claims are results partly of the interactions of the scientists with an independently existing reality, partly of the social situation of the scientists in their culture and community. This view is compatible with critical scientific realism: the truth or truthlikeness of a scientific claim depends on its correspondence with reality, but the discovery and justification of such a claim may be intertwined with features of the social situation. While the *norm* of objectivity precludes that personal or social wishes and interests are explicitly presented as reasons for accepting or rejecting hypotheses ("I find this theory acceptable, since its truth would give me advantage ..."), the real-life scientists as fallible and limited human beings may in fact be influenced in this task by their non-scientific values. However, critical self-reflection and openness to objections by other scientists helps to make science, as Peirce put it, a self-corrective enterprise.

7. THE DECONSTRUCTION OF GENDER RELATIVISM

Feminist *postmodernism* (see Nicholson, 1990) has challenged the standpoint theory by arguing that it is a "universalizing mistake" to speak in the name of an abstract "woman". White middle-class heterosexual women of North America and Western Europe cannot have the voice of African-American, Asian, Native American or lesbian women. Each female scholar should individually locate or position herself (cf. the discussion in Heinämaa, 1994).

When Harding (1986, 1993) admits that these divisions lead to "different feminisms" that "inform each other", gender relativism is lost and transformed to Protagorean subjectivism: there will then be literally as many feminisms as feminists, since eventually smaller and smaller class divisions end with singletons of one member only.

Postmodernism deconstructs gender relativism also in the sense that it questions all universal generalizations about the nature of women. Even Luce Irigaray (1985), who invites us to rethink the masculine–feminine distinction, has been accused of "essentialism" (cf. the discussion in Heinämaa, 1994). Judith Butler challenges the sex–gender distinction. One might wonder whether it is still appropriate to speak of "feminism" without some opposites of this kind. Feminist postmodernists see themselves as continuing a project "with plural and complexly constructed conceptions of social identity" and with "a patchwork of overlapping alliances, not one circumscribable by an essential definition" (Fraser and Nicholson, 1990).

To complete the deconstruction of feminist epistemology (in the traditional sense), the postmodernists argue that epistemology, as a project of finding ultimate and firm grounds for knowledge claims, is tied to the historical development of the ideology of modern Western culture. In this view, standpoint theories belong to the past. What remains is the successor project of epistemology, the endless discourse or conversation on women and femininity.

Social constructivism has been recommended for feminist philosophy (Gergen, 1988). But the relativist idea that men and women "construct" their own "realities" has also been effectively criticized by Grimshaw (1986).

Social constructivism is a natural ally of postmodernism: science is embedded in its social situation, the relation of scientific claims to nature is bracketed (Collins, 1981), the reality is treated as resulting from negotiations in a laboratory, and a moratorium on epistemological explanations of scientific activities is declared (Latour and Woolgar, 1986).

Elizabeth Potter (1993) has tried to illustrate the gender-relative nature of "social negotiations" in natural science by a case study of Boyle's Law of Gases. The knowledge that the air has spring and weight, she argues, was "influenced by class and gender considerations". Robert Boyle, a Puritan and an opponent of attempts to liberate women from the domestic sphere, supported the mechanistic natural philosophy. He defeated the organicist or hylozooist view, associated with Hermes, Paracelsus, and Campanella, and advocated by the radical men and women of the "mob". Potter concludes that Boyle's work had "direct implications for women of that period".

It should not surprise anyone that scientific knowledge may be highly *non-neutral* in society, since it often corrects popular everyday conceptions and prevailing doctrines. If there were women in Britain who were devoted to what turned out to be the mistaken theory, Boyle's rejection of that theory and the step toward a more truthlike account of gases certainly had "direct implications" for these women. What Potter fails to show, I think, is that gender considerations played any role in Boyle's evaluation of the experimental evidence available at his time.

It seems to me that the common mistake of postmodernism and social constructivism is in their belief that anti-foundationalism about science entails anti-realism. Fallibilist realism provides an alternative that they both ignore.

8. FEMINIST EMPIRICISM AND REALISM

Harding defines *feminist empiricism* as the view that "sexism and androcentrism (in science) are social biases correctable by stricter adherence to the existing methodological norms of scientific inquiry" (Harding, 1986, p. 24). This is perhaps not the best formulation, since the leading feminist empiricists, like Longino (1990) and Nelson (1990), hardly think that we already possess a ready and finished pattern of methodological norms. Reasonable empiricists will agree with Richard Boyd (1984) that the progress of science takes place both in its content and in its methods and methodology. While an empiricist thinks that the self-corrective methods of research are the best means to eliminate whatever "sexist bias" there may be in the *content* of scientific knowledge, there is another kind of meta-level social bias in the *attitudes and institutions*, and its correction can take place by ethical norms and measures of science policy (rather than by methodological norms). (The difference between these two levels is emphasized by Harris, 1992.)

We may also add that of course an important part of feminist writing will belong to *philosophy and the critique of ideology*, which has taken its model from the emancipatory social science. Conceptual and normative questions are here based upon philosophical reflections and arguments. *Pace* Quine, epistemology and ethics cannot be completely naturalized, i.e., reduced to empirical disciplines like psychology, cognitive science, and sociology.

There is a weak sense in which almost all philosophers of science are empiricists: scientific hypotheses are tested against observations and experiments. But there is a stronger sense in which empiricism contrasts with scientific realism: it claims that theories should not go beyond the domain of the observable, or that this theoretical part of scientific theories is merely an instrument for systematizing observational statements, it is not referring to real entities, it does not have a truth value, or its truth value is irrelevant for the purposes of science. Ernst Mach, the Vienna Circle, W. V. O. Quine, Bas van Fraassen, and Larry Laudan represent different variants of empiricism in this strong sense (cf. Leplin, 1984).

When Lynn Hankinson Nelson (1990) bases her feminist empiricism on Quine's philosophy, she associates her view with empiricism in the strong sense. Quine's account of language, truth, and the web of belief has a strong leaning towards relativism (see, e.g., the criticism in Harris, 1992). In spite of her rejection of sociological relativism, Nelson's feminist epistemology leaves the realism–relativism issue largely untouched and unsettled.

Realism and relativism are of course not the only relevant alternatives in the philosophy of science – for example, Laudan (1990) is both an anti-realist and an anti-relativist. Here the feminist writers have not very much used the relevant literature. Cassandra Pinnick (1994) complains that, to the extent they have done so, their views are not up-to-date and appeal to "discredited philosophical ideas" (such as Kuhn, Quine's underdetermination thesis, the Strong Programme).

In particular, there is surprisingly little discussion on realism within feminist philosophy. Linda Alcoff (1987) has rightly emphasized the importance of this question to the feminist movement. When she contrasts the correspondence theory of truth with the "constructive" notion of truth (Foucault, Gadamer), I don't understand why the former is claimed to make truth "abstract" and "universal": Tarskian truth can be applied to singular and temporal statements; and not only languages, but their interpretations and thus the correspondences between languages and the world as well, have to be created by human activities (cf. Niiniluoto, 1994b). Foss-Fridlizius (1990), inspired by Roy Bhaskar (1978), has suggested that feminism might accept scientific realism. But still I have not detected anyone to use the term *feminist realism*.

It is important to distinguish realist and instrumentalist versions of fallibilist epistemology. There are some recent articles on the relations between pragmatism and feminism, but instead of Peirce's realist fallibilism, they are more concerned with Dewey's instrumentalism (see Rooney, 1993). A similar emphasis can be seen in the statement that "postmodern-feminist theory would be pragmatic and fallibilistic" (Fraser and Nicholson, 1990, p. 35).

In my view, critical fallibilist scientific realism would give the best epistemological background for feminist research within Women's Studies, understood as an empirical interdisciplinary attempt to find new descriptive knowledge about nature, mind, culture, and society. It would preserve all the advantages that feminist empiricism has over standpoint theories and postmodernism. It helps us to understand how new research on issues important to feminism may complement and correct the earlier biased views. It insists that truth is not relative to persons or gender, but accepts that different communities have used varying methods of justifying knowledge claims – and that these methods can still be improved. It can combine the ideas that scientific inquiry is always socially situated and cognitively progressive, without falling into the traps of radical relativism. And it could include in the research agenda such theories that refer to realistically interpreted theoretical entities like "feminine consciousness" (cf. Niiniluoto, 1994a).

9. FEMINIST APPLIED RESEARCH

Finally, I wish to show that the realist view of science is not incompatible with a positive program of feminist politics.

Feminist inquiry (besides the philosophical and descriptive lines mentioned above) could take its model from *applied social sciences*, like social policy studies or peace research. I don't know any attempts in this direction, however. This may be due to the misconception that applied science cannot serve critical functions. That this is not the case can be seen from the following consideration (cf. Niiniluoto, 1993).

A typical "design science" studies conditional recommendations or *technical norms* of the form

(TN) If you want G, and believe you are in situation B, then you ought to do Z.

Such statements can be in principle supported in a *value-neutral* way by showing that doing Z in situation B most likely brings about the goal G. Thus, TN is a statement that has a truth value in the realist sense, and so it can be a result of scientific inquiry. On the other hand, TN is *conceptually value-laden* in the sense that it contains a description of a goal G. In applied social science, G may range from the maintenance of *status quo* to radical and even utopian aims. G could also be a goal which expresses the interests and political purposes of a particular social group, like women or some marginalized minority (e.g., Black feminists, lesbians). In this case, TN represents *research for* that group. When descriptive social studies show that the actual situation is of type B, then the normative conclusion "You ought to do Z!" is *true for* the members of this group (Niiniluoto, 1985).

For example, "If you wish to improve the academic position of women scholars, and you live in the present situation in Finland, you ought to do *Z*" and "If an African–American mother wishes to improve her condition of life in the United States, she should do *W*". The questions of what the content of *Z* and *W* might be, I am willing to leave to feminist research.

This is, I think, an attractive model of genuinely *feminist* research, since all the values and political goals of this movement can be packed into the goal *G*. The traditional principle of value-neutrality is nevertheless respected, as long as commitment to *G* does not influence the assessment of the causal connections between the means *Z* and the end *G* in situation *B*.

Perhaps feminist applied science is not found appealing, as it is based upon *instrumental* or means–ends *rationality*, which "gynocentric" writers often view as typically "masculine". It may also seem to be bound to a reformist ideal of "social engineering". I should rather think that the concept of technical norm TN is in fact commonly used in everyday male and female reasoning. Moreover, TN as such is neutral with respect to reformist and revolutionary strategies: which one is appropriate, depends on the situation *B*. The Scandinavian welfare state, with its progressive program of sex equality, was built by following the reformist strategy. But whether other kinds of programs are needed in a different time or place, is another story. Indeed, *B* in TN shows the historical and social context-dependency of rational recommendations of action.

Richmond Campbell (1994) has attempted to illuminate the prospects of feminist science by considering the notion of confirmation, but his argument is clearly insufficient for linking research and political action. The model of feminist applied research, outlined in this section, shows in a concrete way how the principles of empirical science can be combined with feminist politics.

Department of Philosophy,
P.O. Box 24,
00014 University of Helsinki,
Finland

ACKNOWLEDGEMENT

I am grateful to Sara Heinämaa and Ritva Ruotsalainen for suggestions and advice concerning the material discussed in this paper, and for valuable comments on its earlier version. Mistakes and misunderstandings are solely my responsibility.

REFERENCES

Alcoff, L.: 1987, 'Justifying Feminist Social Science', *Hypatia*, **2:3**, 85–103.
Alcoff, L. and Potter, E. (eds): 1993, *Feminist Epistemologies*. Routledge, London.
Bernstein, R.: 1983, *Beyond Objectivism and Relativism*. Blackwell, Oxford.
Bhaskar, R.: 1978, *A Realist Theory of Science*. Harvester Press, Sussex.
Boyd, R.: 1984, 'The Current Status of Scientific Realism', in Leplin, pp. 41–82.
Brown, J. R. (ed.): 1984, *Scientific Rationality: The Sociological Turn*. D. Reidel, Dordrecht.

Bowles, G. and Klein, R. D.: 1983, *Theories of Women's Studies*. Routledge & Kegan Paul, London.

Campbell, R.: 1994, 'The Virtues of Feminist Empiricism', *Hypatia*, **9:1**, 90–115.

Collins, H. M.: 1981, 'Stages in the Empirical Programme of Relativism', *Social Studies of Science*, **11**, 3–10.

Du Bois, B.: 1983, 'Passionate Scholarship: Notes on Values, Knowing and Method in Feminist Social Science', in Bowles and Klein, pp. 105–116.

Feyerabend, P.: 1987, 'Notes on Relativism', in *Farewell to Reason*. Verso, London, pp. 19–89.

Foss-Fridlizius, R.: 1990, 'Relativism och Realism i Feministisk Vetenskapsfilosofi', *Kvinnovetenskaplig Tidskrift*, **3**, 54–64.

Fraser, N. and Nicholson, L. J.: 1990, 'Social Criticism without Philosophy: An Encounter between Feminism and Postmodernism', in Nicholson, pp. 19–38.

Gergen, K. J.: 1988, 'Feminist Critique of Science and the Challenge of Social Epistemology', in M. McCanney Gergen (ed.), *Feminist Thought and the Structure of Knowledge*. New York University Press, New York, pp. 27–48.

Griffiths, M. and Whitford, M. (eds): 1988, *Feminist Perspectives in Epistemology*. Macmillan Press, Basingstoke.

Grimshaw, J.: 1986, *Feminist Philosophers: Women's Perspectives on Philosophical Traditions*. Wheatsheaf Books, Brighton.

Harding, S.: 1986, *The Science Question in Feminism*. Cornell University Press, Ithaca.

Harding, S. (ed.): 1987, *Feminism and Methodology*. Indiana University Press, Bloomington.

Harding, S.: 1993, 'Rethinking Standpoint Epistemology: What is "Strong Objectivity"?', in Alcoff and Potter, pp. 49–83.

Harding, S. and Hintikka, M. B. (eds): 1983, *Feminist Perspectives on Epistemology, Metaphysics, Methodology and Philosophy of Science*. D. Reidel, Dordrecht.

Harris, J. F.: 1992, *Against Relativism: A Philosophical Defence of Method*. Open Court, LaSalle.

Hartsock, N.: 1983, 'The Feminist Standpoint: Developing the Ground for a Specifically Feminist Historical Materialism', in Harding and Hintikka, pp. 283–310. Reprinted in Harding (1987), pp. 157–180.

Heinämaa, S.: 1994, 'The Rhetoric of Positioning in Women's Studies', *Nora*, **2:2**, 83–94.

Hintikka, J.: 1974, *Knowledge and the Known: Historical Perspectives in Epistemology*. D. Reidel, Dordrecht.

Hollis, M. and Lukes, S. (eds): 1982, *Rationality and Relativism*. Blackwell, Oxford.

Irigaray, L.: 1985, *Speculum of the Other Woman*. Cornell University Press, Ithaca.

Keller, E. F.: 1985, *Reflections on Gender and Science*. Yale University Press, New Haven.

Klaus, G.: 1958, *Einführung in die formale Logik*. VEB, Berlin.

Latour, B. and Woolgar, S.: 1986, *Laboratory Life: The Social Construction of Scientific Facts*, 2nd edn. Princeton University Press, Princeton.

Laudan, L.: 1990, *Science and Relativism*. University of California Press, Berkeley.

Leplin, J. (ed.): 1984, *Scientific Realism*. University of California Press, Berkeley.

Lloyd, G.: 1984, *The Man of Reason: "Male" and "Female" in Western Philosophy*. Methuen, London.

Longino, H.: 1990, *Science as Social Knowledge*. Princeton University Press, Princeton.

Margolis, J.: 1991, *The Truth about Relativism*. Blackwell, Oxford.

Meiland, J. W. and Krausz, M. (eds): 1982, *Relativism: Cognitive and Moral*. University of Notre Dame Press, Notre Dame, Indiana.

Merchant, C.: 1983, *The Death of Nature: Women, Ecology and the Scientific Revolution*. Harper & Row, San Francisco.

Nelson, L. H.: 1990, *Who Knows?: From Quine to a Feminist Empiricism*. Temple University Press, Philadelphia.

Nicholson, L. J. (ed.): 1990, *Feminism/Postmodernism*. Routledge, London.

Niiniluoto, I.: 1984, *Is Science Progressive*? D. Reidel, Dordrecht.

Niiniluoto, I.: 1985, 'Truth and Legal Norms', in N. MacCormick, S. Panou, and L. L. Vallauri (eds), *Conditions of Validity and Cognition in Modern Legal Thought*. Frantz Steiner Verlag, Stuttgart, pp. 168–190.

Niiniluoto, I.: 1991a, 'Realism, Relativism, and Constructivism', *Synthese*, **89**, 135–162.

Niiniluoto, I.: 1991b, 'What's Wrong With Relativism?', *Science Studies*, **4:2**, 17–24.

Niiniluoto, I.: 1993, 'The Aim and Structure of Applied Science', *Erkenntnis*, **38**, 1–21.

Niiniluoto, I.: 1994a, 'Scientific Realism and the Problem of Consciousness', in A. Revonsuo and M. Kamppinen (eds), *Consciousness*. Lawrence Erlbaum Associates, Hillsdale, NJ, pp. 33–54.

Niiniluoto, I.: 1994b, 'Defending Tarski against his Critics', in B. Twardowski and J. Wolenski (eds), *Sixty Years of Tarski's Definition of Truth*. Philed, Cracow, pp. 48–68.

Pietilä, V.: 1981, *Social Praxis and the Development of Science*. Yhteiskuntatieteiden tutkimuslaitos, Series A: 54, University of Tampere, Tampere.

Pinnick, C. L.: 1994, 'Feminist Epistemology: Implications for Philosophy of Science', *Philosophy of Science*, **61**, 646–657.

Potter, E.: 1993, 'Gender and Epistemic Negotiation', in Alcoff and Potter, pp. 161–186.

Preston, J.: 1992, 'On Some Objections to Relativism', *Ratio* (New Series), **5:1**, 57–73.

Putnam, H.: 1990, *Realism with a Human Face*. Harvard University Press, Cambridge, MA.

Redlow, G. *et al.*: 1971, *Einführung in dem dialektischen und historischen Materialismus*. Dietz Verlag, Berlin.

Roll-Hansen, N.: 1989, 'The Practice Criterion and the Rise of Lysenkoism', *Science Studies*, **2:1**, 3–16.

Rooney, Ph.: 1993, 'Feminist–Pragmatist Revisionings of Reason, Knowledge, and Philosophy', *Hypatia*, **8:2**, 15–37.

Sandkühler, H. J. (ed.): 1975, *Marxistische Erkenntnistheorie*. Fischer Athenäum, Frankfurt am Main.

Seller, A.: 1988, 'Realism versus Relativism: Towards a Politically Adequate Epistemology', in Griffiths and Whitford, pp. 169–189.

Siegel, H.: 1987, *Relativism Refuted: A Critique of Contemporary Epistemological Relativism*. D. Reidel, Dordrecht.

Spender, D.: 1982, *Man Made Language*. Routledge & Kegan Paul, London.

Stanley, L. and Wise, S.: 1983, *Breaking Out: Feminist Consciousness and Feminist Research*. Routledge & Kegan Paul, London.

PART III

FEMINIST PHILOSOPHY OF SCIENCE AND THE SOCIOLOGY OF KNOWLEDGE, SOCIAL CONSTRUCTIVISM, AND THE DEBATE OVER SCIENCE STUDIES

KAREN BARAD

MEETING THE UNIVERSE HALFWAY: REALISM AND SOCIAL CONSTRUCTIVISM WITHOUT CONTRADICTION

Because truths we don't suspect have a hard time
making themselves felt, as when thirteen species
of whiptail lizards composed entirely of females
stay undiscovered due to bias
against such things existing,
we have to meet the universe halfway.
Nothing will unfold for us unless we move toward what
looks to us like nothing: faith is a cascade.
The sky's high solid is anything
but, the sun going under hasn't
budged, and if death divests the self
it's the sole event in nature
that's exactly what it seems.
[From the poem "Cascade Experiment", by Alice Fulton (Fulton, 1990)]

1. INTRODUCTION

The morning after giving an invited lecture on the socially constructed nature of scientific knowledge, I had the privilege of watching as a STM (scanning tunneling microscope) operator zoomed in on a sample of graphite, and as we approached a scale of thousands of nanometers ... hundreds of nanometers ... tens of nanometers ... down to fractions of a nanometer, individual carbon atoms were imaged before our very eyes. The experience was so sublime that it sent chills through my body – and I stood there, a theoretical physicist who, like most of my kind, rarely ventures into the basements of physics buildings experimental colleagues call "home", conscious that this was one of those life moments when the amorphous jumble of history seems to crystallize in a single instant. How many times had I recounted for my students the evidence for the existence of atoms? And there they were – just the right size and grouped in a hexagonal structure with the interatomic spacings as predicted by theory! "If only Einstein, Rutherford, Bohr, and especially Mach, could have seen this!" I found myself exclaiming. And as the undergraduate students operating the instrument (that they had just gotten to work the day before by carefully eliminating sources of vibrational interference – we're talking nanometers here!), disassembled the chamber which held the sample so that I could see for myself the delicate positioning of the probe above the graphite surface, expertly

L. H. Nelson and J. Nelson (eds), Feminism, Science, and the Philosophy of Science, 161–194.

cleaved with a piece of scotch tape, I mused outloud that "seeing" atoms will quickly become routine for students (as previous generations in turn found the examination of cells by visual light microscopes to be and then the structure of molecules by electron microscopes so) and that I was grateful to have been brought up in a scientific era without this particular expectation.

At this point in my story, I imagine there will be scientific colleagues who will wonder whether this presented a moment of intellectual embarrassment for your narrator who had on the previous night insisted on the socially constructed nature of scientific knowledge. In fact, although I was profoundly moved by the event I had just witnessed, standing there before the altar of the efficacy of the scientific enterprise, I was unrepentant. For as social constructivists have tried to make clear, empirical adequacy is not an argument that can be used to silence charges of constructivism. The fact that scientific knowledge is socially constructed does not imply that science doesn't "work", and the fact that science "works" does not mean that we have discovered human-independent facts about nature. (Of course, the fact that empirical adequacy is not proof of realism is not the endpoint, but the starting point for constructivists, who must explain how it is that our constructions work – an obligation that seems all the more urgent in the face of increasingly compelling evidence that the social practice of science is conceptually, methodologically, and epistemologically allied along particular axes of power.[1])

On the other hand, I stand in sympathy with my scientific colleagues who want science studies scholars to remember that there are cultural *and* natural/material causes for knowledge claims. While most social constructivists go out of their way to attempt to dispel the fears that they are either denying the existence of a human-independent world or the importance of material factors in the construction of scientific knowledge, the bulk of the attention has been on cultural factors. To be fair, this is where the burden of proof has been placed: social constructivists have been responding to the challenge to demonstrate the falsity of the worldview that takes science as the mirror of nature. Nonetheless, as both the range and sophistication of constructivist arguments have grown, the charge that they embrace an equally extreme position – that science mirrors culture – has been levied against them with increasing vigor. While few constructivists actually take such an extreme position, we would be remiss in simply dismissing this charge as a trivial over-simplification and misunderstanding of the varied and complex positions that come under the rubric of constructivism. For the anxiety being expressed, though admittedly displaced, touches upon the legitimate concern about the privileging of epistemological issues over ontological ones in the constructivist literature. Ontological issues have not been totally ignored, but they have been overshadowed.

The ontology of the world is a matter of discovery for the traditional realist. The assumed one-to-one correspondence between scientific theories and reality is used to bolster the further assumption that scientific entities are unmarked by the discoverers: that is, nature is taken to be transparently given. Acknowledging the importance of Cartwright's (1983) philosophical analysis decoupling these assumptions and her subsequent separation of scientific realism into two independent positions – realism about theories and realism about entities – Hacking (1982), like Cartwright, advocates realism towards entities. Shifting the traditional emphasis in science studies away from theory construction to the examination of experimental practice, Hacking grounds his position on the ability of the experimenter to manipulate entities in the laboratory. Galison (1987) also centers experimental practice in his constructivist analysis comparing three different periods of twentieth-century physics experimentation, wherein he generalizes Hacking's criterion for the reality of entities by underlying the importance of the notions of stability (i.e., invariances of results under changing experimental conditions, rather than the narrower category of manipulation) and directness (i.e., epistemologically, but not necessarily logically, non-inferential). There are other constructivist approaches which go further in interrogating the transparency of our representations of nature. Latour (1993) prioritizes stability as well, posing it as one variable of a two-dimensional geometry whose other axis connects the poles of Nature and Society. Essence then becomes the trajectory of stabilization within this geometry that is meant to characterize the variable ontologies of quasi-objects. In contrast, Haraway (1988) emphasizes instability: it is the instability of boundaries defining objects that is the focal point of her explicit challenge not only to conceptions of nature that claim to be outside of culture, but also to the separation of epistemology from ontology. Interestingly, the instability of boundaries and Haraway's insistence that the objects of knowledge are agents in the production of knowledge, feature her notions of cyborgs (1985) and material-semiotic actors (1988) which strike up dissonant and harmonic resonances with Latour's hybrids and quasi-objects (1993). Moving to what some consider the opposite pole of the traditional realist position is the poststructuralist position. To many scientists as well as science studies scholars, Derridian forms of poststructuralism that disconnect sign from signified seem to be the ultimate in linguistic narcissism. While insisting that we are always already in the "theater of representation", Hayles (1993) takes exception to extreme views that hold that language is groundless play, and while she does not provide us with access to the real she does attempt to place language in touch with reality by reconceptualizing referentiality. Hayles' theory of constrained constructivism (1993) relies on consistency (in opposition to the realist notion of congruence) and the semiotic notion of negativity to acknowledge the importance of constraints offered by a reality that cannot be seen in its positivity: as she puts it, "Although there may be no outside that we can know, *there is a boundary*" (p. 40, original emphasis).

These attempts to say something about the ontology of our world are exceptions rather than the rule in the constructivist literature. There is a need to elaborate further upon the crafting of ontologies. We need to understand the technologies by which nature and culture interact. Does nature provide some template that gets filled in by culture in ways that are compatible with local discourses? Or do specific discourses provide the lenses through which we view the layering of culture upon nature? Does the full "texture" of nature get through or is it partially obliterated or distorted in the process? Is reality an amorphous blob that is structured by human discourses and interactions? Or does it have some complicated irregular shape that is differently sampled by varying frameworks that happen to "fit" in local regions like coincident segments of interlocking puzzle pieces? Or is the geometry fractal so that it is impossible for theories to match reality even locally? At what level of detail can any such question be answered, if at all? And what would it mean? Is it possible to take any of these questions seriously within the academy, in the U.S., in the late twentieth century? Won't this still sound too much like metaphysics to those trained during the various states of decay of positivist culture? And if we don't ask these questions what will be the consequences? For as Donna Haraway reminds us, "what counts as an object is precisely what world history turns out to be about" (1988, 588). I seek some way of trying to understand the nature of the interplay of the material and the cultural in the crafting of an ontology. Consequently, I will place considerably more emphasis on ontological issues than is common in science studies, although I will not ignore the epistemological issues either, since like Haraway's material-semiotic actors, the ontology that I will offer is not outside of epistemology.

Upon articulating a new ontological and epistemological framework, I will own up to its realist tenor. After a resurgence of interest in scientific realism in the 1980s, its popularity seems to have waned once again, if not the result of the deathknell sounded by Fine's (1984) clever according of the metatheoretical failure of arguments for realism, then at least by the commonplace tendency on the part of constructivists to present scientific realism as naive, unreflexive, and politically invested in its pretense to assume an apolitical posture. In fact, the pairing of social constructivism with some form of antirealism has come to seem almost axiomatic: if we acknowledge the cultural specificity of scientific knowledge construction, are we not obligated to relinquish the hope of constructing theories that are true representations of independent reality? For example, in offering a concrete case of the underdetermination thesis, Cushing (1994) argues that the fact that distinctive theories can account for the same empirical evidence means that realists are hard-pressed to make an argument for theoretical access to the actual ontology of our world.[2] For the most part, social constructivists have expressed either outright disdain for or at least suspicion towards realism, and have explicitly adopted antirealist positions, or they have refused the realism–antirealism debate altogether either because they feel limited by this very opposition (see for example Fine, 1984; Pickering, 1994) or they have thought it more fruitful to focus on other issues. As an admitted social constructivist, I must confess to having sympathy with all of these positions, but I do not want to deny my own realist tendencies or the realist features of the framework I present. While I acknowledge that realism has been invoked to support both oppressive and liberatory positions and projects, my hope is that at this historical juncture, the weight of realism – the serious business and related responsibility involved in truth hunting – can offer a possible ballast against the persistent positivist scientific culture that too easily confuses theory with play (see Barad, forthcoming).

Realizing the multiplicity of meanings that realism connotes, at this juncture I want to clarify how I take realism in the first instance. As a starting point, I follow Cushing's lead:

I assume, perhaps unreasonably, that a scientific realist believes successful scientific theories to be capable of providing reliable and understandable access to the ontology of the world. If one weakens this demand too much, not much remains, except a belief in the existence of an objective reality to which we have little access and whose representation by our theories is nebulous beyond meaningful comprehension. In such a situation, is it worth worrying about whether or not one is a realist? (Cushing, 1994, 270, fn. 26).

Although I will ultimately add substantive qualifications to this definition, I do not intend to weaken what I take to be the spirit of this demand, and I therefore have selected this starting point to clarify the sense of realism with which I mean to engage, as separate from some other more general uses in the science studies literature, including discussions that oppose realism to relativism, or realism to linguistic monism, or realism to subjectivism. My first concern is not with realism in these senses: I grant that there are forms of antirealism that are not relativist, that do not deny the existence of an extralinguistic reality, and that are compatible with various notions of objectivity. That is, in the spirit of Cushing's query, I want to limit the elasticity of the meaning of realism for my initial purposes. Science studies scholars have labored long and hard to articulate moderate social constructivist positions that reject the extremes of objectivist, subjectivist, absolutist, and relativist stances, but it is perhaps inappropriate to label these as realist on just such bases alone. That is, I do not want to turn these accomplishments aside by setting up realism as the foil to the entire family of apparitions, including some that scientists find most haunting. For example, feminist science studies scholars in particular have overwhelmingly rejected the specter of epistemological relativism, with an intensity shared by scientists (a fact which may come as a surprise to scientists who have not studied the feminist literature). Seeing epistemological relativism as the mirror twin of objectivism, and both as attempts to deny the embodiment of knowledge claims, feminist theories of science including Haraway's theory of situated knowledges (1988), Harding's strong objectivity (1991), Keller's dynamic objectivity (1985), and Longino's contextual empiricism (1990) articulate nonrelativist constructivist positions. Consequently, although my discussion of realism in this paper is concerned with the sense in which access to the ontology of our world is possible, additionally I will also attempt to satisfy the high standards that have already been set by specifying the ways in which the new form of realism that I propose rejects these other extreme oppositions. I use the same label, "agential realism", for both the new form of realism and the larger epistemological and ontological framework that I propose. (My motivation for using an adjectival form of agency as the modifier will be clarified later.)

2. AGENTIAL REALISM: AN OVERVIEW

The inspiration for agential realism comes from my reading of Niels Bohr's philosophy-physics. (I use this hyphenated structure, instead of the usual "philosophy of physics", to emphasize Bohr's unwillingness to think of these interests as distinctive in any sense, contrary to the sharp disciplinary boundaries that are important to contemporary physics culture (Barad, 1995).) Bohr's philosophy-physics provides a fruitful starting point because it involves a critical examination of observation/measurement processes: in contrast to the inconsequential role of the observer in Newtonian physics, Bohr argued that quantum physics requires a new logical framework that takes the observation processes into account. Measurement is a potent moment in the construction of scientific knowledge – it is an instance where matter and meaning meet in a very literal sense. For example, in the context of studies of the practice of experimental high-energy physics, science studies scholars have emphasized the role of detectors as sites for making meaning (Traweek, 1988; Galison, 1987; Pickering, 1984). My focus here is on the embodiment of culture within *theory*. That is, I read Bohr's philosophy-physics as an argument for the necessity of including practice within theory: that, contrary to traditional views of physical theory that take the actual practice of measurement to be outside of theory, and according to the logical positivist/empiricist program which assumes that measurements transparently adjudicate among theories, Bohr situates practice within theory, since to ignore practice is to get the theory wrong. This is not to suggest that all is reduced to theory, but that theory, as a matter of principle, must itself be embodied in practice and cannot abstract itself from these issues.[3] While I fully suspect postmodern readers to be readily suspicious of theoretical moves that elevate practical issues to the realm of principles, I will show that this implicit universality amounts to the common constructivist assertion that all knowledges are local knowledges. That is, I will indicate how this theoretical analysis of measurement can be understood as the literal embodiment of objectivity in the sense of Haraway's theory of situated knowledges (1988; see also Barad, 1996).

Now I am quite aware that the ubiquitous appropriation of quantum theory makes it dangerous material to handle these days, and the addition of feminist theory to my list of concerns seems to be quite enough to detonate the explosive mixture, so a few preliminary words of caution are in order. In a sense, to accomplish my task I need to "rescue" quantum theory from both its overzealous advocates and its unreflective practitioners. In the popular literature quantum physics is often positioned as the scientific path leading out of the West to the metaphysical garden of Eastern mysticism. Paralleling these popular renditions, one can find suggestions in the feminist literature that quantum physics is inherently less androcentric, more feminine, and generally less regressive than the masculinist tendencies found in Newtonian physics. But those who naively embrace quantum physics as some exotic Other that will save our weary Western souls forget too quickly that quantum physics underlies the workings of the A-bomb, that particle physics (which relies on quantum theory) is the ultimate manifestation of the tendency towards scientific reductionism, and that quantum theory in all its applications continues to be the purview of a small group of primarily Western-trained males. It is not my intention to contribute to the romanticizing or mysticizing of quantum theory. On the contrary, as a physicist, I am interested in engaging in a rigorous dialogue about particular aspects of specific discourses on quantum physics and the implications. Similarly, I do not make any claims here about Niels

Bohr being an unappreciated or closet feminist. Nor is my aim to critique physics by holding it up to some fixed notion of gender. On the contrary, the feminist analysis I present here simultaneously interrogates the notions of identity and science.[4]

On the other hand, I part company with my neo-positivist physics colleagues who believe that philosophical concerns are superfluous to the real subject matter of physics. Indeed, I am sympathetic to Bohr's view that philosophy is integral to physics. Niels Bohr was one of the most important physicists of the twentieth century, and his "philosophical" writings span a period of approximately four decades.[5] Bohr is considered to be the primary author of the so-called Copenhagen interpretation of quantum mechanics.[6] Although alternative interpretations have been advanced ever since the formulation of quantum theory during the mid-1920s, the physics community claims allegiance to the Copenhagen interpretation.[7] Unfortunately, the vast majority of physicists have no more than a passing interest in the philosophical issues, and prefer to focus on the powerful tools that the quantum formulation provides for purposes of calculation. This avoidance has had its cost: the foundational issues of this fundamental physical theory remain unresolved and the culture of physics is such that unreflective attitudes and approaches are rewarded. While I will not make any arguments about the superiority of Bohr's approach to quantum physics, the simultaneous centrality and marginality of his approach makes it particularly interesting.[8]

Bohr often makes reference to the epistemological lessons of quantum theory, and he sees the framework that he offers for quantum physics as having general relevance beyond physics (Folse, 1985). So it is not at all inappropriate that attention has been given to the larger philosophical implications of Bohr's philosophy-physics, leaving specific issues surrounding the interpretation of quantum theory aside. My approach will be to draw out the specifics of a consistent Bohrian framework, grounding the analysis in the physics, and then to consider the larger implications.

The first task is necessary since there is much disagreement in the secondary literature about how to interpret Bohr. For example, Bohr has been called a positivist, an idealist, an instrumentalist, a (macro)phenomenalist, an operationalist, a pragmatist, a (neo)Kantian, and a realist by various authors. One of the difficulties in assigning a traditional label to Bohr's interpretative framework is the fact that Bohr is not specific about his ontological commitments. In order to fill this crucial gap, I propose an ontology that I believe to be consistent with Bohr's views, although I make no claim that this is what he necessarily had in mind. That is, as a result of Bohr's inattentiveness to concerns of ontology, there will be vast differences in opinion about this matter, and I am less interested in trying to figure out what Bohr was actually thinking than what makes sense in the context of what Bohr does tell us. My approach is to use Bohr's writings as the context for my thinking about these issues; I do not take them as scripture (see Methodological Interlude). Using this analysis of Bohr's philosophy-physics as inspiration, I introduce agential realism as a framework that ties together the epistemological and ontological issues.

I then show how agential realism can be used to address particular concerns that social constructivist approaches to science make apparent, including some of the ones enumerated in the previous section. I diverge from Bohr in strategy here, but not in spirit. Bohr's methodological approach was to draw out the epistemological lessons of quantum theory for other fields of knowledge by essentially trying to guess what the relevant Complementary variables would be in each arena. This analogic strategy often failed: both because he proposed a set of variables that turned out not to be Complementary, and because the implications drawn on this basis watered down the complexity and richness of the "epistemological lessons".[9] My approach will be to examine specific implications by directly taking on a different set of epistemological and ontological commitments. That is, I will not use the notion of Complementarity as a springboard; instead I directly interrogate particular philosophical background assumptions that underlie specific concerns.

Finally, I want to make explicit the distinction between my approach and a host of analogical (mis)appropriations of quantum theory that are more common in the literature than physicists would wish. I will not put forward any argument to the effect that quantum theory of the microworld is analogous to situations that interest us in the macroworld – be they religious, spiritual, psychological, or even those encountered in science studies. My focus is on the development of widely applicable epistemological and ontological issues, that can be usefully investigated by a rigorous examination of measurement processes as explicated by Bohr's understanding of quantum physics. To ask whether it is not suspect to apply arguments made specifically for microscopic entities to the macroscopic world is, in this case, to mistake the approach as analogical. The epistemological and ontological issues are not circumscribed by the size of Planck's constant (see note 12). That is, I am not interested in mere analogies but rather widely applicable philosophical issues such as the conditions for objectivity, the appropriate referent for empirical attributes, the role of natural as well as cultural factors in scientific knowledge production, and the efficacy of science.

3. MEASUREMENT MATTERS

Often the development of physics has taught us that a consistent application of even the most elementary concepts indispensable for the description of daily experience, is based on assumptions initially unnoticed, the explicit consideration of which is, however, essential if we wish to obtain a classification of more extended domains of experience as clear and as free from arbitrariness as possible .... This development has contributed to the general philosophical clarification of the principles underlying human knowledge (Bohr, 1937, 289–290).

Classical epistemological and ontological assumptions, such as the ones found to underlie Newtonian physics, include an autonomously existing world that is describable independently of our experimental investigations of it. This accounts for the fact that the process of measurement is transparent and external to the discourse of Newtonian science. It is assumed that objects and observers occupy physically and conceptually separable positions. Objects are assumed to possess well-defined intrinsic attributes, and it is the job of the scientist to cleverly discern these inherent characteristics, obtaining the values of the corresponding context-independent variables through some benignly invasive measurement procedure. The reproducibility of measured values under the methodology of controlled experimentation is used in support of the objectivist claim that what has been obtained is an objective representation of intrinsic properties that characterize the objects of an uncontrolled, independent reality.[10] The transparency of the measurement process in Newtonian physics is a root cause of its value to and prestige within the Enlightenment culture of objectivism.

The two basic assumptions of measurement transparency underlying Newtonian physics that were challenged by Bohr's interpretation of quantum physics are:[11]

(1) Measurements involve continuous, determinable interactions; that is, an unambiguous, inherent, Cartesian-like cut between knower and known delineates object from observational apparatus.
(2) The applicability of conceptual schema is independent of measurement processes; concepts are abstractable, universal, definite, and context-independent.

The hallmark of Newtonian physics is its determinism, its proclaimed ability to predict and retrodict the full set of physical states of a system for all time: given the initial conditions (i.e., the simultaneous specification of position and momentum at one instant in time), entire particle trajectories can be calculated. In the Newtonian picture, the position and momentum of an object can be determined by a time-of-flight measurement, for example, in which light impinges on the object and the scattered light is collected at a detector. Although light has momentum and energy, the process of illuminating the object can either be made to impart negligible disturbance on the object (intuitively, by continuously lowering the intensity) or the disturbance can be determined and subtracted out, thereby yielding the desired values of the position and the momentum of the object as they would have been had the measurement not been performed. According to Niels Bohr, this objectivist account of the measurement process rests on false assumptions.[12]

Quantum physics is based on an empirically verified discreteness or discontinuity (the quantum of action = Planck's constant = $h \neq 0$) in measurement interactions, initially observed in experiments probing atomic phenomena at the turn of the century. The lack of continuity means that measurement interactions cannot be reduced to the point of being negligible and, therefore, determination of the properties of an independent object are contingent upon subtraction of the effects of the measurement interaction. Bohr argued that subtraction is impossible; that the measurement interaction cannot be precisely specified without intervening in such a way as to disrupt the purpose of the intended measurement. Furthermore, he argued that *the indeterminable discontinuity undermines the separability of the "object" and the "agencies of observation"*.

Bohr's argument for the indeterminability of measurement interactions is based on his insistence that *concepts are defined by the circumstances required for their measurement*, and therefore mutually exclusive experimental arrangements would need to be employed simultaneously (which is impossible) in order to determine all the features of the measurement interaction. For example, in a time-of-flight measurement used to determine the initial conditions, the momentum imparted by the light impinging on the object would need to be subtracted out. But a measurement of the momentum requires an apparatus with movable parts (i.e., the concept 'momentum' is necessarily defined by reference to an apparatus with movable parts[13]), which would then exclude the equally necessary measurement of the position since position measurements require an apparatus with fixed parts (i.e., the concept 'position' is necessarily defined by reference to a fixed apparatus). Therefore, observation is only possible on the condition that the interaction is indeterminable (i.e., it cannot be subtracted out). Consequently, since observations involve an indeterminable discontinuous interaction, *as a matter of principle, there is no unambiguous way to differentiate between the "object" and the "agencies of observation" – no inherent/naturally occurring/fixed/universal/Cartesian cut exists.* Hence, *observations do not refer to objects of an independent reality*. Bohr's interpretation of quantum theory provides profound challenges to both of the assumptions of measurement transparency underlying the Newtonian framework. In fact, Bohr's philosophy-physics undermines a host of Enlightenment notions, requiring him to construct a new logical framework (see especially, Folse, 1985), including a new epistemology, for understanding science.

Bohr moves away from reference to the classical notion of 'disturbance' in his later writings and emphasizes "*quantum wholeness*", or *the lack of an inherent/Cartesian distinction between the "object" and the "agencies of observation*", as the central feature of his new descriptive framework. For Bohr, "*object" and "agencies of observation" form a nondualistic whole in the sense that it is conceptually incoherent to refer to an inherent distinction between the two.* "*Descriptively*, there is a *single* situation, no part of which can be abstracted out without running into conflict with other such descriptions (namely, those of complementary situation). The object cannot be ascribed an 'independent reality in the ordinary physical sense'" (original italics; Hooker, 1972, 156). This is a central notion in Bohr's philosophy-physics and he uses the term "*phenomenon*" (in his later writings) *to designate particular instances of wholeness*: "While, within the scope of classical physics, the interaction between object and apparatus can be neglected or, if necessary, compensated for, in quantum physics this *interaction thus forms an inseparable part of the phenomenon*. Accordingly, *the unambiguous account of proper quantum phenomena must, in principle, include a description of all relevant features of the experimental arrangement*" (my italics, Bohr, 1963c, 4). Furthermore, "[t]he essential wholeness of a proper quantum phenomenon finds logical expression in the circumstance that any attempt at its well-defined subdivision would require a change in the experimental arrangement incompatible with the appearance of the phenomenon itself" (Bohr 1963b, 72).

If a cut delineating the "object" from the "agencies of observation" is not inherent, what sense, if any, should we attribute to the notion of observation? Bohr suggests that "by an experiment we simply understand an event about which we are able in an unambiguous way to state the conditions necessary for the reproduction of the phenomena" (quoted in Folse, 1985, 124). The specification of these conditions is tantamount to *the introduction of a constructed/agentially positioned/ movable/local/"Bohrian" distinction between an "object" and the "agencies of observation"*.[14] That is, although no inherent distinction exists, every measurement involves a particular choice of apparatus, providing the conditions necessary to give definition to a particular set of classical variables, at the exclusion of other essential variables, and thereby placing a particular constructed cut delineating the "object" from the "agencies of observation". *This particular constructed cut resolves the ambiguities only for a given context; it marks off and is part of a particular instance of wholeness, that is, a particular phenomenon.*

For example, consider once again an experiment in which light is scattered from a particle. The scattered light may be directed towards a photographic plate rigidly fixed in the laboratory and therefore used to record the position, or the light may be directed towards a piece of equipment with movable parts used to record the momentum of the scattered light. The first case essentially describes the process of taking a picture of a particle with a flash camera. In that case, the light is part of the measuring apparatus. In the latter case, the light's momentum is being measured and hence it is part of the object in question.[15] (This example nicely illustrates the Bohrian assertion that observation is possible only on the condition that the measurement interaction is indeterminable: since at least one particle of light, or photon, is required to make a mark on the film recording the position (illustrating the "quantum discontinuity"), and since the effect that this photon has on the particle cannot be accounted for unless the photon's momentum is simultaneously determined, and given that both variables ("position" is definable in the context of an apparatus with a fixed photographic plate, and "momentum" is definable in the context of photographic plate on a movable platform) cannot be unambiguously defined using one particular choice of measuring apparatus, the observation entails an indeterminable interaction.) Therefore, the measurement of unambiguously defined quantities is possible through the introduction of a constructed cut which serves to define "object" and "agencies of observation" *in a particular context*.[16]

Especially in his later writings, Bohr insisted that quantum mechanical measurements are "objective".[17] Since Bohr also emphasized the essential wholeness of phenomena, he cannot possibly have meant that measurements reveal objective properties of independent objects. Rather Bohr's use of the term "objectivity" is tied to the fact that "[n]o explicit reference is made to any individual observer" (Bohr, quoted in Murdoch, 1987, 99). "Objective" means reproducible and unambiguously communicable – in the sense that "permanent marks ... [are] left on bodies which define the experimental conditions":

> Common to the schools of so-called empirical and critical philosophy, an attitude therefore prevailed of a more or less vague distinction between objective knowledge and subjective belief. By the lesson regarding our position as observers of nature, which the development of physical science in the present century has given us, a new background has, however, been created just for the use of such words as objectivity and subjectivity. From a logical standpoint, we can by an objective description only understand

a communication of experience which does not admit of ambiguity as regards the perception of such communications (Bohr, quoted in Folse, 1985, 15).

Clearly, Bohr's notion of "objectivity", which is not predicated on an inherent/Cartesian distinction between "objects" and "agencies of observation", stands in stark contrast to any Newtonian sense of "objectivity" denoting observer independence.

Bohr's term "agencies of observation" is evocative of the central role of agency in the new epistemological and ontological framework that I will introduce later in this paper. "Agencies of observation", instead of the more common term "observer", already signals the inseparability of the material and semiotic apparatuses. That is, in my reading, a pivotal point in Bohr's analysis is that *the physical apparatus (existing in the realm of classical, macroscopic, everyday, direct experience) marks the conceptual subject–object distinction: the material and semiotic apparatuses form a non-dualistic whole.* In other words, classical descriptive concepts obtain their *meaning* by reference to a particular physical apparatus which in turn marks the placement of a constructed cut between the "object" and the "agencies of observation". Finally, the point of reference for unambiguous communication is "from permanent marks – such as a spot on a photographic plate, caused by the impact of an electron – left on the bodies which define the experimental conditions" (Bohr, 1963c, 3). Therefore, "bodies which define the experimental conditions" serve as both the endpoint and the starting point for meaningful observation. For Bohr, measurement and description entail one another (though not in a narrowly operationalist sense but in the sense of epistemological indistinguishability).

Quite atypical of the writings of theoretical physicists, Bohr's writings often include very detailed drawings of experimental apparatuses. As Honner points out "Bohr insisted on providing elaborate drawings of mechanical devices used for observing quantum events [in many of his discussions of complementarity], as if to emphasize the connection between descriptive concepts and classical apparatus" (Honner, 1987, 119). Though Bohr was a theoretical physicist, he was obsessed with the details of measurement and was not satisfied to deal with abstract concepts – for Bohr *meaning is tied to the experiential world.*[18] (There is historical evidence that Rutherford, whom some regard as the greatest experimental physicist of this century, had a profound influence on Bohr, who was a postdoc under Rutherford.)

The question remains: what is the referent for a given objective property (as unambiguously defined in reference to a given constructed cut) that is obtained by a given measurement process? Since there is no inherent distinction between object and instrument, the property which is determined cannot be meaningfully attributed to either an abstract object or an abstract measuring instrument. That is, the measured value is neither attributable to an observation-independent object, nor is it a property created by the act of measurement (which would belie any sensible meaning of the word "measurement").[19] My reading is that the *measured properties refer to phenomena*, remembering that the crucial identifying feature of phenomena is that they are particular instances of wholeness, that "the unambiguous account of proper quantum phenomena must, in principle, include a description of all relevant features of the experimental arrangement" (Bohr, 1963c, 4).

Implicit in our classical descriptive concepts is a subject–object distinction, and since phenomena entail the placement of a constructed subject–object distinction, it is consistent to use classical concepts to describe phenomena. In fact, Bohr strengthens the claim for the appropriateness of our use of classical concepts to describe phenomena to one of necessity. The following detailed mapping of the relationship between classical concepts and phenomena can be given to provide justification for this move: since by their very definition classical descriptive concepts entail a particular subject–object distinction, as specified by the circumstances required for their measurement, and since phenomena include a constructed subject–object distinction, namely the one in question that gives definition to a particular classical concept, it follows that these particular classical concepts are just the ones that are useful in describing phenomena.[20] That is, phenomena are necessarily described using concepts conditioned by particular subject–object distinctions. Another way to appreciate the necessity of this condition is that unambiguous communication necessarily refers to "permanent marks ... left on bodies", that is macroscopic bodies, that in a particular context are defined as the "apparatus", and since the "apparatus" in turn specifies the circumstances required for the definition of particular classical concepts (derived from everyday experience in the macroscopic world and therefore premised on an object–subject distinction), it follows that phenomena, which include the particular constructed cut in question, are necessarily described using classical concepts appropriate to the given context. Again, *reference must be made to bodies in order for concepts to have meaning.*

"While in the [classical] mechanical conception of nature, the subject-object distinction was fixed, room is provided for a wider description through the recognition that the consequent use of our concepts requires different placings of such a separation" (Bohr, 1963b, 92). In fact, according to Bohr's Principle of Complementarity all possible ways of drawing the subject–object distinction must be considered to obtain the maximal accounting of our investigations. That is, mutually exclusive constructed cuts constituting mutually exclusive experimental circumstances, thereby agentially manifesting mutually exclusive phenomena serve to denaturalize the nature of the observational process.

Bohr's epistemological and descriptive framework is radically different from that associated with Newtonian physics. For Bohr, measurement, far from being external to the discourse of scientific theories, must play a prominent role in scientific theorizing: that is, Bohr situates practice within theory. As a result, method, measurement, description, interpretation, epistemology, and ontology are not separable considerations. These connections are explored in the sections following the Methodological Interlude.

4. METHODOLOGICAL INTERLUDE

Einstein once remarked of Bohr, "He utters his opinions like one perpetually groping and never like one who believes to be in possession of definite truth" (Einstein, quoted in Pais, 1982, 417).

Many of the philosophers, historians, and the few physicists who have tried to read Bohr's works have commented on the difficulty of this task. Bohr's style is atypical of most science writing. His writing reflects a self-conscious regard of his own descriptive process, which is consistent with his thorough-going examination of the role of description in scientific knowledge production, fundamental to his approach to understanding quantum physics. In like manner, I have tried to remain attentive to my own descriptive/interpretative process in my reading of Bohr. Consequently, I make no claims here to have discovered what Bohr was actually thinking or intending, as separate from my own interpretative apparatus; rather I attempt to provide a consistent reading within the context of particular ways of resolving ambiguities. (Recall that for Bohr descriptions refer to phenomena, not to some independent reality.) There are clear parallels between this methodology and feminist and other located-knowledges methodologies. This is not mere coincidence but, as will become clear later, a reflection of a common critical reflexivity.

My presentation of the major features of Bohr's post-Newtonian framework and corresponding epistemology come from more than a decade of extensive study of Bohr's writings. Interpretative questions about quantum theory plagued me as a graduate student in theoretical particle physics. (It may seem peculiar to non-scientists to discover that physics graduate school is not the appropriate context for engaging such questions.[21]) By the time I was an assistant professor of physics, my focus broadened to include the larger philosophical issues in Bohr's post-Newtonian framework.

The ideas as I have presented them so far are in considerable agreement with individual features of many of the standard secondary texts on Bohr's philosophy of physics, including the work of Feyerabend (1962), Hooker (1972), Bohm (1985), Folse (1985), Petersen (1985), Honner (1987), and Murdoch (1987). It is important to point out that the views of these scholars are widely divergent on many crucial points. I do not agree *in toto* with the views presented in any of these other accounts, though as I read through the primary texts time and again from the perspective of a theoretical particle physicist, various aspects of these works have been and continue to be helpful to me while I formulate my own evolving views on Bohr's philosophy-physics.

As a measure of the disagreement among Bohr scholars, consider the question of the nature of Bohr's interpretative framework. Most Bohr scholars, and many other scholars who have not studied Bohr, attribute some form of antirealism to Bohr, who has been called a positivist, an idealist, an instrumentalist, a (macro)phenomenalist, a relativist, a pragmatist, and a (neo)Kantian. Folse has been one of the strongest proponents of the minority view that sees Bohr as a realist. As I indicated in the Introduction, one of the difficulties in resolving the ambiguities in Bohr's position is that Bohr focuses on epistemological issues in his writings and he never spells out his ontological commitments. Consequently, it is difficult to discern the nature of any correspondence he may hold between theory and reality. Without a clear-cut presentation of a coherent Bohrian ontology, the task of determining what kind of realist or antirealist position is consistent with Bohr's philosophy-physics seems doomed. In the next section, I present an ontology I believe to be consistent with Bohr's views, and I then address the question of a correlative interpretative stance.

I will argue that Bohr's philosophy-physics can be understood to be consistent with a particular form of realism, that I label "agential realism". But at I noted from the outset, my aim is not so much to provide a faithful representation of Bohr's philosophy-physics, as to propose a framework for thinking about critical epistemological and ontological issues, particularly in science studies. In addressing these issues in the remainder of the paper it would be just as dishonest to attribute the full development of this framework to Bohr as it would be to deny that my thinking about Bohr's philosophy-physics is everywhere present in my formulation.

5. AGENTIAL REALITY AND AGENTIAL REALISM

Bohr has often been badly misunderstood, I believe, because his readers have insisted on reading the classical ontological and epistemological assumptions into ... [his] remarks ... it presupposes some autonomously existing atomic world which is describable independently of our experimental investigation of it. There is no such world for Bohr.... There is no godlike approach possible to the physical world whereby we may know it as it is "absolutely in itself"; rather we are able to know only as much of it as can be captured in those situations which we can handle conceptually – that is, those situations where unambiguous communication of the results is possible. ... This is in complete contrast to the classical realist metaphysics and epistemology where the world is concerned as being the way classical theory says it is, independently of our experimental exploration of it. ... (Clifford A. Hooker, 1972, 155–6)

The realism–antirealism distinction is often drawn on the basis of questions about belief in a correspondence theory of truth, which is rooted in a subject–object / culture–nature / word–world dualisms. The separation of epistemology from ontology is a reverberation of these dualism. Bohr's philosophy clearly contests a Cartesian (inherent, fixed, universal) subject–object distinction, and I will argue here that this undermines conceptions which see reality as either prior to or outside of language. What is being described is our participation *within* nature.

Aage Petersen, in an article entitled "The Philosophy of Niels Bohr", writes:

Traditional philosophy has accustomed us to regard language as something secondary, and reality as something primary. Bohr considered this attitude toward the relation between language and reality inappropriate. When one said to him that it cannot be language which is fundamental, but that it must be reality which, so to speak, lies beneath language, and of which language is a picture, he would reply "We are suspended in language in such a way that we cannot say what is up and what is down. The word 'reality' is also a word, a word which we must learn to use correctly"[22] (Petersen, 1985, 302).

In my effort to provide a consistent Bohrian meaning to the term 'reality', I turn to a very important passage from Bohr's writings: a passage from his response to the 1935 paper of Einstein, Podolsky, and Rosen, the so-called "EPR paper", wherein Bohr directly challenges the EPR definition of "physical reality".[23] Many scholars have pointed out that the argument Bohr articulates in this passage is pivotal to his attempt to discredit the analysis of EPR and to resolve the "EPR paradox" once and for all. I say this both to highlight the fact that I have not chosen some obscure or arbitrary passage from Bohr's writings, but the one in which Bohr has the most at stake in being careful with the presentation of his ideas on the notion of reality, and also to express my surprise that none of the scholarship that I have read on Bohr emphasizes the *positive* feature of this passage – that Bohr offers his own definition of physical reality in the final sentence:

From our point of view we now see that the wording of the above-mentioned criterion of physical reality proposed by Einstein, Podolsky, and Rosen contains an ambiguity as regards the meaning of the expression "without in any way disturbing the system". Of course there is in a case like that just considered no question of a mechanical disturbance of the system under investigation during the last critical stage of the measuring procedure. But even at this stage there is essentially the question of *an influence on the very conditions which define the possible types of predictions regarding the future behaviour of the system.* Since these conditions constitute an inherent element of the description of any phenomenon to which the term "physical reality" can be properly attached, we see that the argumentation of the mentioned authors does not justify their conclusion that quantum-mechanical description is essentially incomplete (original italics, Bohr, 1935, 700).

In discussing Bohr's use of the word 'phenomenon' earlier, I pointed out that the conditions which define the possible types of predictions constitute an inherent element of the description of any phenomenon. Therefore, the first phrase of the last sentence is consistent with Bohr's use of the term phenomenon.[24] The last sentence then indicated that the term 'physical reality' can properly be attached to the phenomenon. *Phenomena are constitutive of reality.* Reality is not composed of things-in-themselves or things-behind-phenomena, but things-in-phenomena.

This interpretation is consistent with the following point made by von Weizsäcker:

The fact that classical physics breaks down on the quantum level means that we cannot describe atoms as "little things". This does not seem to be very far from Mach's view that we should not invent "things" behind the phenomena. But Bohr differs from Mach in maintaining that "phenomena" are always "phenomena involving things", because otherwise the phenomena would not admit of the objectification without which there can be no science of them. For Bohr, the true role of things is that they are not "behind" but "in phenomena" (quoted in Honner, 1987, 15).

Or as Honner puts it:

The term [phenomenon] was not intended to signify the uninterpreted appearance of the object of experience itself. Nor was Bohr trying to follow the Kantian distinction between the thing-in-itself and our perception of it. If one wanted to talk about such "things", then they were as Weizsäcker put it, to be found *in* the phenomena rather than behind it (Honner, 1987, 68).

The nature of this relationship is a point of contention among Bohr scholars. My own studies of Bohr's writings brought me to a conclusion similar to von Weizsäcker's before I ever started reading any of the secondary texts, and in spite of subsequent readings of the many different interpretations offered, it has always seemed very clear to me that this is the only interpretation that respects the complex intention of the Bohrian notion of 'phenomena'.[25] The point is that phenomena constitute a non-dualistic whole so that it literally makes no sense to talk about independently existing things as somehow behind or as the causes of phenomena. A Bohrian ontology does not entail some fixed notion of being that is prior to signification (as the classical realist assumes), but neither is being completely inaccessible to language (as in transcendental idealism) nor completely of language (as in linguistic monism) – what is being described is our participation *within* nature, what I term "agential reality".

Bohr often refers to the fact that we are in nature: "In our own century the immense progress of the sciences has ... given us an unsuspected lesson about *our position as observers of that nature of which we are part ourselves*" (my emphasis, Bohr, 1963c, 8). The introduction to the collection *Essays 1933–1957 on Atomic Physics and Human Knowledge* begins:

> The importance of physical science for the development of general philosophical thinking rests not only on its contributions to our steadily increasing *knowledge of that nature of which we ourselves are part*, but also on the opportunities which time and again it has offered for examination and refinement of our conceptual tools (my emphasis, Bohr, 1963b, 1).

The passage from Bohr's response to EPR continues:

> On the contrary, this description, as appears from the preceding discussion, may be characterized as a rational utilization of all possibilities of unambiguous interpretation of measurements, compatible with the finite and uncontrollable interaction between objects and the measuring instruments in the field of quantum theory. In fact, it is only the mutual exclusion of any two experimental procedures, permitting the unambiguous definition of complementary physical quantities, which provides room for new physical laws, the co-existence of which might at first sight appear irreconcilable with the basic principles of science. It is just this entirely new situation as regards the description of physical phenomena that the notion of *complementarity* aims at characterizing (original italics).

Notice that in this last sentence we are told that scientific theories describe physical phenomena. Since phenomena constitute agential reality, and it is phenomena that scientific theories describe, it follows that scientific theories describe agential reality. Were it not for the crucial adjective "agential", emphasizing the non-objectivist nature of Bohrian ontology, as I've described it here, the conclusion of this syllogism would sound like the proclamation of a die-hard realist who is advocating a classical correspondence theory of truth. However, the correspondence in question is between theories and *agential reality*, not an observer-independent reality. Hence, I conclude that Bohr's framework is consistent with a particular notion of realism, which I label "agential realism". Agential realism is compatible with the point I made earlier in this section that any notion of realism that is consistent with Bohr's philosophy must not be parasitic on subject–object / culture–nature / word–world distinctions.[26]

That Bohr subscribed to some sort of realism is also supported by his practice of science. A particularly poignant example of how different philosophical positions guided the efforts of different segments of the physics community during the 1920s is given by considering a range of reactions to the notion of "wave/particle duality". These reactions constitute the twentieth-century contribution to a long historical debate about the nature of light.

To say that light consists of particles is to insist that light consists of localized object that occupy a given location at each moment in time. On the other hand, to say that light consists of waves is to insist that light consists of objects with extension in space, occupying more than one position at any moment of time, like ocean waves that move along a stretch of beach; and furthermore, different waves can overlap and occupy the same position at any moment of time, unlike particles. Obviously, the concepts of "wave" and "particle" are mutually exclusive: an object is either localized or it is extended, it can't be both. And yet, early twentieth-century experiments seemed to indicate that light behaves as a wave under certain experimental conditions, and as a particle under a mutually exclusive set of experimental conditions. This result was surprising since in the latter part of the nineteenth century the wave theory of light was well confirmed by both theoretical (Maxwell's electromagnetic theory) and experimental (diffraction and interference effects) considerations. Hence, a community-wide struggle ensued to resolve this paradox.

Classical realists hoped to resolve the paradox by finding some unifying explanation. Could it be that all objects are ultimately waves, but that on certain scales they look like particles? Another type of response came from the positivists/instrumentals who, like Heisenberg, put their faith in the mathematical formalism itself and saw the efforts to assign appropriate visualizable concepts to the mathematics as specious. While this seemed to be a neat and pragmatic solution to some physicists, others were not so willing to give up on interpretation and meaning.

Bohr's affinity for some kind of realistic interpretative stance led him to continue to seek out a solution to this paradox. Bohr participated with a tenacious passion in the debate. If Bohr had adopted an antirealist attitude it is doubtful that he would have found it necessary to develop an entirely new approach for understanding the role of descriptive concepts in science, which became the basis for Complementarity, and ultimately the so-called Copenhagen interpretation of quantum mechanics. Clearly, interpretative stances matter in the construction of scientific theories.

A few more historical details may be illuminating here. In 1924, Bohr wrote a paper with Kramers and Slater that put forth the radical conjecture that perhaps the most sacred principle in all of physics – the conservation of energy – would have to be sacrificed at the atomic level in order to find a satisfying resolution of the wave/particle duality paradox. Surely an instrumentalist or a die-hard antirealist would not have gone this far in attempting to explain the applicability of dual representations. The trio quickly retracted this proposal as soon as contrary empirical evidence came to light, but Slater never forgave Bohr for convincing him to go along with such a radical proposal. Bohr then adopted a new approach that entailed the examination of the circumstances under which these characteristics are manifest (they only appear under mutually exclusive circumstances), and consequently to an examination of the context-dependence of descriptive concepts.[27] Complementarity's development was contingent on certain realist commitments on Bohr's part. Otherwise, Bohr would have been content with the use of alternative descriptions (wave and particle) as evidenced by Heisenberg's instrumentalist stance.

Furthermore, there is important historical evidence that shows that Bohr strongly disagreed with Heisenberg about the importance and interpretation of wave/particle duality. Bohr and Heisenberg went off on separate vacations and developed the framework of Complementarity and the Uncertainty Principle, respectively; upon returning to Copenhagen, Bohr passionately criticized Heisenberg's derivation of the Uncertainty Principle for its gross neglect of the centralness of wave/particle duality for an appropriate analysis.[28]

Bohr's interpretative framework deviates in a unique and nontrivial fashion from classical correspondence or mirroring theories of science. For Bohr, the paradox is resolved as follows: "wave" and "particle" are classical descriptions that refer to different mutually exclusive phenomena, and not to independent physical objects. He emphasized that this saved the theory from inconsistencies since it was impossible to observe particles and wave behaviors simultaneously since mutually exclusive experimental arrangements are required.

Ambiguity and paradox do not find a Newtonian/Cartesian resolution in this post-Newtonian framework. No final unifying reductionistic explanation is offered; only contextual understanding, located knowledges are obtained from the multiple contestations of the assumption of an inherent/fixed/universal/Cartesian subject–object distinction. The ambiguity is only temporarily, contextually decided, and therefore, descriptive characterizations do not signify properties of abstract objects or observation-independent beings, but rather describe the "between of our intra-actions" as it is marked by particular constructed delineations. (Since there is no sense of two things to interact, I have introduced the term "intra-action" to avoid reinscription of the contested dichotomy.) In other words, measurements of the values of the well-defined variables are attributable to the phenomenon as a particular instance of wholeness, the fully contextual be-in' where the matter and meaning meet.

6. AGENTIAL REALISM: THE FRAMEWORK

Throughout the field of meanings constituting science, one of the commonalities concerns the status of any object of knowledge and of related claims about the faithfulness of our accounts to a "real world", no matter how mediated for us and no matter how complex and contradictory these worlds may be (Haraway, 1991, 197)

In addition to the question of interpretative stances in science studies, agential realism provides a framework for addressing broad epistemological and ontological issues. In this section I develop a few key points that are relevant to the issues I will address in the next section:[29] (1) agential realism grounds and situates knowledge claims in local experiences: objectivity is literally embodied; (2) agential realism privileges neither the material nor the cultural: the apparatus of bodily production is material-cultural, and so is agential reality; (3) agential realism entails the interrogation of boundaries and critical reflexivity; and (4) agential realism underlines the necessity of an ethics of knowing.

(1) *Agential realism grounds and situates knowledge claims in local experiences: objectivity is literally embodied.*

On the one hand, feminists and other Enlightenment critics have expressed skepticism towards objectivism, especially

[t]he idea of a basic dichotomy between the subjective and objective; the conception of knowledge as being a correct representation of what is objective; the conviction that human reason can completely free itself of bias, prejudice, and tradition; the ideal of a universal method by which we can first secure firm foundations of knowledge and then build the edifice of a universal science; the belief that by the power of self-reflection we can transcend our historical context and horizon and know things as they really are in themselves (Bernstein, 1983, 36).

In the post-Kuhnian era in which we live, the arguments against objectivism have been robust and extensive, reaching across disciplinary boundaries and out into the world beyond the academy, so that few scholars currently find it tenable to subscribe to the set of Enlightenment doctrines outlined above. Enlightenment defenders are hard-pressed to show how objectivism can bootstrap its way out of the murky waters of spacetime contingencies. Ironically, mainstream anti-Enlightenment theorists, including Derrida, Foucault, and Lyotard, have ignored crucial social markers such as gender and race in their critiques of the universalizing tendencies characteristic of the Enlightenment project. However, it is not only the limitations of these critiques that have concerned feminists, but their thoroughgoing rejection of the entire set of Enlightenment goals as well. Feminist theorists have taken exception with anti-Enlightenment scholarship that abandons the possibility of positive epistemologies in their embrace of interpretationism, relativism, and strong social constructivism.[30]

Haraway's theory of situated knowledges presents a direct challenge to the objectivist "view from nowhere", the "godtrick" of infinite passive vision, and the equally irresponsible relativist "view from everywhere", posing embodied sight – the view from somewhere, along with the responsibility that that entails – as the key to feminist objectivity. According to Haraway:

There is no unmediated photograph or passive camera obscura in scientific accounts of bodies and machines; there are only highly specific visual possibilities, each with a wonderfully detailed, active, partial way of organizing worlds. ... Understanding how these visual systems work, technically, socially, and psychically, ought to be a way of embodying feminist objectivity (1988, 583).

Agential realism gives us a technology of embodiment (Barad, 1996). Recall that concepts obtain their meaning by reference to a particular apparatus marking the placement of a constructed boundary between the "object" and the "agencies of observation". And in turn, the point of reference for objective description of phenomena is "from permanent marks ... left on bodies which define the experimental conditions." Therefore, bodies which define the experimental conditions serve as both the endpoint and the starting point for objective accounts of our intra-actions. In other words, *objectivity is literally embodied.* According to agential realism, knowledge is always a view from somewhere – objective knowledge is situated knowledge.

(2) *Agential realism privileges neither the material nor the cultural: the apparatus of bodily production is material–cultural, and so is agential reality.*

While theoretical constructs are not to be understood as representing transparently given observation-independent properties possessed by independent material objects/beings as they exist in isolation from all observational interactions, neither are we to interpret these constructs as artifacts of the observational process, purely discursive gestures imprinted on the blank slate of passive matter. As Bohr tell us:

> These problems were instructively commented upon from different sides at the Solvay meeting.... On that occasion an interesting discussion arose also about how to speak of the appearance of phenomena.... The question was whether, as to the occurrence of individual effects, we should adopt a terminology proposed by Dirac, that we were concerned with a choice on the part of "nature" or, as suggested by Heisenberg, we should say that we have to do with a choice on the part of the "observer" constructing the measuring instruments and reading their recording. Any such terminology would, however, appear dubious since, on the one hand, it is hardly reasonable to endow nature with volition in the ordinary sense, while, on the other hand, it is certainly not possible for the observer to influence the events which may appear under the conditions he [*sic*] has arranged. To my mind, there is no other alternative than to admit that, in this field of experience, we are dealing with individual phenomena and that our possibilities of handling the measuring instruments allow us only to make a choice between the different complementary types of phenomena we want to study (Bohr, 1949, 223).[31]

There are three important points that we can take from this passage: (i) nature has agency, but it does not speak itself to the patient, unobtrusive observer listening for its cries – there is an important asymmetry with respect to agency: we do the representing, and yet (ii) nature is not a passive blank slate awaiting our inscriptions, and (iii) to privilege the material or the discursive is to forget the inseparability that characterizes phenomena.

As evidenced in the above quote, when Bohr and other physicists engaged in dialogue about quantum theory they spoke about the "choice made on the part of the experimenter," as if the experimenter is a liberal humanist actor of individual will.[32] There is no reference to the social dimensions of scientific knowledge production. (It is interesting to note though that Bohr does acknowledge the role of linguistic constraints.) However, without intending any anachronistic projections, it must be the case that material–semiotic apparatuses are fully cultural (i.e., social, linguistic, historical, political, etc.) frameworks, not the result of individual will, since reproducibility and unambiguous communication are the criteria for objectivity. That is, scientists make meanings within specific communities, they do not do so autonomously. Therefore, according to agential realism, the apparatus that is theorized must be a multidimensional material–cultural framework.

Furthermore, agential realism provides an account of the simultaneously material and cultural nature of the ontology of the world. Saying that something is socially constructed doesn't mean that it isn't real – on the contrary, according to agential realism, reality is itself material-cultural.[33] There is no opposition here between materiality and social construction: constructedness does not deny materiality. *The materiality of the body is not dissipated by its constructedness since reality is constituted by the "between", the inseparability of nature-cultural / world–word / physical–conceptual / material–discursive.* Culture does not displace or replace nature, but neither do things exist outside of culture. Phenomena are material–cultural be-in's. Haraway makes a similar point, I think, in designating objects as "material–semiotic actors". She uses this term "to portray the object of knowledge as an active, meaning-generating part of the apparatus of bodily production, without ever implying the immediate presence of such objects.... Boundaries are drawn by mapping practices; 'objects' do not preexist as such. Objects are boundary projects" (Haraway, 1988, 595). In other words, the apparatus of bodily production, *qua* agencies of observation, are not separable from phenomena.

(3) *Agential realism entails the interrogation of boundaries and critical reflexivity.* Wholeness, according to agential realism, does not signify the dissolution of boundaries. On the contrary, boundaries are necessary for making meanings. Theoretical concepts are only defined within a given context, as specified by constructed boundaries. Wholeness is not about the prioritizing of the innocent whole over the sum of the parts; wholeness signifies the inseparability of the material and the cultural. Wholeness requires that delineations, differentiations, distinctions be drawn; differentness is required of wholeness. Utopian dreams of dissolving boundaries are pure illusion since by definition there is no agential reality without constructed boundaries. There are two common ways to attempt to deny responsibility for boundaries: (1) claim that they are natural, or (2) claim that they are arbitrary partitionings of a holistic oneness, existing outside of human space and time. In contrast, agential realism explicitly shows that boundaries are interested instances of power, specific constructions, with real material consequences. There are not only different stakes in drawing different distinctions, *there are different ontological implications*.

Furthermore, boundaries are not fixed. Productive and creative tensions are set up in consideration of different possible placements of agentially situated cuts. Consideration of mutually exclusive intra-actions, constituting opposing shifts in the conceptual terrain, reminds us that descriptive concepts do not refer to an observer-independent reality, but to phenomena. In fact, descriptions reflect back upon the specification of boundaries, since descriptions refer to phenomena and boundaries are in phenomena (i.e., the conceptual scheme is tied to the physical apparatus and the descriptions refer to the phenomenon, which by definition includes the apparatus; therefore the description refers back to the constructed conceptual scheme). The placement of the boundary becomes part of what is being described: human conceptual schema are part of the quantum wholeness. Descriptions of phenomena are reflexive, and the shifting of boundaries constitutes a meta-critique.

The acknowledgement and interrogation of context is common to many feminist epistemologies. For example, both Longino's theory of contextual empiricism and Harding's theory of strong objectivity call for a critical examination of background assumptions. Harding writes:

> In an important sense, our cultures have agendas and make assumptions that we as individuals cannot easily detect. Theoretically unmediated experience, that aspect of a group's or an individual's experience in which cultural influences cannot be detected, functions as part of the evidence for scientific claims. Cultural agendas and assumptions are part of the background assumptions and auxiliary hypotheses that philosophers have identified. If the goal is to make available for critical scrutiny all the evidence marshaled for or against a scientific hypothesis, then this evidence too requires critical examination within scientific research processes (1991, 149).

Agential realism includes practice within theory: theory is epistemologically and ontologically reflexive of context. Contrary to traditional views of theory that take the actual practice of measurement to be outside of theory, and according to the logical positivist/empiricist program which assumes that measurements transparently adjudicate among theories, Bohr's philosophy-physics entails a reconceptualization of science that places the discourse on science into scientific discourse. That is, *phenomena are the embodiment of cultural practices within theory*. I suspect that the reflexive implications are a root cause of Bohr's marginalization within the physics community (see Barad, 1995 for more details).

(4) *Agential realism underlines the necessity of an ethics of knowing.*

According to agential realism, reality is not independent of our explorations of it – both epistemologically *and* ontologically speaking. Focusing on the ontological as well as the epistemological is crucial to intra-acting responsibly within the world. Knowledge projects entail the drawing of boundaries, the production of phenomena which are material–cultural intra-actions. That is, our constructed knowledges have real material consequences. And therefore, agential realism calls for direct accountability and responsibility. It is to remind us of this fact that the adjectival form of the word "agency" modifies and specifies the form that realism takes here, in defiance of traditional forms of realism that deny any active participation on the part of the knower. Agency is a matter of intra-acting, that is, agency is an enactment, it is not something someone has.

We need to understand the technologies by which particular social constructions have real material consequences. According to agential realism, the full apparatus of bodily production must be theorized as well – the consideration of acontextual variables will give inadequate results. Think again of the existence of wave phenomena in the context of a particular apparatus of bodily production; particle phenomena are tied to a mutually exclusive apparatus. Quantum physics can account for the phenomenon that exists in a particular context if and only if the apparatus of bodily production is included in the calculation. Agential realism provides an understanding of the possible dynamical intra-actions of nature–culture as ontological be-in's, thus helping us to theorize the material consequences of constructing particular apparatuses of bodily production. Knowing involves denaturalizing, multiply contesting and destabilizing the existing apparatus to refigure boundaries. This will have real material consequences, so that agential realism underlines the requirement for an ethics of knowing.

7. AGENTIAL REALISM AND SCIENCE STUDIES

The notion of complementarity, Bohr also wants to say, can be seen to arise out of the nature of our consciousness of what is "other" to us, out of the unresolvable tension between content and form, between reality and concept, and between theory and experience. Our representations of reality do not so much involve a privileged mental mirroring of external reality, in which object and subject are absolutely distant from each other, as a successful compromise between language and activity.... Yet for Bohr the relationship between word and world is not seen as entirely relative, with the implication that our words have no anchorage in world; instead given the nature of our consciousness of what is demonstrably "other" to us, a relationship between word and world is accepted as necessarily denying complete resolution (Honner, 1987, 103).

As a scientist I have been very interested in feminist science studies in part because the scholars in this field, many of whom are scientists as well, have resisted the polarization often found in contemporary discussions about the nature of science as posed by the more traditional and monodisciplinary approaches. Evelyn Keller identifies two noncommunicating discourses about science,

> ... one an increasingly radical critique that fails to account for the effectiveness of science, and the other a justification that draws confidence from that effectiveness to maintain a traditional, and essentially unchanged, philosophy of science. What is needed is a way of thinking and talking about science that can make sense of these two very different perspectives – that can credit the realities they each reflect and yet account for their differences in perception (Keller, 1985, 6).

I think these tensions are quite productive and, in my opinion, Keller's challenge marks one of the most important issues for contemporary science studies.

If the "discovery model" of science, that sees the production of scientific knowledge as a one-actor show – nature at center stage with a passive audience of observers patiently looking on – is no longer acceptable, and neither is some extreme version of social constructivism that presents science as an arbitrary compendium of power-laden rhetorical moves, then is it possible to give a detailed understanding of the interaction of nature and culture in the production of scientific knowledge? Agential realism provides a framework that can be useful for retheorizing a range of issues generated by reliance on classical epistemologies and ontologies. In this section, I will explore the implications of agential realism for science studies. I have in mind the following questions: How can we reconcile the claim of science studies scholars that scientific knowledge is a socially constructed product that is conceptually, methodologically, and epistemologically allied along particular axes of power with both the liberatory and oppressive interventions that are possible because of the reliability of empirically adequate scientific knowledges? What, if anything, can be said about the ontology of our world through our investigations of it? Is there a notion of realism that is consistent with the assertion that scientific knowledge claims are culturally specific?

The scientific method, which was our Enlightenment birthright, promised to serve as a giant distillation column, removing all cultural influences, and allowing patient practitioners to collect the pure distillate of Truth. The transparency of Newtonian physics to the process of measurement grew out of and helped reinforce this cultural milieu of objectivism that made the successes of science unparadoxical: science works because scientists are able to obtain the facts about the world as it exists independently of us human beings. The Enlightenment notion of science is premised on a separation between knowing subjects and observation-independent objects. Agential realism challenges this conceptualization of science on epistemological and ontological grounds.

According to agential realism, scientific concepts obtain their meaning by reference to a particular physical apparatus marking the placement of an agentially constructed cut between the "object" and the "agencies of observation". In turn, the point of reference for objective description of phenomena is "from permanent marks ... left on the bodies which define the experimental conditions"[34] (Bohr,

1963c, 3). Therefore, bodies serve as both the endpoint and starting point for objective accounts of our intra-actions. In other words, agential realism gives us an embodied account of objectivity.

Scientific results are not reproducible because we are able to measure the observer-independent properties of an independent reality. Reproducibility is possible because scientific investigations are embodied, grounded in experience, in praxis. Reproducibility means the possibility of the reproduction of phenomena, and phenomena are written on the "body"; phenomena are the place where matter and meaning meet. Reproducibility of phenomena does not require or serve as proof for access to the transcendent. "The overall force of Bohr's argument is that we are without absolute foundation in our participation in the world, despite the acceptance that our language works by being anchored in everyday experience of reality" (Honner, 1987, 222). Reproducibility of phenomena is not innocent – it depends upon the choice of some constructed cut for which the ambiguity is only temporarily, contextually decided in such a way as to lend meaning to certain concepts, at the exclusion of others. Reproducibility is not a filter for shared biases; the apparatus of bodily production is culturally situated. The scientists marking off the boundaries are marked by the cultural specificities of race, history, gender, language, class, politics, etc. In stark contrast to the classical framework, there is a sense of agency and therefore accountability. Since reproducibility is the cornerstone of Western science, in the context presently under discussion, science has meaning, but not in any classical sense.[35] According to agential realism, *science is movement between meanings and matter, word and world, interrogating and redefining boundaries, a dance not behind or beyond, but in "the between", where knowledge and being meet.*

Scientific knowledge is not an arbitrary construction independent of "what is out there", since it is not separate from us; and given a particular set of constructed cuts, certain descriptive concepts of science are well-defined and can be used to achieve reproducible results. However, these results cannot be decontextualized. Scientific theories do not tell us about an independent reality; scientific concepts are not simple namings of discoveries of objective attributes of an independent Nature with inherent demarcations. Scientific concepts are not innocent or unique. They are constructs which can be used to describe "the between", rather than some independent reality. (Why would we be interested in such a thing as an "independent reality" anyway? We don't live in such a world.) The point is that phenomena constitute reality. That is, reality itself is material–cultural. And according to agential realism, scientific knowledges are situated knowledges describing *agential* reality. My revision of an important quote by Niels Bohr goes like this: "It is wrong to think that the task of physics is to find out how nature is. Physics concerns what we can say about [our intra-actions within] nature." *We are in reality, we must be in our theories.* In other words, scientific theories describe agential reality – which is just what we are interested in (we don't live in a transcendent reality). For scientific theories to be able to describe agential reality, scientific knowledge must take material–cultural factors into account since they are in agential reality, otherwise we would not expect scientific knowledge to produce empirically adequate accounts of our intra-actions within nature. Reliability is not premised on access to the transcendent, but on the grounding of practice within theory. (The nonclassical epistemology and ontology have removed the paradox of the classical position which sees the reliability of scientific theories as contingent upon objective discoveries of an independent reality.) Consideration of mutually exclusive sets of concepts produce crucial tensions and ironies which underline the point that *it is the fact that scientific knowledge is socially constructed that leads to reliable knowledges about reproducible phenomena – which is just what we are interested in.* Therefore, the understanding that science as a social practice is conceptually, methodologically, and epistemologically allied along particular axes of power can indeed be reconciled with the fact that scientific knowledge is empirically adequate, that it provides effective interventions which may be used towards either regressive or liberatory purposes.

It is not that we attempt to view nature through the lens of culture with an optics that has varying degrees of transparency or opaqueness. We do not try to fit our theories to reality by probing the fixed boundary between nature and culture. Phenomena constitute our ontology. And since scientific concepts can be used to describe phenomena and phenomena are not "out there", but are material-cultural be-in's, agential realism provides us with a form of realism that is compatible with social constructivism. Agential realism is a form of social constructivism that is not relativist, does not reduce knowledge to power plays or language, and does not reject objectivity.

8. CONCLUSIONS

> So, I think my problem and "our" problem is how to have *simultaneously* an account of radical historical contingency for all knowledge claims and knowing subjects, a critical practice for recognizing our own "semiotic technologies" for making meanings, *and* a no-nonsense commitment to faithful accounts of a "real" world, one that can be partially shared and friendly to earth-wide projects of finite freedom, adequate material abundance, modest meaning in suffering, and limited happiness (Haraway, 1991, 187).

Agential realism denies the innocence of naive realism; instead, it entails a conscious, critical reflexivity. Dualisms, binary oppositions, dichotomies, and other demarcations are not secured with natural status as Cartesian cuts which form the foundation of all knowledge – not even in physics. The lines drawn are power-laden epistemological moves with stakes in a given conceptual scheme. This doesn't mean that we can't justify drawing lines, or that crafted conceptual schemes are unusable. Just because science is exposed as being socially constructed doesn't mean that it doesn't work. And empirical adequacy is not an argument that can be used to silence charges of constructivism. But neither is constructivism a proof of epistemological relativism. I have argued that reliable theories about our intra-actions are necessarily socially constructed theories with real material consequences. We need knowledge systems that are both reliable and accountable guides to action. Agential realism creates an alternative to objectivist accounts of knowledge production that deny the situated nature of knowledges and social constructivist accounts that do not address the effectiveness of knowledge systems.

Agential realism is not a call for feminists and others to bow down once again to the hegemony of science in finding a new epistemology. On the contrary, agential realism undermines the hegemony of science (though not its effectiveness). Agential realism insists that science incorporate a reflexive critical discourse, like all other human endeavors. Bohr argued that quantum physics, considered by many to be the most highly esteemed field of science, requires a new framework for understanding the role of descriptive concepts in scientific knowledge production. The notions of wave and particle deconstruct one another, exposing the limitations of the classical framework. There is irony, though perhaps little surprise, in the fact that our interactions with light – oh light! that ever resilient metaphor for knowledge illuminating the dark terrain of ignorance – plays a central role in undermining the hegemony of Newtonian physics, that bright star of the En-lightenment, deconstructing the objective–subjective and nature–culture dualisms that have plagued many attempts to understand the nature of scientific knowledge.

What I am proposing is not some holistic approach in which subject and object reunite into some apolitical relativized whole, but a theory which insists on the importance of constructed boundaries and also the necessity of interrogating and refiguring them. The intra-action involving the subject–object problematizes natural, pure, and innocent separations, but not in a way which reaches for the rapid dissolution of boundaries. Boundaries are not our enemies; they are necessary for making meanings, but this does not make them innocent. Boundaries have real material consequences – cuts are agentially positioned and accountability is mandatory. The shifting of boundaries often helps bring to the surface questions of power which the powerful often try to submerge. Agential realism insists that mutually exclusive, shifting, multiple positionings are necessary if the complexity of our intra-actions are to be appreciated.[36] Multiple contestations of agentially positioned boundaries keep concepts alive, and protects them from reification and petrification. Our goal should not be to find less false boundaries for all spacetime, but reliable, accountable, located temporary boundaries, which we should anticipate will quickly close in against us. Agential realism will inevitably be a casualty of its own design, but I suggest that there is power there presently for some of our purposes.[37] Agential realism involves located or situated knowledges (Haraway, 1988), knowledges that reject transcendental, universal, unifying master theories in favor of understandings that are embodied and contextual.

Who are the agents in agential realism? The history of science parallels the history of knowledge in other arenas: the powerful effectively portray their own knowledge systems as universal, denying their own agency. Within this tradition, agency has been an issue quite separate from authorship. Rivalries over primary authorship are common in the history of science, but what is at stake is cleverness and ingenuity; what is "discovered" is presumed unmarked by its "discoverer". The claim is that the well-prepared scientist can read the universal equations of Nature that are inscribed on G–d's blackboard: Nature has spoken. The paradox is that the objects being studied are given all the agency, even and most especially when they are seen as passive, inert things, culture-free and existing outside of human space and time, moving aimlessly in the void. Completing this Enlightenment scenario, are the passive human observers who are without agency. The overdetermination of Enlightenment discourse is revealed in the juxtaposition of this mythology with the liberal humanist story that provides man with individual will and dominion over nature.

The nature–culture and object–subject dualisms are constructed cuts passed off as inherent and fixed in the service of this legacy. Agential realism makes other moves: shifting and destabilizing boundaries. Here knowledge comes from the "between" of nature–culture, object–subject, matter–meaning. The Cartesian split between the agencies of observation and the object is a classical illusion. Agency cannot be designated as residing in one or the other in isolation. The observer does not have total agency over passive matter – not any representation of reality will do – since not any result one can think of is possible: the world "kicks back". Neither does the object have total agency, whispering its secrets, mostly through the language of mathematics, into the ear of the attentive scientist – knowledge is not so innocent; it doesn't "just come out that way" all by itself. Nature is neither a blank slate for the free-play of social inscriptions, nor some immediately present, transparently given "thingness". Agential realism acknowledges the agency of both subjects and objects without pretending that there is some utopian symmetrical wholesome dialogue, outside of human representations. Science is not the product of some interaction between two well-differentiated entities: nature and culture, since it flies in the face of any matter–meaning dichotomy, like an electron that tunnels through boundaries set up to confine its motion. Meaning and matter are more like interacting excitations of non-linear fields – a dynamic, shifting dance we call science.[38]

Phenomena are the intra-actions of knowledge and being, word and world, culture and nature. Phenomena are material-cultural be-in's. Agential realism relies on a non-classical ontology. The material is not fixed and prior to discursive signification, but in it. Jeanette Winterson writes in her recent novel *Written on the Body*: "That is how I know you. You are what I know" (Winterson, 1992, 120). Intra-acting is an activity that theorizes the mechanics of an embodied objectivity. In our attempt to understand we actively participate within reality. *Realism is not about representations of an independent reality, but about the real consequences, interventions, creative possibilities, and responsibilities of intra-acting within the world.*

Finally, materiality matters: there are social and material reasons for knowledge claims – the intra-actions of the material and the discursive are the technologies of embodied objectivity – and socially constructed knowledges have real material consequences. These conceptions of materiality are opposed to the immediacy of matter in naive realist accounts and its neglect in some social constructivist accounts. It seems to me that giving up on realism would be as hasty as giving up on objectivity. Feminists have interrogated, redefined, and retheorized objectivity; agential realism is an attempt to formulate a feminist notion of realism. Agential realism goes beyond the recognition that there are material and cultural reasons for knowledge claims, beyond the reconceptualization of description in knowledge systems, to providing us with a positive sense of the ontology of our world and some important clues as to how to intra-act responsibly and productively within it.

Judy Grahn suggests that: "To understand, to get to the basis, the root or hidden meaning, is the wrong tool to bring" to our own work. "Perhaps interstand [or better yet intra-stand] is what we do, to engage with the work, to mix with it in an active engagement, rather than 'figuring it out'. Figure it in" (Grahn, 1989, 39). Knowledges are not innocent representations, but intra-actions of natures–cultures: knowledge is about meeting the universe halfway.

Pomona College,
Claremont, CA

ACKNOWLEDGEMENTS

The original version was presented at the Pew Gender and Science Workshop, Lake Arrowhead, CA, September 1991. Successive versions were presented at numerous conferences and public lectures. I am grateful to Ruth Wilson Gilmore, Donna Haraway, Sandra Harding, Martin Krieger, Helen Longino, Lynn Hankinson Nelson, Jennifer Rycenga, Sharon Traweek, and Roanne Wilson for their insightful comments and support.

NOTES

[1] A less obvious point is that the success of scientific theories is not automatic for realists either, as Laudan (1981) and Fine (1984) argue.

[2] Cushing asserts that "realism is in double jeopardy", in the sense that although Bohm's interpretation is realist, he tags Bohr's interpretation of quantum physics as antirealist, and furthermore, the existence of this concrete example of underdetermination means that it would be very difficult to make the case for realism. Although I will be arguing here for a realist interpretation on the part of Bohr, this divergence in and of itself does not weaken the underdetermination aspect of Cushing's argument. (There are a few independent issues however. One is the fact that the empirical equivalence of these theories depends upon the resolution of the measurement problem for the Copenhagen interpretation (see fn. 6) since rigorously speaking without such a resolution the Copenhagen interpretation does not offer definite predictions (see Albert, 1992). And of course, it still remains to be seen whether Bohm's theory and the Copenhagen theory are empirically coincident in all respects.)

[3] This is not a circularity. As I will explain later, it is indicative of a critical reflexivity.

[4] The destabilization of liberal humanist conceptions of identity that follow from the framework of agential realism will not be my focus here. My focus here will be primarily on science. For more details on agential realism and identity see Barad, forthcoming.

[5] The collected volumes of Bohr's writings have been made available thanks to Rosenfeld, (1972–).

[6] While physicists talk of *the* Copenhagen interpretation, in one sense there are really many Copenhagen interpretations, or to put it another way there is no well-defined, coherent, and complete Copenhagen interpretation. This is due to the fact that the physicists who are seen as the contributors had strong philosophical/interpretative differences, so that what is taken to be *the* Copenhagen interpretation is actually a superposition of the views of Bohr (complementarity), Heisenberg (uncertainty), Born (probability), and Von Neumann (collapse), to name a few of the key players.

[7] For more details see Cushing, 1994. (Although Bohr's philosophy-physics is not a primary focus for Cushing, I note that my reading of Bohr diverges substantially from Cushing's. As I specify in more

detail later, my reading has much more overlap with interpretations presented by a number of Bohr scholars.)

[8] There are an increasing number of quantum textbooks that do not mention any of Bohr's contributions to the field (except for reference to his pre-quantum theory atomic model). That is, there is often no mention of his principle of correspondence and the role it played in the development of the quantum theory, or Complementarity and its importance to an understanding of quantum theory.

[9] For Bohr, Complementary means simultaneously necessary and mutually exclusive (as explained in detail in the next section. NB: I capitalize 'complementary' when it is used in Bohr's sense of the term). See Bohr (1963b) for examples for this approach. An attempt by Bohr to resolve the vitalism–mechanism debate in biology failed because he assumed, from his limited technological perspective, that the conditions for examining the underlying mechanics of life processes and the conditions for maintaining the life of the specimen under investigation were mutually exclusive.

[10] Although one is free to give antirealist interpretation of Newtonian physics, the "classical realist" one articulated here is particularly seductive to our Enlightenment intuitions, and I have heard variations of this classical realist tenet espoused time and again to students in undergraduate physics classes. (It is of course ironic to attribute a realist stance towards the physics of one who was unwilling to feign any hypothesis, but not many students would pick up on this since physics courses overwhelmingly lack any overt discussion of the different interpretative stances with regard to science. See Barad (1995) for the pedegogical implications of this widespread inattention to metatheoretical issues and the lack of critical reflexivity.)

[11] A re-visioning of the nature of light is common to both of the major conceptual revolution of twentieth century physics: special relativity and quantum theory. Special relativity will not be considered in this paper, but a few words to distinguish some of the more popular implications of this theory from quantum mechanics may be helpful to some readers. The special theory of relavity is based on the empirically verified invariance and finiteness of the speed of light (1/speed of light = 1/constant = $1/c \neq 0$). Einstein transformed Galilean relativity into a new theory of relativity, redesignating certain previously held invariant quantities as relative and vice versa. (Einstein had thought of calling this theory "the theory of invariances", and he may have been better off doing so given the political climate in Europe during the first half of the 20th century.) The theory of relativity gives the measurement process some limited visibility: concepts such as "time" and "length" are defined relative to a particular frame of reference in which the measurements are performed. (It is not that time appears to slow down; time is what you measure with a clock.) The theory of relativity may have undermined the universality of certain concepts, but the assumption that measurements are continuous and determinable is never questioned. That is, according to the special theory of relativity there still is a well-defined separation of object and measuring instrument, i.e., a clear subject/object distinction is preserved. The properties measured are attributable to an independent object as measured relative to a particular frame of reference. (The frame of reference simply specifies which "time" we are talking about, that is, what we mean by "time" in each case.)

[12] It is important to note that the fact that Newtonian physics "works" in the macroscopic domain does not mean that the assumptions of measurement transparency are true in that domain. On the contrary, this simply explains why the assumptions lay hidden for centuries. That is, the fact that Newtonian physics makes predictions that are approximately the same as those made by quantum theory in the macroscopic domain is due to the fact that in that regime the ratio of Planck's constant to the mass of the particle is smaller than the accuracy required of the macroscopic situation in question – but it is *not* zero. This is why Bohr refers to the general epistemological lessons of quantum theory.

[13] A rough, intuitive picture is the following: think of catching a ball, the relative amount by which your arm moves back is an indication of the momentum of the ball.

[14] I have chosen to use the term 'constructed' instead of Bohr's term 'arbitrary' for two reasons. First of all, 'arbitrary' is misleading since the cut is not totally arbitrary in that the cut must be made in such a way that the measuring device is always macroscopic (this is necessary since the use of classical concepts is predicated on a subject/object split). Secondly, the term 'arbitrary' carries misleading connotations such as the inappropriate associations of relativism. The point that I think Bohr is out to emphasize in using the term 'arbitrary' is that since there is no inherent/Cartesian distinction some non-inherent distinction still must be drawn. Since the choice of a conceptual apparatus necessarily draws

this distinction, I will use the term 'constructed' in the hope that this term will connote agency. The full contrast is that classical physics is premised on an inherent/naturally occurring/fixed/universal/Cartesian distinction while quantum physics requires constructed/agentially positioned/movable/local/"Bohrian" cuts, positioning Bohr as Descartes counterpart.

[15] Another way of expressing this quantum quandary is by noticing that this means that the act of measuring can be accounted for only if the measuring device is itself treated as an object, defying its purpose as a measuring instrument.

[16] See Barad (1995) for a more detailed discussion of this example.

[17] "Bohr's provocative tendency, especially in earlier writings, to 'emphasize the subjective character of all experience' (Bohr, 1963a, 1) brought his entire interpretation of quantum theory into peril" (quoted in Honner, 1987, 65), parallel to terminological choices made by some science studies scholars early on that proved equally rhetorically disadvantageous.

[18] Bohr interchanges the phrases "language of everyday experience" and "language of classical physics". The connection for Bohr is that everyday experiences take place within the macroscopic realm to which the language of classical physics is applied. However, he does suggest that the language of everyday experience may include "suitable generalizations" of the language of classical physics.

[19] Bohr: "It is just arguments of this kind which recall the impossibility of subdividing quantum phenomena and reveal the ambiguity in ascribing customary physical attributes to atomic objects" (Bohr, 1963b, 51).

[20] This detailed mapping of the relationship between classical concepts and phenomena is meant to clarify Bohr's position with respect to the necessity of using classical concepts in the description of quantum phenomena. Confusion about this issue is widespread in the literature. Many physicists trying to understand Bohr's interpretative framework have accused Bohr of conservatism with respect to the future development of physical theories: why they asked should we limit our descriptive concepts in this way? It is also not uncommon to find philosophers describing this aspect of Bohr's theory as Kantian. I hope that I have clearly communicated here why I think it is that Bohr was not denying the possibility for future creative developments in physics, nor was he advocating transcendental idealism in his insistence on the use of *empirically grounded* classical concepts *even* within a *nonclassical* framework (which is already admitting the possibility of the evolution of ideas). An important related fact is that Bohr offers the observation that "everyday" languages are based on subject–predicate forms (a point that he unfortunately makes without qualification); that is, everyday languages structurally assume subject–object distinctions. I believe that this a contributing factor to what is commonly described as the obscurity of Bohr's writings, since he uses many circumlocutions to try to talk about things that are not inherently structured along this distinction.

[21] See Keller's "Anomaly of a Woman in Physics" (1977) for one telling of a graduate school experience in the US that is typical in its discouragement of reflexivity and contemplation of interpretative questions in physics, though specific in its gendering.

[22] This quote is from Petersen, 1985, 302. Petersen goes on to say that Bohr had no use for an ontology. Perhaps Bohr didn't feel the need to articulate one, but this is not to say that he held a thorough-going pragmatic or positivist view. In fact, I will argue later in this section that Bohr had a realistic attitude towards wave-particle duality, for example, though his views diverged dramatically from classical realism. Honner (1987) also argues for a realistic interpretation and against pragmatic or positivist perspectives, although the version of realism that Honner ascribes to Bohr does not address the issue of a reference for our representations. Folse (1985) also advocates for an interpretation which sees Bohr as a realist, but Folse seems to take phenomena as the result of an underlying reality.

[23] Einstein *et al.*, 1935 and Bohr, 1935. In an article entitled "Discussion with Einstein on Epistemological Problems in Atomic Physics", for a volume honoring the epoch-making contributions of his long time friend Albert Einstein, Bohr quotes extensively from this particularly important passage of his 1935 paper (see Bohr, 1949, 234).

[24] I have presented what may seem like a pedantic analysis of Bohr's use of the term 'phenomenon' in this passage, but I do so because as of 1935 his use of this term was still somewhat inconsistent, and it is therefore crucial to justify by the context of his usage of this term that it indeed is consistent with the specific signification he assigns to it in his later writings. In fact, Bohr's usage of 'phenomenon'

to signify the wholeness in the interaction between "objects of investigation" and "agencies of observation" is consistent throughout this particular 1935 article.

25 This fact motivates my introduction of the term 'intra-action' at the end of this section since phenomena are the instantiation of intra-actions.

26 This criterion would apply as well to any suggestion of a Bohrian notion of anti-realism. In particular, realists cannot expect to rely on an independent external reality, but also antirealists would be hard pressed to argue against realism on the basis of some postulated inaccessible independent reality.

27 This mutual exclusivity highlights the problematics of an instrumentalist stance for Bohr. How does the instrumentalist account for the non-arbitratriness of this feature? (What auxiliary criterion must be applied?)

28 After a few weeks of intensive discussion, Heisenberg finally acquiesced to Bohr's point of view and added a postscript to his article on the uncertainty principle in which he states: "In this connection Bohr pointed out to me that I have overlooked essential points in some of the discussion of this work. Above all the uncertainty in the observation does not depend exclusively on the occurrence of discontinuities, but is directly connected with the necessity of doing justice simultaneously to the different experimental data which are expressed in the corpuscular theory on the one hand and the wave theory on the other [i.e., wave–particle duality]" (quoted in Murdoch, 1987, 51) Recent papers in quantum optics (e.g., see Scully *et al.*, 1991) give empirical evidence in support of Bohr's interpretation of the uncertainty principle over the one given by Heisenberg which is not consistent with these findings. NB: it is Heisenberg's analysis (without Bohr's corrections) that is taught to physics students. See Barad (1995) for more details. The divergence of Bohr's and Heisenberg's interpretations of the uncertainty principle highlights their philosophical (realist and instrumentalist, respectively) differences. The construction of scientific theories is influenced by philosophical attitudes.

29 Other pivotal aspects of the framework of agential realism are developed in Barad, forthcoming. In particular, there is a more in-depth discussion of the issues of agency and identity. The fact that agential realism can be used to think about rather disparate issues from the destabilization of identity to the destabilization of science is not a matter of a more parallelism, but different instances of the same epistemological and ontological issues.

30 For a more detailed discussion see Harding (1990) and other articles in *Feminism/Postmodernism*, ed. Nicholson.

31 The positions that Heisenberg and Dirac articulate here are consistent with the former's instrumentalist leanings and the latter's traditional realist leanings.

32 In a related fashion I have stayed away from Bohr's term "Complementarity" because of the associated connotations of liberal humanist conceptions of choice. Take the example of the complementary (intended here in the colloquial sense of the word) theory of genders where essentialized differences between men and women are theorized on a level playing field, denying the unequal power relations represented by unequal material conditions. Matrix theory (not the mathematical kind but the kind that comes from social theory) and other nonessentializing analytic moves fully deconstruct such liberal conceptualizations.

33 To assert that we only get to study nature through the distorting lense of culture is to reinstate the privileged position of the transcendent once again, resulting in further classical epistemological astigmatism.

34 Bohr makes direct note of this point himself: "the description of atomic phenomena has in these respects a perfectly objective character, in the sense that no explicit reference is made to any individual observer and that therefore ... no ambiguity is involved in the communication of information" (Bohr, 1963c, 3).

35 Notice that experiments in some fields, like high energy physics, are rarely repeated due to constraints imposed by limited resources or other community priorities, but the issue here is not whether or not the results have actually been reproduced, the issue is the possibility of reproducibility due to the literal embodiment of objectivity. Also, note that reproducibility is still an issue for scientists studying chaotic systems (which are highly sensitive to initial conditions) in the sense that chaotic systems do not behave differently for different observers (it is just that it is very difficult, to start an experiment with the very same initial conditions, but simulations of chaotic systems are often reproduced). Other criteria delineat-

ing science (read "Western science") from nonscience have been offered (see Harding, 1993). This project of delineation is of course part and parcel of Western imperialism's focus on distinguishing "us" from "them". Nonetheless, this same distinction along these lines is extremely common in the science studies literature and it is therefore useful in this context.

[36] Kondo (1989) and Sandoval (1991) make a similar point. Anzaldúa (1987) theorizes the constructed nature of boundaries.

[37] Feminist scientists, economists, political scientists, historians, psychologists, geographers and literary critics are among those who have expressed seeing the utility of agential realism for their projects.

[38] "Tunneling" is a quantum phenomenon whereby classically confined particles escape. This is the result of the uncertainty principle and it explains many different physical phenomena such as nuclear decay, transistors, etc.

REFERENCES

Albert, David Z.: 1992, *Quantum Mechanics and Experience*. Cambridge, MA: Harvard University Press.

Anzaldúa, Gloria: 1987, *Borderlands/La Frontera*. San Francisco: Spinsters/Aunt Lute Book Co.

Barad, Karen: 1995, 'A Feminist Approach to Teaching Quantum Physics', in *Teaching the Majority: Breaking the Gender Barrier in Science, Mathematics, and Engineering*, edited by Sue V. Rosser. NY: Teachers College Press.

——: 1996, 'Getting Real: Technologies of Embodiment and Materiality', keynote address, Gender, Technology, Place Conference, Rutgers University, March 30, 1996.

——: forthcoming, *Meeting the Universe Halfway*, Book Manuscript.

Bernstein, Richard J.: 1983, *Beyond Objectivism and Relativism: Science, Hermeneutics, and Praxis.* Phila: U. of Penn. Press.

Bohm, David.: 1985, 'On Bohr's Views Concerning Quantum Theory', in *Niels Bohr: A Centenary Volume*, ed. French, A. P. and Kennedy, P. J. Cambridge, MA: Harvard U. Press.

Bohr, Niels: 1935, 'Can Quantum-Mechanical Description of Physical Reality Be Considered Complete?', *Physical Review*, **48**, 696–702.

——: 1937, 'Causality and Complementarity', *Philosophy of Science*, **4**.

——: 1949, 'Discussion with Einstein on Epistemological Problems in Atomic Physics', in *Albert Einstein: Philosopher-Scientist*, ed. P. A. Schilpp. Evanston: Northwestern U. Press.

——: 1963a, *The Philosophical Writings of Niels Bohr, Vol. I: Atomic Theory and the Description of Nature*. Woodbridge, Conn: Ox Bow Press.

——: 1963b, *The Philosophical Writings Of Niels Bohr, Vol. II: Essays 1933–1957 on Atomic Physics and Human Knowledge*. Woodbridge, Conn: Ox Bow Press.

——: 1963c, *The Philosophical Writings of Niels Bohr, Vol. III: Essays 1958–1962 on Atomic Physics and Human Knowledge*. Woodbridge, Conn: Ox Bow Press.

Cartwright, Nancy: 1983, *How the Laws of Physics Lie*. New York: Clarendon Press.

Cushing, James T.: 1994. *Quantum Mechanics: Historical Contingency and the Copenhagen Hegemony*. Chicago: University of Chicago Press.

Einstein, A., Podolsky, B., and Rosen, N.: 1935, 'Can Quantum-Mechanical Description of Physical Reality Be Considered Complete?' *Physical Review*, **47**, 777–780.

Feyerabend, P. K.: 1962, 'Problems in Microphysics', in *Frontiers of Science and Philosophy*, ed. Robert G. Colodny. Pittsburgh: U. of Pitt.

Fine, Arthur: 1984, 'The Natural Ontological Attitude', in *Scientific Realism*, ed. by J. Leplin. Berkeley: University of California Press.

Folse, Henry: 1985, *The Philosophy of Niels Bohr: The Framework of Complementarity*. NY: North-Holland.

Fulton, Alice: 1990, *Powers of Congress*. Boston: David R. Godine.

Galison, Peter: 1987, *How Experiments End*. Chicago: The University of Chicago Press.

Grahn, Judy: 1989, *Really Reading Gertrude Stein*, Freedom, CA: Crossing Press.

Hacking, Ian: 1982, 'Experimentation and Scientific Realism', *Philosophical Topics*, **13**, 71–87.

Haraway, Donna: 1985, 'Manifesto for Cyborgs: Science, Technology, and Socialist Feminism in the 1980s', *Socialist Review*, **80**, 65–108, (reprinted in Haraway, 1991).

——: 1988, 'Situated Knowledges: The Science Question in Feminism as a Site of Discourse on the Privilege of Partial Perspective', *Feminist Studies*, **14(3)**, 575–599, (reprinted in Haraway, 1991).

——: 1991, *Simians, Cyborgs, and Women: The Reinvention of Nature*. NY: Routledge Press.

Harding, Sandra: 1990, 'Feminism, Science, and the Anti-Enlightenment Critiques', in *Feminism/Postmodernism*, ed. Linda J. Nicholson. NY: Routledge.

——: 1991, *Whose Science? Whose Knowledge?: Thinking from Women's Lives*. Ithaca: Cornell University Press.

—— (editor): 1993, *The "Racial" Economy of Science: Towards A Democratic Future*. Bloomington: Indiana University Press.

Hayles, N. Katherine: 1993, 'Constrained Constructivism: Locating Scientific Inquiry in the Theater of Representation', in *Realism and Representation: Essays on the Problem of Realism in Relation to Science, Literature, and Culture*, ed. by George Levin. Madison: The University of Wisconsin Press.

Honner, John: 1987, *The Description of Nature: Niels Bohr and the Philosophy of Quantum Physics*. Oxford: Clarendon Press.

Hooker, Clifford A.: 1972, 'The Nature of Quantum Mechanical Reality', in *Paradigms and Paradoxes*, ed. R. G. Colodny, Pittsburgh: U. of Pittsburgh Press.

Keller, Evelyn Fox: 1977, 'The Anomaly of a Woman in Physics', in *Working It Out*, ed. Ruddick and Daniels.

——: 1985, *Reflections on Gender and Science*. New Haven: Yale U. Press.

Kondo, Dorinne: 1989, *Crafting Selves: Power, Gender, and Discourses of Identity in a Japanese Workplace*. Chicago: Chicago Univ. Press.

Latour, Bruno: 1993, *We Have Never Been Modern*. Cambridge, MA: Harvard University Press.

Laudan, Larry: 1981, 'A Confutation of Convergent Realism', *Philosophy of Science*, **48**, 19–48.

Longino, Helen: 1990, *Science as Social Knowledge*. Princeton: Princeton Univ. Press.

Murdoch, Dugald: 1987, *Niels Bohr's Philosophy of Physics*. NY: Cambridge U. Press.

Pais, Abraham: 1982, *'Subtle is the Lord…': The Science and Life of Albert Einstein*. NY: Oxford Univ. Press.

Petersen, Aage: 1985, 'The Philosophy of Niels Bohr', in *Niels Bohr: A Centenary Volume*, ed. French, A. P., and Kennedy, P. J. Cambridge, MA: Harvard U. Press.

Pickering, Andrew: 1984, *Constructing Quarks: A Sociological History of Particle Physics*. Chicago: University of Chicago Press.

——: 1994, 'After Representation: Science Studies in the Performative Idiom'. Talk given at the Philosophy of Science Association Annual Meeting New Orleans, 14–16 October, 1994.

Rosenfeld, Leon (general ed.): 1972– , *Collected Works*. Amsterdam: North-Holland.

Sandoval, Cheyla: 1991, 'U.S. Third World Feminism: The Theory and Method of Oppositional Consciousness in the Postmodern World.' *Genders*, **10**.

Scully, Marlan O., Englert, Berthold-Georg, and Walther, Herbert: 1991, 'Quantum Opical Tests of Complementarity', *Nature*, **351**, 111–116.

Traweek, Sharon: 1988, *Beamtimes and Lifetimes: The World of High Energy Physicists*. Cambridge, MA: Harvard U. Press.

Winterson, Jeanette: 1992, *Written on the Body*. NY: Vintage Books.

JOSEPH ROUSE

FEMINISM AND THE SOCIAL CONSTRUCTION OF SCIENTIFIC KNOWLEDGE

INTRODUCTION

Feminist science studies and the sociology of scientific knowledge have emerged within the past twenty years as explicit challenges to the epistemological individualism that still predominates within most philosophy of science. The "Strong Programme" for the sociology of scientific knowledge was put forward first, to provide distinctively sociological explanations for the diversity of human beliefs about the natural world.[1] Whereas earlier programs in the sociology of knowledge had exempted the natural sciences and mathematics from their purview, and the dominant Mertonian approaches to the sociology of science had confined their studies to scientific institutions, motivations, and organizational norms, the new sociologists proposed to explain the content of scientific knowledge in the same way that they would explain any other systems of beliefs and practices. To do otherwise, they often argued, would invoke a scientifically unjustifiable *a-priori* decision to exempt the sciences from empirical sociological investigation. Indeed, they called for a methodological commitment to some form of epistemological relativism to prevent centuries of cultural admiration for and epistemic deference to science from prejudicing sociological inquiry.

Alternative conceptions of scientific knowledge as a social achievement have emerged from recent feminist scholarship. An initial concern of feminist science studies was to examine and criticize the ways in which biological, psychological, and social scientific studies of women and men, and of sex and gender more generally, have been androcentric. The aim was to develop an explicitly feminist *science* alongside feminist transformations of scholarship in the humanities and social sciences. Scientific and philosophical resistance to these initial feminist criticisms of sexism in the guise of science has encouraged more general feminist reconceptions of scientific inquiry which would recognize feminist criticism as a constructive contribution to science.[2] In the resulting reconceptions of the sciences as politically engaged social practices, sex and gender have become aligned with race, colonialism or imperialism, sexual orientation, and other politically significant categories; the concern to challenge sexism specifically has remained a powerful motivation for feminist science studies, but sex and gender are no longer privileged or isolable analytic categories.

In this paper, I compare these two traditions, as challenges to philosophical orthodoxies and as constructive proposals for a *social* understanding of science. In juxtaposing feminism and sociology of science, however, my principal aim is to

L. H. Nelson and J. Nelson (eds), Feminism, Science, and the Philosophy of Science, 195–215.

clarify the significance of feminism for philosophy of science, and thereby to contribute to this volume's proposed dialogue. Feminist science studies are often thought to seek an intermediary position between traditional philosophy of science and the sociology of scientific knowledge: whereas the sociologists reject any normative account of the objectivity, rationality, or truth of scientific claims or methods, feminists would revise norms of rationality or objectivity rather than abandon them. I shall challenge this interpretation, despite its familiarity and initial plausibility. The supposedly sharp differences between normative philosophy and descriptive sociology presuppose a shared conception of knowledge, which I characterize as *epistemological.* Feminist scholars, I argue, are developing a different ontology of knowing, whose articulation displays and challenges the continuity between epistemological philosophies and sociologies of science.

Before initiating this comparison, I offer two significant caveats. First, my discussion deliberately overlooks contested issues within the feminist and sociological traditions in order to accentuate some shared conceptions that frame those internal disputes. My aim is not to minimize or suppress the disagreements, which in other contexts remain vital, but only to recognize some interwoven themes connecting otherwise disparate approaches. Some feminist theorists may well dissent from my presentation of these themes, but my hope is that dissenters will still find it useful in clarifying such differences. Even for my purposes, we should recognize that some recent work in the sociological tradition, most notably by Bruno Latour, Leigh Star, Andrew Pickering, Karin Knorr-Cetina, Michael Lynch, and others, challenges orthodox sociology of scientific knowledge in ways that encourage rapprochement with feminist studies of science. Yet the significance of such a rapprochement may also become clearer in light of my contrast between feminist and sociological approaches to scientific knowledge.

The second caveat is that my comparison brackets the most obvious difference between feminist and sociological studies of science, namely that gender and the sex/gender distinction play almost no role in sociologists' empirical case studies and methodological reflections. The absence of sociological attention to gender in science should be surprising, given the importance of sex and gender as social categories, and especially given the historical predominance of men within the sciences and the widespread use of gendered imagery to characterize both science and scientific interpretations of nature. Even if the relative unimportance of sex and gender for understanding science were an outcome of empirical research in sociology, one might expect that this conclusion would be highlighted, and supported by extensive argument. No such argument is apparent in the sociological literature, and its absence ought to be regarded *prima facie* as a serious problem for the sociological tradition. Yet despite the importance of this issue, there are good reasons to bracket it for my purposes. Sociologists' inattention to gender, if mistaken, might reflect only a faulty application of their theoretical frameworks; perhaps the frameworks themselves leave ample room for a full appreciation of the significance of gender relations as a social explanans for the content of scientific knowledge. Moreover, bracketing questions of sex and gender can usefully emphasize that feminist science studies are not limited to studies of gender in science. Feminist scholars have called for a much more far-reaching reconception of the sciences as social, political, and cultural practices, a reconception that would thereby *enable* critical studies of gender to occupy a more prominent place in the culture of science.

The differences I shall identify between the sociological and feminist traditions emerge from a background of three important shared themes. First, both traditions understand scientific knowledge as a collective or social achievement. We may still reasonably speak of individuals as knowers, but such attributions of knowledge to individuals are derivative from or dependent upon the social achievement or authorization of knowledge. Second, both traditions argue that important aspects of traditional epistemologies and philosophies of science are not merely false, but ideological. Sociologists of scientific knowledge argue that philosophical explications of the rationality or verisimilitude of scientific methods and practices misconstrue the actual practice of science as revealed by empirical sociological studies, and thereby unjustifiably legitimate the cultural and political authority of the sciences. Moreover, they argue, the exceptional or asymmetrical treatment accorded to the sciences by such philosophical accounts obscures important similarities between the sciences and other less authoritative practices and belief systems. Many feminists, meanwhile, have argued that familiar conceptions of scientific knowledge as achieved or possessed by individuals, as disinterested and apolitical, or as rational and cognitive in ways that exclude the affective and embodied aspects of human experience, have simultaneously served to distort our understanding of science, to rationalize male dominance in science and elsewhere, and to reinforce the alienation or exclusion of many women from effective participation in science.

This insistence upon the ideological character of individualist and rationalizing interpretations of scientific knowledge points to the third common theme between feminist and sociological studies of science: both are politically engaged projects. Political commitment has been more readily apparent in feminism, since feminist science studies have always been projected as part of a larger political and cultural movement to criticize sexism and empower women. Yet beneath the sociologists' frequent insistence upon the need for methodological detachment and a symmetrical treatment of all belief systems has also been a broad political commitment, to a humanism that is effectively articulated by Collins and Yearley (1992a):

> The effect of SSK has been to show that the apparent independent power of the natural world is granted by human beings in social negotiation. Because the special power and authority of natural scientists comes from their privileged access to an independent realm, putting humans at the center removes the special authority.... Symmetry between the true and the false requires a human-centered universe (pp. 310–11).

Thus, both feminists and sociologists of science argue that important aspects of science are contingent and alterable, even though traditional epistemologies present them as natural and immutable, and that recognizing their contingency makes a political difference.

THE SOCIOLOGY OF SCIENTIFIC KNOWLEDGE AS EPISTEMOLOGICAL

From this shared background, however, the two traditions importantly diverge. Many of their differences result from sociologists' commitment to an *epistemological* conception of their project, or so I shall argue, whereas many feminist science studies begin to develop a philosophy of science that rejects an epistemological conception of knowledge. But what do I mean by an "epistemological" understanding of science? The intelligibility of epistemology as a field of study presumes that "knowledge" demarcates a coherent, surveyable domain of inquiry. Michael Williams has recently challenged the commonplace philosophical acceptance of that presumption:

> It is tempting to use "human knowledge" and "our knowledge of the external world" as though it were obvious that such phrases pick out reasonably definite objects of study. But it isn't obvious, or shouldn't be. We can talk of "our knowledge of the world," but do we have any reason to believe that there is a genuine totality here and not just a loose aggregate of more or less unrelated cases? (Williams, 1991, p. 102).

The corresponding temptation in philosophy and sociology of science would be to presume that *scientific* knowledge is a "reasonably definite object of study," either on its own or parasitic upon its subsumption within knowledge in general.[3] To see why the program for a sociology of scientific knowledge remains committed to epistemology in this sense (and why most feminist science studies are not), we need to consider the epistemological project in more detail. Williams succinctly summarized the characteristic forms in which this epistemological commitment has been deployed within philosophy:

> The traditional philosophical examination of knowledge ... points to four central ideas: an *assessment* of the *totality* of our knowledge of the world, issuing in a judgment delivered from a distinctively *detached* standpoint, and amounting to a verdict on our claim to have knowledge of an *objective* world (Williams, 1991, p. 22).

Sociologists of scientific knowledge have given a distinctive spin to each of these ideas, but all four remain important to their project.

The ideal of detachment has clearly played a central role in the recent sociological tradition; indeed, sociological criticisms of mainstream philosophy of science often focus upon philosophers' insufficient detachment from the norms and cultural familiarity of the natural sciences. The classic statement of sociologists' conception of an appropriately detached standpoint to study science is David Bloor's programmatic demand that the sociology of scientific knowledge be "impartial with respect to truth or falsity, rationality or irrationality, success or failure, ... and symmetrical in its style of explanation ... [of] true and false beliefs" (Bloor, 1991, p. 7). But relativism has not been the only methodological commitment that has resulted from sociologists' desire for a detached standpoint from which to understand the sciences. Latour and Woolgar deliberately construct their account of Latour's participation in a neuroendocrinology laboratory from the perspective of a relative *stranger* to the culture of science:

We regard it as instructive to apprehend as strange those aspects of scientific activity which are readily taken for granted... . The uncritical acceptance of the concepts and terminology used by some scientists has had the effect of enhancing rather than reducing the mystery which surrounds the doing of science... . For us, the dangers of "going native" outweigh the possible advantages of ease of access and rapid establishment of rapport with participants (Latour and Woolgar, 1986, p. 29).

More recently, Collins and Yearley (1992a) have suggested an alternative image of detachment from epistemic norms and commitments. The sociologist of science must be able to *alternate* between competing beliefs, methods, and epistemic ideals, understanding each from the "inside" but without the insider's epistemic commitment. Moreover, the sociologist must combine this "promiscuous" epistemic alternation with a "meta-alternation" which also permits the sociologist to accept naively the categories and norms of everyday life and of sociology itself. From Collins' and Yearley's perspective, Latour and Woolgar's methodological estrangement is thus not detached enough – ultimately, what is called for is a disciplinary *antagonism* between natural and sociological realisms:

We provide a prescription: stand on social things – be social realists – in order to explain natural things. The world is an agonistic field; others will be standing on natural things to explain social things. That is all there is to it (Collins and Yearley, 1992b, 382)

Whichever form of sociological detachment is preferred, the result is to situate the sciences within a larger epistemic *totality*. Claims to scientific knowledge are to be regarded as examples of a more general kind, which can be surveyed and assessed as a whole. Sociologists characterize this more general object of study variously, as beliefs (Barnes, Bloor), order (Latour and Woolgar, Collins, Shapin and Schaffer), discourse (Gilbert and Mulkay), inscriptions (Latour and Woolgar), or representation (Woolgar), but underlying these differences is a shared commitment to subsuming the sciences under a more general analytical category. Of course, the new sociologists of science rightly pride themselves on the wealth of detailed case studies that exemplify their methodological and theoretical commitments. Steven Shapin's (1982) prominent review already listed an impressive range of such sociological studies, which has since greatly expanded. Yet the supposed *significance* of these studies is oddly disconnected from their empirically rich particularity. MacKenzie and Barnes recognized this disconnection in reflecting upon the implications of their own study of early 20th-century controversies between Mendelians and biometricians:

The general point is not that the goal-directed character of scientific judgment implies its relationship to any particular contingency, or to external factors, or political interests; what is implied is that any such contingency *may* have a bearing on judgment and that contingent sociological factors of some kind *must* have (MacKenzie and Barnes, 1979, p. 205).

What matters is that all scientific practices and achievements exemplify a more general kind that can be accounted for by appeal to distinctively social contingencies.

These general categories that sociologists of science have used to define their object of inquiry have been perhaps surprisingly continuous with the "semantic ascent" that so often characterizes philosophical reflection upon science.

Sociologists, too, have shifted from accounting for scientists' interaction with the world to explaining scientists' beliefs, accounts, inscriptions, or representations, or at still further remove, the *content* of their beliefs or representations. Undoubtedly the sociologists have multiple reasons for their semantic ascent. The Strong Programme was explicitly defined in opposition to Mannheim's and Merton's sociological approaches, which had exempted the content of knowledge from sociological study. The sociologists also intended to challenge normative philosophical accounts on their own turf, and that challenge might be facilitated if the contested turf were described commensurably. Finally, a sociology of science needs a sociologically accessible domain. Collins and Yearley emphasize this concern in challenging Michel Callon's (and Bruno Latour's) attempts to recognize natural objects as "actants":

> There is only one way we know of measuring the complicity of scallops [or other natural objects that scientists work with] and that is by appropriate scientific research. If we are really to enter scallop behavior into our explanatory equations, then Callon must demonstrate his [natural] scientific credentials (1992, p. 316).

If scientific knowledge can instead be characterized in terms of "collectively accepted systems of belief," consensus, discourse, order, representation, or other forms of mediation, then sociological credentials may regain their relevance. Whatever the reason, however, the new sociology of science defined its domain of inquiry in ways that significantly abstract from and generalize over the diverse forms of scientific practice.

The sociology of scientific knowledge may nevertheless initially seem to diverge from epistemological philosophy of science by rejecting the normative perspective from which philosophers would assess the rationality or verisimilitude of scientific knowledge. Sociologists have hoped to explain why scientists accept some beliefs (representations, accounts, etc.) rather than others, in ways that are carefully severed from whatever justification they might have for accepting those beliefs themselves. As Barnes and Bloor put the point,

> the incidence of all beliefs without exception calls for empirical investigation and must be accounted for by finding the specific, local causes of this credibility. Whether the sociologist evaluates a belief as true or rational, or as false and irrational, he must search for the causes of its credibility (1982, p. 23).

But this shift from a classically normative epistemological stance to an empirical and explanatory account is a move within epistemology rather than against it; the sociological accounts thereby make common cause with the increasingly widespread philosophical commitment to *naturalized* epistemology.[4] Moreover, such naturalized (or socialized) accounts of knowledge do not thereby lose all normative force. Explanations of the acceptance or credibility of beliefs may then reinforce or undermine one's own inclination to accept them. Knowing the causes of certain beliefs might well increase confidence in them, by displaying reliable causal connections to aspects of their intended objects. Adherents of the Strong Programme's commitment to explanatory symmetry typically go in the opposite direction: they suggest that the causes for the acceptance of scientific beliefs are disconnected from any reason for the analyst to believe them, and hence that "all beliefs [including scientific beliefs] are on a par with one another with respect to the causes of their credibility" (Barnes and Bloor, 1982, p. 23). To that extent, the most prominent adherents of the sociology of scientific knowledge do engage in the epistemological project as Williams described it: they assess the totality of scientific beliefs *as* claims to knowledge of an objective world, and find them to have no greater (but also no less) justification than any other collectively accepted systems of belief. The new sociology has not abandoned justification; rather, it has engaged in the quite general epistemological project of showing why standard philosophical defenses of the rationality or truth of scientific claims should instead be regarded as *ex post facto* rationalizations. In Latour and Woolgar's words, "'reality' cannot be used to explain why a statement becomes a fact, since it is only after it has become a fact that the effect of reality is obtained" (1986, p. 180).

The distinctively epistemological character of the new sociology of science can be seen clearly in another way. Despite their rejection of realist philosophy of science, few of the new sociologists of science espouse a metaphysical anti-realism. Barnes and Bloor are especially clear about this point, for they are metaphysical *realists*. They only insist that truths about the natural world, whatever those truths are, are useless to explain the diversity of conflicting beliefs about that world. Latour and Woolgar, whose views differ in so many important ways from Barnes's or Bloor's nevertheless concur in distinguishing sociological from metaphysical anti-realism: "we do not wish to say that facts do not exist or that there is no such thing as reality… . Our point is that 'out-there-ness' is the *consequence* of scientific work rather than its *cause*" (1986, pp. 181–2; N.B. the quotation marks, indicating that the supposed consequence of scientific work is not reality itself, but the *effect* of reality, i.e., what can *count* as reality for social beings like us). The sociologists separate their commitment to an epistemological relativism from a metaphysical anti-realism by arguing that the world's effects upon our beliefs or representations are always socially mediated to an extent that makes beliefs referentially opaque. Social practices, interests, or interactions, and the holistic interconnections among beliefs (and among the concepts in which they are framed and the experimental and observational practices that provide them evidential support) always intervene *between* us and the world, in much the way that strict empiricist anti-realists take *experience* to mediate the world opaquely. The sociologists' commitment to relativism is a natural consequence of conjoining the social mediation of belief or representation with acknowledgement of diverse social practices among humans. Sociological relativism is thus epistemologically parallel to the relativism strict empiricists would presumably accept if there were compelling evidence that human beings differed significantly in their sensory modalities and capacities.

There remains one last revealing feature of the new sociology of science that further displays its continuity with the epistemological tradition. Among David Bloor's four tenets of the Strong Programme is the call for a *reflexive* account of one's own claims to sociological knowledge (Bloor, 1991, p. 7). Subsequent theorists in the sociological tradition, most notably Woolgar and Ashmore, argue that neither Bloor nor the ensuing sociological tradition have taken reflexivity seriously enough, while others (Collins and Yearley, 1992a) respond that there must be limits upon reflexive criticism if the sociological project is to proceed at all. Yet these disagreements conceal an unexamined presumption that the important question to ask reflexively is epistemological. Bloor makes very clear that what is at stake in reflexivity are internal consistency and completeness: the sociology of scientific knowledge should neither be self-refuting nor should it accept arbitrary limits upon its scope. Woolgar (1988) and Ashmore (1989), by contrast, sometimes characterize their reflexive concern as rhetorical. Yet the central rhetorical problem they pose is that the authoritative authorial voice of sociological narrators and the apparent transparency of their referential prose are in pragmatic contradiction with the explicit content of their "findings" about the social construction of all accounts and representations. Woolgar and Ashmore are far too sophisticated to ask that sociological rhetoric accurately represent the findings it expresses, but they do ask that sociologists adopt literary forms that disrupt and interrogate readers' too-easy acceptance of their claims as transparent representations *of* the social construction of scientific knowledge. The "other voices" that Woolgar hopes would be included in sociologists' texts are there not because of their own need to be heard, but only to satisfy the sociologists' need not to be believed too readily. Collins and Yearley object to Woolgar's project primarily because they do, after all, want their own accounts to be believed straightforwardly.

FEMINIST RECONCEPTIONS OF KNOWING

Feminists' insistence that scientific knowledge is socially constructed appears in a new light when juxtaposed against my subsumption of the sociology of scientific knowledge within the epistemological tradition. Feminist science studies scholars most evidently differ from the new sociologists in their opposition to relativism, their normative stance toward particular scientific claims, and their willingness to retain and employ suitably revised conceptions of evidence, objectivity, and a distinction between belief and knowledge. Yet in many cases, these familiar differences are a consequence of feminist scholars working toward *post-epistemological* conceptions of knowledge, evidence, justification, and objectivity, and thereby opposing a framework *shared* by traditional philosophies of science and the new sociologies of scientific knowledge.

There are five ways in which I shall elaborate such feminists' transcendence of epistemology. First, these feminist science studies shift their primary object of study, from sociologists' focus upon the semantic "content" of knowledge or belief, to a concern with *relationships* among knowers and known. Second, these feminist studies take up a participatory stance toward scientific practices and scientific knowledge, rather than trying to explain or assess scientific knowledge as a totality. Third, such feminist science studies have a different temporal orientation than either the sociology or classical philosophy of science, as their primary concern is less with the present state of knowledge than its future possibilities. Fourth, many feminist reconstructions of the concept of "objectivity" in science and science studies dissolve any sharp conceptual distinction between epistemic and political criticism, a transformation which also prohibits reducing one category to the other. Finally, some feminists develop a more adequate conception of reflexivity and its epistemic, rhetorical, and political significance for science studies.

My first point concerns the ways many feminists have conceptualized scientific knowledge as an object of study. Feminist science studies have typically worked with a conception of knowledge that is less austere and abstract than the various forms of semantic ascent that characterize much recent philosophy and sociology of science. My claim is not that feminist science studies are an epistemological analogue to an ethics of care,[5] but rather that feminist scholars conceive of "knowing" as concretely situated, and as more interactive than representational. Knowledge is not merely a propositional attitude (belief or acceptance) toward some ideal or abstracted propositional content, but a *relationship* between knower and known, a *situation* that guides what knowers do and how the known responds and can be understood. Evelyn Fox Keller pointed clearly in this direction when she suggested that:

> Although scientific theories cannot be understood as faithful reflections of either culture or nature, perhaps they can be understood as good enough reflections of the forms of interaction that speaking and desiring social actors seek to implement with that mute but nonetheless responsive world of actors we call nature (1992, p. 95).

Yet even Keller's formulation still preserves an unnecessary vestige of the representational idiom; it would be consistent with her overall argument to omit altogether any talk of "reflections," and to say that theorizing is indispensable to the "forms of interaction" that she takes as central to science.

Undoubtedly this emphasis upon knowledge as a concrete relationship to its intended object gains its centrality and its urgency from initial feminist concern with the human sciences. Feminist theorists saw early on that knowledge claims about women, or about human beings in general, which excluded or diminished the humanity or rationality of women, are not simply *instruments* that could be used to oppress women or justify that oppression, but are integral to patterns of domination. These concerns with scientific knowledge as itself a form of action were reinforced by parallel reflections upon the politics of the human sciences within post-colonialist anthropology, where Western ethnographers' interactions with and writings about the people they study are increasingly seen to embody power and not merely to serve it. Observing, writing, and reading are not merely proposing or accepting the content of certain beliefs, but are themselves actions with consequences (one must consider to whom one writes, in what language, available to whom, drawing upon what patterns of interaction, using what narrative conventions and authorial stances, and who is permitted or enabled to respond, with what effects).[6] Yet feminist science theorists apply these lessons more generally. As Donna Haraway noted:

> The agency of people studied itself transforms the entire project of producing social theory, ... but the same point must apply to the other knowledge projects called sciences. A corollary [for] ... the sciences

as a heterogeneous whole, and not just in the social sciences, is granting the status of agent/actor to the "objects" of the world. Actors come in many wonderful forms. Accounts of a "real' world do not, then, depend on a logic of "discovery," but on a power-charged social relation of conversation (Haraway, 1991, p. 198).

Feminist science studies have as a consequence often been explicitly concerned with *different* ways in which knowers might interact with the objects of knowledge. One way of manifesting this concern has been criticism of the sexual politics embedded in some epistemic models and practices. Feminist scholars have directed critical attention, for example, to Bacon's vision of scientific mastery over a feminized nature, to Boyle's articulation of a distinctively masculine modesty as an epistemically constitutive virtue, to molecular biologists' invocation of their project as "a calculated assault on the secret of life," and to the sadism that structured the scientific vision, narrative structure, and experimental practices of Harry Harlow's Wisconsin Primate Laboratory.[7] Lynn Hankinson Nelson reminds us that these issues

are not about language or mysterious metaphysical agendas. The commitments to linear and hierarchical relationships, to executives and controllers, and to laws that phenomena obey are incorporated in our theories, methodologies, and models, and a commitment to "dominating nature" is incorporated in many of our scientific practices (Nelson, 1990, pp. 213–14).

Feminists' concern for different ways of knowing have also led to the articulation of possibly more constructive relationships between knowers and knowns. Most familiar, perhaps, are Evelyn Fox Keller's biographical evocation of Barbara McClintock's "feeling for the organism," and many feminist scholars' defense of holistic and interactive explanatory models in various scientific fields.[8] But these projects do not exhaust feminist reconstructions of knowing. Haraway has described her own erotic response to some "rigorously analytical and biotechnical" procedures in cell biology as a "knowing love [that] took shape in quite particular, historical–social intercourse, or 'conversation', among machines, people, other organisms and parts of organisms" (Haraway, 1992a, pp. 71, 72). She has also argued for a feminist reconstrual of vision as an epistemic model:

The "eyes" made available in modern technological sciences shatter any idea of passive vision; these prosthetic devices show us that all eyes, including our own organic ones, are active perceptual systems, building in translations and specific *ways* of seeing, that is, ways of life (Haraway, 1991, p. 190).

Feminist defenses of holistic models, then, are best regarded not as *constitutive* of feminist reconstructions of knowing, but as one specific group of contestable moves within a more widely shared concern to understand knowledge as embedded within specific ways of engaging the world. Thus, Haraway crucially captures this feature of feminist science studies as a critical project: "the point is not new representations, but new *practices*, other forms of life rejoining humans and not-humans" (1992a, p. 87).

Sociologists of scientific knowledge might reasonably object that I have overstated the contrast between the feminist and sociological traditions. After all, most recent constructivist sociologists of science emphasize that scientific knowledge is situated within forms of life, while Shapin and Schaffer's influential sociological reconstruction of 17th-century conflicts over experiment perhaps goes further in claiming that "solutions to the problem of knowledge *are* solutions to the problem of social order" (1985, p. 332, my emphasis). Differences reemerge, however, upon considering the place of feminist and sociological *accounts* of scientific knowledge with respect to the forms of life within which knowing is to be situated. The crucial difference, marking the second of my five main points, is located in the sociologists' aspiration to *explain* the content of scientific knowledge, and the attempt to achieve a *detached* standpoint from which such explanations could be launched. Feminist science studies, by contrast, have generally eschewed detachment, and the explanatory project, in favor of a participatory stance. Feminist science studies belong to the culture of science, and most feminists have been concerned to have an effect upon scientific knowledge, and to legitimate the specific effects they hope to bring about.

Feminists' opposition to the sociologists' explanatory detachment has multiple motivations. Latour and Woolgar's hesitation to "go native" among the scientists, for fear of being taken in, ironically contrasts to feminists' recognition that for women, being taken in*to* science has too often not been an option. Adopting a stance of estrangement or antagonism toward the culture of science would risk *endorsing* an exclusion of feminist concerns from that culture which would dangerously reenact its history of excluding women. A thoroughgoing constructivist detachment from science threatens to be just "one more excuse for not learning any post-Newtonian physics and one more reason to drop the old feminist self-help practices of repairing our own cars (they're just texts anyway, so let the boys have them back)" (Haraway, 1991, p. 186, my parentheses), along with one more rationalization for those who would take feminist criticism to signify that women do not belong in science.

Perhaps more fundamentally, however, feminists often regard the sociologists' aspiration to a detached and totalizing *explanation* of the content of scientific knowledge to be objectionably androcentric. Feminist theorists have been suspicious of attempts to escape (metaphorically, methodologically, or theoretically) from the concrete particularity of bodies and social relationships.[9] Sociologists of scientific knowledge may initially seem to share feminists' concern, and to respond by insisting that knowledge claims are always situated within particular forms of social life. Yet their explanatory aspirations require that these particular forms be surveyable as a totality: as determined by interests, as the outcome of negotiations, as the manipulation of inscriptions, etc. As Bruno Latour has since put this point, such explanations are attempts to act at a distance, and thereby to exercise power (Latour, 1987, Ch. 6; 1988). The "strategic position" from which such explanatory abstractions can account for the totality they survey (in this case, for scientific knowledge in its diverse manifestations) is precisely what Nancy Hartsock (1984, pp. 240–7) once critically characterized as the standpoint of abstract masculinity, and what other feminists have since characterized as the suppression of differences. It is not simply the desire for "explanatory power" in its dual political and epistemic senses to which feminists object, but the specific connections between this aspiration to explanatory detachment and the androcentrism that feminists have been specifically concerned to criticize.

Feminists have also objected to the specific forms that sociological detachment has taken, for they cannot afford to follow Bloor in his "impartiality to truth and falsity." Whereas the political commitment of the sociology of scientific knowledge has been to challenge the supposedly unjustified cultural hegemony of natural scientific knowledge in general, feminist science studies were initially concerned with the adequacy of particular scientific projects and proposals that are damaging to women. Feminists have been concerned to substitute scientific approaches and accounts that are not harmful, but it matters to them that those proposals provide not merely more congenial beliefs, but more adequate knowledge. Feminists certainly hope to change important aspects of the way the world is, but they want the political struggles on behalf of their utopian aspirations to be responsive to their actual situation, and that calls for more reliable knowledge. A feminist understanding of scientific knowledge thus requires not detachment or neutrality,[10] but a reflective and self-critical participation in the assessment of particular scientific projects and knowledge claims.

The explanatory ambitions of some recent sociologists of scientific knowledge are problematic for feminist science studies for another reason. Sociologists have often aspired to an explanation of scientific knowledge in a very stringent sense that would require the (social) explanans to be independently variable from or constitutive of the explanandum, so that a sociological explanation of the content of scientific knowledge would require that the relevant sociological categories not be dependent upon or interdependent with the categories whose application is to be explained (recall Collins' and Yearley's call, cited above, for sociologists to "stand upon social things to explain natural things"). Yet feminist science studies have neither sought nor provided such independence from the terms of their interpretations. Elizabeth Potter's (forthcoming) study of Robert Boyle, for example, does not try to *explain* the content of Boyle's scientific project by reference to a pregiven conception of gender, but instead interprets Boyle's work as itself a linked reformulation of received conceptions of both gender and natural philosophy: Boyle was simultaneously "making gender [and] making science." Haraway has insisted upon the importance of this point more generally, in articulating her "nervousness about the...appropriationist logic of domination built into the nature/culture binarism and its generative lineage, including the sex/gender distinction" (1991, p. 198) while Butler has trenchantly argued that feminists need to problematize the discursively conditioned experience through which sex and gender emerge as possible categories:

> Whether gender or sex is fixed or free is a function of a discourse which seeks to set certain limits to analysis....The locus of intractability, whether in "sex" or "gender" or in the very meaning of "construction," provides a clue to what cultural possibilities can or cannot be mobilized through any further analysis (Butler, 1990, p. 9).

I shall return to the concern Butler expresses to understand the effects and limits of feminist analysis itself, when considering sociological and feminist conceptions of reflexivity.

These differences between feminist and sociological approaches to science are also manifest in their *temporal* orientation, which is my third main point. The sociologists' explanatory project would account for the present state of scientific knowledge as an outcome of a social history and the present social situation. An important aim of the sociological project is to display the contingency of scientific beliefs and practices to counteract the appearance of their natural or rational necessity. The sociology of scientific knowledge does not by itself point toward specific changes in scientific belief or practice, but instead opens a space of contingency within which human agency can be exercised more freely: "scientific choice is in principle irreducible and open, [even though] historically, options are foreclosed according to the opportunities perceived for future practice" (Pickering, 1984, p. 405). As I noted earlier, the underlying cultural politics of the classical sociology of scientific knowledge is thus a *humanism* that would allow broader scope to human freedom in constructing views of the world and practices within it.

The temporal orientation of feminist science studies is more specifically futural and transformative. The aim of feminist science studies is not to expose scientific knowledge as *in general* contingent and alterable if "we" choose, but rather to show it as in need of alteration in specific respects, and as potentially open to changes responsive to that need. Feminist science studies are thus specifically oriented toward a "successor science" in a way that the sociology of scientific knowledge has not been.[11] To some extent, this concern has a specifically utopian dimension that would encourage envisioning new ways of organizing specific scientific fields and the cultural politics in which they are situated.[12]

This difference in orientation emerges especially clearly in contrasting *uses* of a superficially similar rhetorical and argumentative strategy. Both feminists and sociological constructivists frequently display alternatives to well-established historical or contemporary scientific programs, and argue that there was or is no *compelling* epistemological justification for a choice many scientists actually made. Yet when Pickering, for example, argues for the contingency and underdetermination of high-energy physicists' commitment to the Weinberg–Salam model of electroweak interactions in the face of apparently contrary experimental results (1984, especially Ch. 10), the point is emphatically not to advocate the resurrection of the contingently defeated points of view.

Feminists use comparable juxtapositions to quite different ends: when Longino (1990, Ch. 7) compares the evidence for and against linear hormonal and neural selectionist explanations of higher brain functions, Keller (1985, Ch. 8) argues for the underdetermination by evidence of biologists' preference for pacemaker cell models of aggregation in cellular slime molds, or Haraway (1989, Ch. 8, 14, 15) recontextualizes debates over the human evolutionary models of "man-the-hunter" and "woman-the-gatherer," their aims are specifically to encourage reconsideration of the merits of the less dominant view, and to show how "one story is not as good as another" (Haraway, 1989, p. 331). Feminist science studies have been specifically concerned to criticize androcentrism and sexism in the development and acceptance of scientific work,[13] to envision less constraining and differently revealing scientific practices, and to enhance recognition of the contribution feminist inquiry can make to science and the culture of science.

Feminists resurrect ideals and norms of objectivity in the context of this concern to reconstruct science, and to secure a place for feminist inquiry in that ongoing reconstruction. This continuing endorsement of objectivity marks the fourth main point of feminist transcendence of epistemology. Yet this endorsement is more often read as an attempt to find an intermediate or compromise position between a strongly internalist claim that scientific knowledge and practice are (or ought to be) fully determined by reason and evidence alone, and a thoroughgoing sociological constructivism whose explanations of knowledge as the outcome of interests, ideologies, or the contingencies of social negotiation would leave no rationalist residue.[14] I think feminist science studies are better understood as attempting to rescue a conception of objectivity from the clutches of *both* epistemology and sociology of knowledge. Epistemological accounts of objectivity or rationality have traditionally been surrogates for realism: if objects of knowledge cannot directly regulate practices of inquiry from "outside," perhaps the *concern* for objective representation can regulate them from within. The new sociology of science would deny even this last vestige of transcendence, but such denial is only *significant* if one accepts the epistemological framing of the question (Rouse, 1996, Introduction, Ch. 7). Feminist science studies would reclaim objectivity not by finding a new route to transcendence of our all-too-human epistemic limitations, but by carefully distinguishing the desire for objectivity from desires for transcendence.

A plausible but mistaken reading of these feminist reconstructions of objectivity might initially characterize them as sociopolitical *rather than* epistemic. The crucial virtue that objectivity would better serve would then be justice rather than truth.[15] The exclusion or marginalization of groups of knowers, whose lives, concerns or needs are thereby prevented from contributing to the critical assessment of knowledge claims, is after all a characteristic failure of objectivity cited by many feminist theorists, and the objectionable consequence seems to be domination mediated by misrepresentation rather than vice versa. Yet by thus contrasting epistemic to sociopolitical virtues and ends, we risk seriously misunderstanding and understating many feminist theorists' interest in better knowledge. For most feminist theorists of science, knowledge is neither external to nor merely instrumental for justice, but is itself a valued end for which justice is integral.

To understand this point, we have to take very seriously my earlier claim that feminist theorists are construing knowledge as multidimensional *relationships* between knowers and knowns, rather than a simple relation of representation and correspondence (or of intertextuality, for those who, like Woolgar (1988), would deny any transcendence of representation). Moreover, these relationships are overlapping, so that scientific knowing always involves complex interrelations among knowers as well as relations to the proximate object of knowledge.[16] The question is how all parties involved can be accorded what is due them. Representation is a power-charged relation, that involves not merely speaking for other people and things, but also the power to shape their circumstances. Feminist reconstructions of objectivity are attempts to hold knowers accountable for what they do, and to determine to whom and to what they need to be held accountable. These attempts take place with the recognition that inquiry and representation are inevitably partial, perspectival, and interested. The demand for justice cannot be an impossible demand for completeness or equal significance, but must instead call for recognition of partiality, openness to criticism and to alternative practices of inquiry and the concerns that motivate them, responsibility for one's actions and position as inquirer and authoritative knower, and accountability for the effects of those actions and that positioning.

This concern to make knowledge more adequately accountable is manifest at multiple levels. Feminist science studies are centrally concerned with questions of evidence, not as a vestige of a theory of confirmation or rational belief, but as an aspect of understanding how knowers should be accountable for what they do. Feminist theorists such as Longino (1990, 1992) and Wylie (forthcoming) contextualize questions of evidence, and focus upon *how* phenomena come to count as evidence and what they can be evidence *for*, what other assumptions, concerns, and practices play a role in constituting evidential relations, and how we can come to a more adequate critical assessment of evidential relations in their many dimensions. Similarly, Haraway's concern to reconceive vision should be understood as a more richly articulated account of the partiality, activity, malleability, and relationality of seeing, and with it a recognition of more complex and far-reaching possibilities for criticism and transformation of what technologies of vision can make evident. In these feminist re-contextualizations of seeing and making evident, "the goal of an epistemology and politics of engaged, accountable positioning remains eminently potent; the goal is better accounts of the world, that is, 'science'" (Haraway, 1991, p. 196). This goal cannot be achieved by subordinating or reducing the epistemic to the political, but instead requires recognition of how familiar conceptions of *both* "domains" are transformed when their boundaries dissolve.

Feminists' moral-epistemology/epistemic-politics also raise questions about how knowledge claims and practices of inquiry become *significant* and authoritative. For example, not all or even most recognized truths about the world count as *scientific* truths. Science as an ongoing practice of inquiry discounts truths that are trivial, marginal, anomalous, arcane, or otherwise "uninteresting," in order to focus resources and attention upon others that are taken to be significantly revealing.[17] Feminist objectivity would incorporate self-critical assessment of judgements about what is interesting or important as well as what is well-confirmed. Not surprisingly, for example, feminists see the coupling of an obsessive interest in researching possible sex-linked differences in cognition with an abiding disinterest in research on diseases afflicting women as simultaneously epistemic and political failings. But the critical assessment of scientific significance extends much further than just the exposure of blatant sexism. The feminist quest to hold judgements of scientific significance accountable extends to critical narrative reconstructions of the cultural significance of what is at issue in whole fields of inquiry, as in Keller's (1992) account of the reorientation of biology around a molecular biological quest for "the secret of life," or Haraway's (1989) examination of how Western primatology came to be focused upon questions of origins.

These questions about how knowledge becomes significant are closely related to feminists' critical examination of how knowers are positioned, a further reminder of why knowing is a multi-dimensional network of relationships from which distinctively epistemic and political judgement cannot be readily disentangled. Feminist assessment of knowledge is directed not merely at what is said on what grounds, but also who gets to speak, who is heard as authoritative, whose concerns and possible responses must be taken into account in constructing knowledge claims, who has access to the material and social resources needed for research, what sustains or compromises these various forms of credibility, and how the resulting authorization of knowers and knowledge changes people's life situation, and constrains or enables their lives. The normative aspirations of feminist science studies are addressed not only to the content of knowledge and justification, but also and inseparably to questions of who knows, with what effects. Yet these aspirations are intertwined without being subordinated to one another. The aim is better knowledge and a better world, together.

This broadening of the normative questions at stake in feminist science studies points to the fifth and final contrast I would draw between feminist science studies and the sociology of scientific knowledge. I maintained earlier that sociological discussions of reflexivity have been guided by epistemological concerns about consistency and completeness. Even Woolgar's and Ashmore's questioning of sociological rhetoric turned out to be focused upon a pragmatic *inconsistency* between what sociologists of science have said and how they have said it. Reflexivity has also incorporated a political dimension for feminist science studies. This contrast does not exempt the rhetoric of science studies from reflexive criticism. Sharon Traweek, for example, criticizes the same rhetorical strategies that concern Woolgar, but to different ends. Traweek notes that, like scientists,

> almost all those writing the newer social studies of science and technology also account for everything and reject all other stories. Almost all these stories, whether about nature, scientists, or science, are narrative leviathans, producing and reproducing all-encompassing stories of cause and effect through the same rhetorical strategies. (Traweek, 1992, p. 430)

In criticizing such rhetoric, Traweek's concern is not to "interrogate" and defamiliarize representational practices generally. From such a feminist perspective, Woolgar's stories comprise yet another narrative leviathan, about how all representations (including his own) are projections of "the Self."[18] Moreover, this conception of "the Self" as encountering only its own constructions is widely recognized in feminist theory as characteristically masculine. For Traweek and other feminist theorists, by contrast, reflexivity discloses partiality and situatedness, not self-enclosure. It exposes the illusion that representation is autonomous and self-projecting; feminists respond that we can never encounter or understand *our-*

selves (and especially not "the Self") except through our interactions with others in partially shared surroundings. If rhetoric is always situated, then reflexive concern for one's own authorship cannot remain internal to texts. The textual self-presentation of the author is subject to reflexive criticism only as part of a larger concern for writing and speaking as forms of action. What do these writings and sayings *do*? To whom and about whom are they expressed? In what ways do they allow for and acknowledge, or foreclose and not hear, the responses of those they speak to, about, or past? Above all, to whom are they accountable? Critical reflection upon knowledge claims is thus always both moral/political and epistemological, and feminist reflexivity would reconfigure the *politics* of science and science studies (including the consequences of the knowledges they produce) in reconstructing their rhetoric. Reflexive attention to one's own practices of speaking and writing would encourage a science, and a political engagement with science, that would be appropriately *modest* and self-critical.[19] Such a reflexive science would be attentive to the effects of its own investigations, including the foreclosing of some questions and concerns by its own theoretical categories and experimental practices.[20]

CONCLUSION

I have been arguing that the contrast between feminist and sociological science studies points toward a *post-epistemological* conception of science and scientific knowledge. My aim in this paper, however, has been to clarify the significance of such a post-epistemological approach, not to develop the more extensive arguments needed to defend it.[21] Williams' characterization of the epistemological project offers a useful way to summarize the contrast. Feminist science studies as I understand them would abandon the epistemological aspiration to a detached assessment of the totality of knowledge (or scientific knowledge) and its relation to an objective world. The alternative is engaged and self-critical participation in the making and remaking of scientific knowledges of the world we live in. Such participation requires an abiding interest in questions of justification, consistency, clarity, and so forth. These questions now arise more locally, however, while drawing upon more wide-ranging considerations. Thus, what is *at issue* in feminist accounts is not scientific knowledge as a totality, but particular scientific practices, projects, and claims, that are understood as ongoing interactions among knowers and the world known. What is *at stake* in feminist participation in science and science studies is at once better knowledge and a better world.

ACKNOWLEDGEMENTS

I am grateful to Lydia Goehr, Jill Morawski, and Lynn Hankinson Nelson for helpful comments on earlier versions of the paper.

Wesleyan University

NOTES

[1] I interpret the scope of the Strong Programme broadly; my discussions of the sociology of scientific knowledge will presume that its central adherents are the Edinburgh School (e.g., Barnes, Bloor, Shapin, Edge, MacKenzie, Pickering), and Bath and York groups (Collins, Pinch, Mulkay, Gilbert, Woolgar, Yearley), the early sociological ethnographers (Latour and Woolgar, Knorr-Cetina) and the social history of Shapin and Schaffer.

[2] Among the more prominent critical assessments of gender in science are Bleier, 1984; Hubbard, 1990; Birke, 1986; Fausto-Sterling, 1985; my discussion, however, will focus upon the broader feminist reconceptions of science found in Haraway, 1989, 1991, 1992a, 1992b; Longino, 1990, 1992, 1993; Nelson, 1990, 1993; Keller, 1985, 1992; Wylie, 1991, 1992, forthcoming; Harding, 1986, 1991, 1992; Alcoff and Potter, 1993; Potter, forthcoming; Addelson and Potter, 1991; Addelson, 1993.

[3] Rouse (1996) argues that the principal research traditions in philosophy of science and sociology of scientific knowledge are committed to epistemology in this sense, and develops in much more detail what a non-epistemological philosophy of science might look like.

[4] Both naturalized epistemologists and sociologists of scientific knowledge emphasize their break from classical normative epistemology, because they seek to explain rather than to justify knowledge. I am emphasizing their underlying continuity in taking 'knowledge' to be a theoretically coherent domain that needs to be accounted for as such, whether the account is an explanation or a justification.

[5] Some feminist studies of science or other ways of knowing may indeed fit under this heading, e.g., Belenky *et al.*, 1986, or the conception of knowledge that Evelyn Fox Keller (1983) attributes to Nobel laureate Barbara McClintock. My point, however, is perhaps best exemplified by theorists like Haraway, for whom feminist reconceptions of how we might engage the world epistemically require *suspicion* of metaphors of organicism, holism, caring, etc., yet do strongly suggest that new forms of knowledge are as much changes in our *practices* as changes in our beliefs or other representations.

[6] Clifford and Marcus (1986), Marcus and Fischer (1986), Rosaldo (1989), among others.

[7] On Bacon, Keller, 1985, Ch. 2; on Boyle, Potter, forthcoming; on the "secret of life," Keller, 1992, part II; on Harlow, Haraway, 1989, Ch. 9.

[8] Keller, 1983; Longino, 1990; Nelson, 1990; Hubbard, 1990, among others, defend specific holistic, interactive, and/or dialectical models as epistemically and politically preferable to more simplified causal accounts of the same domains.

[9] Feminists' concern not to overlook the concrete particularity of bodies and relations does not preclude all generalization, or any feminist *theory*. The objection is rather to a specific kind of theoretical detachment that seeks to *overlook* particularity by appeal to free-floating, ahistorical categories. As Fraser and Nicholson (1990) concluded, "[feminist] theory would be explicitly historical, attuned to the cultural specificity of different societies and periods and to that of different groups within societies and periods. [Theoretical] categories would be inflected by temporality, ... non-universalistic, ... pragmatic and fallibilistic. It would tailor its methods and categories to the specific task at hand, using multiple categories when appropriate and forswearing the metaphysical comfort of a single feminist method or feminist epistemology" (pp. 34–5).

[10] See Proctor, 1991 and Rouse, 1991, on criticisms of recent social constructivist views of science for their detachment and their aspiration to value neutrality, and Harding, 1992 and Haraway, 1991, Ch. 9, on feminist alternatives to neutrality and detachment.

[11] I use the phrase "successor science" here more expansively than some feminist theorists do, to denote any transformation of scientific practice or belief motivated by feminist criticism.

[12] To some extent, this utopian dimension is present in any feminist imagination of alternative ways of organizing scientific culture and practice that would escape or confront the androcentrism or sexism revealed by feminist criticism. It is perhaps most strikingly manifest, however, in Haraway's explicit use of SF literature (1989, Ch. 16; 1991, Ch. 10; 1992b) as a way of posing the utopian imaginative task set by feminist science studies.

[13] As noted above, scholarship has been identified as specifically "feminist" due to its critical focus upon sexism and androcentrism, but feminist scholars have been as much or more concerned to examine racism, heterosexism, colonialism, and other forms of domination and exclusion, *within* feminist politics as well as elsewhere.

[14] McMullin, 1992, p. 22, exemplifies this familiar reading of feminist science studies as "less epistemically radical" than the sociology of scientific knowledge.
[15] In recent political philosophy, "justice" is often used to refer only to fair distribution of rights, goods, and obligations; feminists sometimes criticize not merely particular conceptions of justice, but the focus upon justice as the most central political virtue (e.g., Baier, 1987). Here, I use "justice" in a more expansive sense, as a placemarker for whatever criteria or concerns should be used in the moral and political assessment of actions, and the institutions and relationships that facilitate or sustain them.
[16] Rouse, 1987, Ch. 7 exemplifies this point in discussing the kinds of disciplines that must be imposed to enable laboratory knowledge, and the extension of those disciplines that must accompany the extension of knowledge outside of the local setting of the laboratory.
[17] For further discussion, see Elgin, 1993; Rouse, 1996, Ch. 6.
[18] Woolgar's own capitalization (1988, p. 109).
[19] Haraway's (1989) Chapter 10 discussion of the scientific career of primatologist Alison Jolley provides an illuminating example of how such reflexive modesty might be realized in one very particular setting. Jolley's career and its scientific and cultural setting are unusual in ways that would strongly discourage taking her work as a *model* for politically engaged scientific practice, but it nevertheless illustrates Haraway's conception of a reflexive rhetoric and politics.
[20] Butler (1995) thematizes this concern for the contingent limits of one's own theoretical foundations, but for a natural science the concern must certainly be extended to the material practices at laboratories and field sites.
[21] In Rouse (1996) I articulate and defend my own approach to a post-epistemological philosophy of science, which I take to support and complement the developments in feminist science studies that I characterize here.

REFERENCES

Alcoff, Linda, and Elizabeth Potter, eds: 1993, *Feminist Epistemologies*. New York: Routledge.

Addelson, Kathryn Pyne: 1993, 'Knowers/Doers and their Moral Problems', in Alcoff and Potter, 1993: 265–94.

——, and Elizabeth Potter: 1991, 'Making Knowledge', in Ellen Messer-Davidow and Joan Hartman, eds, *Engendering Knowledge: Feminists in Academe*. Knoxville: University of Tennessee Press, 259–77.

Ashmore, Malcolm: 1989, *The Reflexive Thesis: Wrighting Sociology of Scientific Knowledge*. Chicago: University of Chicago Press.

Baier, Annette: 1987, 'The Need for More Than Justice', in Marsha Hanen and Kai Nielsen, eds, *Science, Morality and Feminist Theory*. Calgary: University of Calgary Press.

Barnes, Barry, and David Bloor: 1982, 'Relativism, Rationalism and the Sociology of Knowledge', in Martin Hollis and Steven Lukes, eds, *Rationality and Relativism*. Cambridge: MIT Press, 21–47.

Belenky, Mary, Blythe Clinchy, Nancy Goldberger, and Jill Tarule: 1986, *Women's Ways of Knowing*. New York: Basic Books.

Birke, Lynda: 1986, *Women, Feminism, and Biology*. Brighton: Wheatsheaf.

Bleier, Ruth: 1984, *Science and Gender: A Critique of Biology and its Theories on Women*. New York: Pergamon Press.

Bloor, David: 1983, *Wittgenstein: A Social Theory of Knowledge*. New York: Columbia University Press.

——: 1991, *Knowledge and Social Imagery*, 2nd edition. Chicago: University of Chicago Press.

Butler, Judith: 1990, *Gender Trouble*. New York: Routledge.

——: 1995, 'Contingent Foundations', in Linda Nicholson, ed., *Feminist Contentions: A Philosophical Exchange*. New York: Routledge, 35–57.

Clifford, James, and George Marcus: 1986, *Writing Culture: The Poetics and Politics of Ethnography*. Berkeley: University of California Press.

Code, Lorraine: 1991, *What Can She Know?* Ithaca: Cornell University Press.

Collins, Harry: 1992, *Changing Order: Replication and Induction in Scientific Practice*, 2nd edition. Chicago: University of Chicago Press.

—— and Steven Yearley: 1992a, 'Epistemological Chicken', in Pickering, 1992: 301–26.

——: 1992b, 'Journey into Space', in Pickering, 1992: 369–89.

Blythe Clinchy, Nancy Goldberger, and Jill Tarule

Elgin, Catherine: 1993, 'Understanding in Art and Science', *Synthese*, **95**, 13–28.

Fausto-Sterling, Anne: 1985, *Myths of Gender*. New York: Basic Books.

Fraser, Nancy, and Linda Nicholson: 1990, 'Social Criticism Without Philosophy: An Encounter Between Feminism and Postmodernism', in Linda Nicholson, ed., *Feminism/Postmodernism*. New York: Routledge: 19–38.

Haraway, Donna: 1989, *Primate Visions: Gender, Race, and Nature in the World of Modern Science*. New York: Routledge.

——: 1991, *Simians, Cyborgs, and Women*. New York: Routledge.

——: 1992a, 'Otherworldly Conversations; Terran Topics; Local Terms', *Science as Culture*, **3**, 64–98.

——: 1992b, 'The Promises of Monsters: A Regenerative Politics for Inappropriate/d others', in Grossberg, Nelson, and Treichler, eds, *Cultural Studies*. New York: Routledge: 295–337.

Harding, Sandra: 1986, *The Science Question in Feminism*. Ithaca: Cornell University Press.

——: 1991, *Whose Science? Whose Knowledge?* Ithaca: Cornell University Press.

——: 1992, 'After the Neutrality Ideal', *Social Research*, **59**, 567–88.

Hartsock, Nancy: 1984, *Money, Sex and Power*. Boston: Northeastern University Press.

Hubbard, Ruth: 1990, *The Politics of Women's Biology*. New Brunswick: Rutgers University Press.

Keller, Evelyn Fox: 1983, *A Feeling for the Organism*. San Francisco: W. H. Freeman.

——: 1985, *Reflections on Gender and Science*. New Haven: Yale University Press.

——: 1992, *Secrets of Life, Secrets of Death: Essays on Language, Gender and Science*. New York: Routledge.

Knorr-Cetina, Karin: 1981, *The Manufacture of Knowledge: An Essay on the Constructive and Contextual Nature of Science*. Oxford: Pergamon.

____ and Michael Mulkay: 1983, *Science Observed: Perspectives on the Social Study of Science*. London: Sage.

Latour, Bruno: 1987, *Science in Action*. Cambridge: Harvard University Press.

——: 1988, 'The Politics of Explanation: An Alternative', in *Knowledge and Reflexivity*, ed. Steve Woolgar. Beverly Hills: Sage: 155–76.

—— and Steve Woolgar: 1986, *Laboratory Life: The Construction of Scientific Facts*, 2nd edition. Princeton: Princeton University Press.

Longino, Helen: 1990, *Science as Social Knowledge*. Princeton: Princeton University Press.

——: 1992, 'Essential Tensions – Phase Two: Feminist, Philosophical, and Social Studies of Science', In McMullin, 1992: 198–216.

——: 1993, 'Subjects, Power, and Knowledge: Description and Prescription in Feminist Philosophies of Science', in Alcoff and Potter, 1993: 101–120.

MacKenzie, Donald, and Barry Barnes: 1979, 'Scientific Judgment: the Biometry–Mendelism Controversy', in *Natural Order: Historical Studies of Scientific Culture*, ed. Barry Barnes and Steven Shapin. Beverly Hills: Sage.

Marcus, George, and Michael Fischer: 1986, *Anthropology as Cultural Critique*. Chicago: University of Chicago Press.

McMullin, Ernan, ed.: 1992, *The Social Dimensions of Science*. Notre Dame: University of Notre Dame Press.

Nelson, Lynn Hankinson: 1990, *Who Knows?: From Quine to a Feminist Empiricism*. Philadelphia: Temple University Press.

____: 1993, 'Epistemological Communities', in Alcoff and Potter, 1993: 121–60.

Pickering, Andrew: 1984, *Constructing Quarks: A Sociological History of Particle Physics*. Chicago: University of Chicago Press.

____: 1992, *Science as Practice and Culture*. Chicago: University of Chicago Press.

____: 1995, *The Mangle of Practice*. Chicago: University of Chicago Press.

Pinch, Trevor: 1986, *Confronting Nature*. Dordrecht: D. Reidel.

Potter, Elizabeth: forthcoming, 'Making Gender/Making Science', in Bonnie Spanier, ed., *Making a Difference: Feminist Critiques in the Natural Sciences*. Bloomington: Indiana University Press.

Proctor, Robert: 1991, *Value-Free Science? Purity and Power in Modern Knowledge*. Cambridge: Harvard University Press.

Rosaldo, Renato: 1989, *Culture and Truth*. Boston: Beacon Press.

Rouse, Joseph: 1987, *Knowledge and Power: Toward a Political Philosophy of Science*. Ithaca: Cornell University Press.

____: 1991, 'Policing Knowledge: Disembodied Policy for Embodied Knowledge', *Inquiry*, **34**, 353–64.

____: 1996, *Engaging Science: How to Understand its Practices Philosophically*. Ithaca: Cornell University Press.

Shapin, Steven: 1982, 'History of Science and its Sociological Reconstructions', *History of Science*, **20**, 157–211.

—— and Simon Schaffer: 1985, *Leviathan and the Air-Pump*. Princeton, New Jersey: Princeton University Press.

Traweek, Sharon: 1992, 'Border Crossings: Narrative Strategies in Science Studies and Among Physicists in Tsukuba Science City, Japan', in Pickering, 1992: 429–465.

Williams, Michael: 1991, *Unnatural Doubts: Epistemological Realism and the Basis of Scepticism*. Oxford: Basil Blackwell.

Woolgar, Steve: 1988, *Science: The Very Idea*. London: Tavistock.

Wylie, Alison: 1991, 'Gender Theory and the Archaeological Record: Why Is There No Archaeology of Gender?', in *Engendering Archaeology: Women and Prehistory*, ed. Joan M. Gero and Margaret W. Conkey. Cambridge: Blackwell: 31–54.

——: 1992, 'The Interplay of Evidential Constraints and Political Interests: Recent Archaeological Research on Gender', *American Antiquity*, **57**, 15–35.

——: forthcoming, 'The Constitution of Archaeological Evidence: Gender Politics and Science', in Peter Galison and David Stump, eds, *The Disunity of Science: Boundaries, Contexts, Power*. Stanford: Stanford University Press.

ELISABETH A. LLOYD

SCIENCE AND ANTI-SCIENCE: OBJECTIVITY AND ITS REAL ENEMIES

INTRODUCTION

As political activist Ti-Grace Atkinson wrote in 1970: "whenever the enemy keeps lobbing bombs into some area you consider unrelated to your defense, it's always worth investigating."[1]

Such an investigation is the primary aim of this essay. There are several interrelated pronouncements that materialize with mystifying but strict regularity whenever "feminism" and "science" are used in the same breath. These include: feminists judge scientific results according to ideological standards instead of truth and evidence, and are recommending that others do the same; feminists are all "relativists" about knowledge, hence they don't understand or don't accept the basic presuppositions of scientific inquiry; feminists – like many historians, sociologists, and anthropologists of science – wish to replace explanations of scientific success that are based on following the methods of science, with explanations purely in terms of power struggles, dominance, and oppression, and to ignore the role of evidence about the real world; in sum, feminists don't believe in truth, they reject "objectivity" as being oppressive, they are hostile to the goals and ideals of scientific inquiry, and they renounce the very idea of rationality itself.

Given that there is a well-known body of feminist work which has systematically articulated *the negation* of each of the above beliefs in black and white, there is certainly a mystery here. In the past four years, a number of scientists and scholars in the studies of science (historians and philosophers), have sounded an alarm about the *anti-scientific* nature of feminist scientists and feminist analyses of science, making highly visible accusations against feminists of the sort just enumerated. Had these critics engaged feminist researchers *other* than those who have made explicit their pro-science commitments, their claims might be interpreted as merely poorly-informed, but this is not the case; the authors they claim to engage are among the most overtly pro-science feminists.

My ultimate goal is to identify and examine central assumptions and loci of concern that play important roles in attempts to discredit feminist contributions to the sciences and to science studies. Because feminists share many goals and analytic tools with other social studies of science, I have found that it is imperative to examine how the critics set up their objections to contemporary science studies. My investigation therefore begins with a display of selected quotations attacking science studies in general (section 1). In section 2, I suggest a framework for analyzing these attacks which involves appreciating the inherent tensions among

L. H. Nelson and J. Nelson (eds), Feminism, Science, and the Philosophy of Science, 217–259.

openness about information, the maintenance of social authority and stability, and scientific and democratic ideals; I think these issues lie at the heart of the conflicts and failures of communication between the critics and the social views of science they attack. In section 3, I analyze several widespread complaints about social and historical studies of science, complaints which our critics raise against feminist approaches as well. I offer diagnoses of the argumentative strategies used, in order to document an important pattern; my goal is to develop conceptual tools adequate to comprehending the origins of the critics' astonishing misrepresentations of feminist science studies.

I review some key features of the relevant set of feminist approaches to the sciences in section 4, and examine several of the standard attacks on them in section 5. I first consider some objections that counterpose scientific and scholarly objectivity and the search for truth with feminists' overtly political goals, and argue that those posing such objections are sorely mistaken about both their simplicity and force. I then illustrate and address claims that certain feminist analyses of science are deeply hostile to the aims and standards of scientific inquiry, and that they appeal ultimately to a rejection of reason itself; I document the distortions necessary to sustain such criticisms. I conclude, in section 6, that the efforts to demonize and dismiss feminist analyses of and contributions to the sciences, amount to illegitimate and unfounded attempts *to exclude participation by fully-qualified colleagues in these aspects of intellectual life solely on the basis of their feminist politics*. Given this dismal conclusion, I must say that I hold no great expectations of convincing the authors I address that such exclusion is a bad idea; rather, I wish to take the opportunity provided by these critics to clarify certain feminist claims regarding the sciences, and to develop the distinctions and frameworks necessary for the rest of us to move on.

1. THE ATTACK

I begin with a selection of quotations in which crucial aspects of science studies are challenged.

Gerald Holton: citing the agenda of the Nobel Conference XXV (October 1989), from the invitation to participants, which presents the issue of whether: "science, as a unified, universal, objective endeavor, is over.... We have begun to think of science as a more subjective and relativistic project, operating out of and under the influence of social ideologies and attitudes – Marxism and feminism, for example. This leads to grave epistemological concerns."[2]

Holton's book is about an "anti-science movement," which he describes as: "a number of different groups which from their various perspectives oppose what they conceive of as the hegemony of science-as-done-today in our culture.... What they ... have in common is that each, in its own way, *advocates nothing less than the end of science as we know it*." He defines the goal of these groups as: "the delegitimation of (conventional) science in its widest sense: a *delegitimation* which extends to science's ontological and epistemological claims." Holton lists the four "most prominent portions of the current counter-constituency, this cohort of delegitimators", at the "intellectually least serious" extreme, he describes the feminist view: "A fourth group, again very different, is a radical wing of the movement represented by such writers as Sandra Harding, who claims that physics today '"is a poor model [even] for physics itself.'" For her, *science now has the fatal flaw of "androcentrism"*; that, together with faith in the progressiveness of scientific rationality, has brought us to the point where, she writes, "a more radical intellectual, moral, social, and political revolution [is called for] than the founders of modern Western cultures could have imagined." One of her like-minded colleagues goes even further, into the fantasy that science is the projection of Oedipal obsessions with such notions as force, energy, power, or conflict.[3] Finally, Holton explains that he has been motivated to write his book because of the social dangers of this "anti-science" movement: "alternative sciences or parasciences by themselves may be harmless enough except as one of the opiates of the masses, but...when they are *incorporated into political movements they can become a time bomb waiting to explode*."[4]

Lewis Wolpert: describing the irrationality that he identifies with *relativism*: "For philosophers of science, and for some sociologists ... the nature of science and the validity of scientific knowledge are central problems.... and *some have even come to doubt whether science is, after all, a special and privileged form of knowledge – 'privileged' in that it provides the most reliable means of understanding how the world works*."[5] In his concern to defend and promote scientific thinking, Wolpert worries about the effects of social studies of science: "emphasis on social processes determining the acceptance of a theory ... can lead one to a relativistic view of science. For if there really is *no rational way of choosing* between rival theories, for choosing between one paradigm or theory and another, then it seems that *science may be a mere social construct* and that a choice of scientific theories becomes like fashion, a matter of taste."[6] In rebuttal, Wolpert insists: "It is not easy to see how the discovery of messenger RNA or the structure and role of DNA could merely be social constructs. They could only appear to be so to someone *ignorant of the complex science involved*."[7] "It is misleading to think ... that *science is really nothing but rhetoric, persuasion and the pursuit of power*."[8]

M. F. Perutz: ridiculing what he describes as "the line laid down by certain social theorists who assert that *scientific results are relative and subjective*, because scientists interpret empirical facts in the light of their political and religious beliefs, and under the influence of wider social and cultural pressures. They allege that instead of admitting their preconceptions, scientists misrepresent their findings as absolute truths in order to establish their power."[9] Perutz also claims that "because references to right and wrong would imply the *existence of objective truth*, they have been eliminated from the vocabulary" of many sociologists of science.[10]

Paul Gross and Norman Levitt: in their book criticizing numerous approaches to understanding the sciences, they concern themselves with those they label the "academic left", i.e., "those people whose doctrinal idiosyncrasies sustain the misreadings of science, its methods, and its conceptual foundations."[11] They are worried about what they see as an "*open hostility toward the actual content of scientific knowledge and toward the assumption, which one might have supposed universal among educated people, that scientific knowledge is reasonably reliable and rests on a sound methodology*."[12] Gross and Levitt claim that this hostility is expressed through adopting "philosophical relativism" and "strong cultural constructivism" about science, which they define as believing that: "science is a highly elaborated set of conventions brought forth by one particular culture (our own) in the circumstances of one particular historical period; *thus it is not*, as the standard view would have it, *a body of knowledge and testable conjecture concerning the 'real' world* ... [instead,] Scientific questions are decided and scientific controversies resolved in accord with the ideology that controls the society wherein the science is done. *Social and political interests dictate scientific 'answers*.'"[13] The central problem with this approach to science is that it "*leaves no ground whatsoever for distinguishing reliable knowledge from superstition*."[14] They explicitly tie this extreme view with feminist approaches to science, asserting that: "Cultural constructivism – in its strong form – is one of the starting points and chief ideological mainstays of the feminist critique of science."[15] Gross and Levitt conclude that the views of science they consider constitute a dire social threat: "What is threatened is the capability of the larger culture, which embraces the mass media as well as the more serious processes of education, to interact fruitfully with the sciences, to draw insight from scientific advances, and, above all, to evaluate science intelligently."[16] In fact, they identify the attitudes of their target social scientific and feminist authors as an "intellectual debility afflicting the contemporary university"; this is extremely dangerous, even genocidal, because the health of our universities is "incalculably important for the future of our descendants and, indeed, of our species."[17]

Themes

There are numerous concerns raised in the above passages; before proceeding, I would like to identify the key issues I shall address in this paper. Most generally, I propose a *way of thinking about the perceived conflicts* between these scientists and those who analyze the sciences within more inclusive social frameworks. In the following section, I suggest and develop a sympathetic interpretation which addresses the *social and political concerns* that are visible in the opening quotations. Having acknowledged that vital socio-political interests are served by maintaining and enhancing the authority of a scientific world-view, I develop an analysis of the tensions and conflicts regarding the control and dissemination of *information* concerning scientific activities – conflicts that arise inevitably when scientific, socio-political, and democratic interests coincide.

The concept of "relativism" seems to be a central obsession of the critics of science studies I quoted above (for brevity, "the critics"). Nevertheless, I will not address the strengths and weaknesses of relativism, *per se*; instead, I shall investigate what these critics believe to be *involved in* a commitment to "relativism" on the part of researchers who examine scientific investigations as social processes. I argue in section 3 that the *threat* of "relativism", as seen by the critics, derives in part from attributing – in error – a simplistic and false dichotomy to science studies research: namely, that social investigations and explanations of scientific processes and products *necessarily exclude and are incompatible with* investigations and explanations which are presented in terms of standards of evidence, theories, testing, and acceptance which compose the framework of "internal", scientific evaluations of knowledge. I consider several specific criticisms which make use of this fundamental dichotomy, concluding with a discussion of the attractions and pitfalls of this particular response to the "demystification" of scientific activities.

2. INFORMATION AND CONTROL

I will not argue that no reason or no argument could be given that might justify a ban on the pursuit or dissemination of social studies of the sciences. In fact, I take it that such an argument could be reconstructed from the chaotic reactions of the alarmists quoted in section 1. One version could go like this: since the collapse of the ruling and controlling power of religious organizations in Western society,[18] the authority of *Science* has taken its place, and for good reasons. Human progress rests on increased knowledge of – not superstition about – the natural world, and how to control it. Accordingly, the awe once inspired by religious ritual is now inspired by feats of insight into and control over nature itself. Science now serves the controlling, inspirational, and authoritative roles formerly belonging to Church-states and their military force.

One consequence of the sciences' central social and political role, as described above, is that a certain *image of science* must be maintained in order to ensure social and economic order: Science must be as believable and trustworthy an authority as possible. Furthermore, some special, authoritative, even mythic stature is psychologically necessary to the maintenance of these essential social roles and functions, and that stature is supported by instantiations of the archetypal story-line known as the "origins myth." In describing a particular picture of science as an origins myth, my aim is *not* to claim that the events recounted did not occur, but rather, to emphasize the cultural and psychological *functions* of the account in question. The focus is on the *role* of a particular picture of science in providing and maintaining specific social goods, such as consensus or agreement (e.g., about public education), peace or non-conflict, and an active and effective democracy.

Let us assume that scientific knowledge and the people who produce it – scientists, technicians, equipment-designers, etc. – play such a central role in maintaining vital social, political, cultural, and economic stability, safety and prosperity, that, therefore, protecting their social *authority* is in all of our best interest. It could therefore appear that feminists and radical critics, when they expose and detail the scientific weaknesses and self-serving interests – whether of class, religious, heterosexual, sex, or racial privilege, usually unconscious – of some scientific approaches and some scientists, are actually destabilizing a great deal: the authority of science and scientists has assumed such a pivotal role in the social order that it can be assaulted only at great peril to the safety and well-being of all.

Let us also accept this social and political perspective on the necessity of maintaining a strong, authoritative role for the sciences in public understandings. I think that such a context creates a tremendous tension between the two primary constructive roles of *studies* of the sciences. One one hand, descriptive accounts of the sciences must serve as the factual basis supporting the myth-level origins stories; on the other hand, studies of science may play vital and detailed roles in the ongoing cultivation and refinement of scientific knowledge and methods.

In essence, one can accept all this and still believe that there is more to scientific investigation and knowledge than a private conversation between Nature and the careful (or bold) and right-minded scientists. Reason and experience tell us that there are good theories, with good supporting evidence, stalled in their tracks for decades because of a favored established theory and its connections to other sciences (e.g., plate tectonics in geology). There are also good theories, with good supporting evidence, rejected for decades because of conflicts – imagined to be irreconcilable – with newer experimental evidence (e.g., the rejection of natural selection as a primary cause of evolution because of breeding experiments). These episodes in the histories of geology and biology, respectively, can be used to teach a variety of lessons, including: true theories don't always look true when we get new evidence; false theories will eventually be rejected in favor of truer ones; being right can pit a scientist against all other established scientists, for life; it's always good to search for more evidence and different kinds of evidence, and so on. I take it that all of these conclusions can be supported by the historical records, even though they may, individually, exemplify conflicting or incompatible lessons about the local workings of scientific inquiry.[19] Hence, there is a great deal of latitude in interest and emphasis when choosing which, if any, of these lessons to spell out using the histories of the sciences. Different lessons are appropriate for different purposes, and the aims of legitimating and maintaining authority can *differ* from the goals of understanding the details of how, when, and why a scientific approach, program, or theory was pursued and then came to be accepted as part of the scientific edifice.

The contemporary situation involving social studies and critiques of science might be better understood by a comparison to military intelligence. A nation interested in commanding global respect for its military might will, quite properly, recount and advertise its successes, the superior training of its personnel, and the efficiency, accuracy, and power of its weapons and war machines. A military commander involved in waging war, however, will find detailed accounts of previous battles – with all of the miscalculations, mistakes, losses, and body counts – to be *essential* to success in actually fighting the war.

To illustrate: suppose that we are currently at war, and suppose further that we have already amassed detailed information that can be analyzed to determine which strategies are most effective against our current enemy, and where our own weaknesses lie. We could, quite reasonably, defend a two-pronged policy regarding this information. Specifically, we might deem it important to disseminate the information and to develop careful analyses of it, for those who can use it to further the goals of our side, i.e., the commanders who make strategic and tactical decisions. At the same time, we might adopt a policy of general secrecy regarding that information and especially our analyses of it, because we gain a strategic advantage through keeping the enemy ignorant of the things we know about them and their operations, and about our own vulnerabilities. In essence, this is the basis of both the value and the secrecy of military intelligence.

Returning to the sciences, we are now in a position to sketch a potential danger of social studies and critiques of science. There are two distinct sets of potential consumers or audiences of such studies and critiques: practicing scientists and their supporters, who already assign substantial authority to the sciences; and everybody else – people who are not directly involved in the development of scientific knowledge but who must, for the sake of order in society, accept the authority of science. While research investigating social aspects of scientific activity – including the foibles, weaknesses, or dead-ends of scientific research – may be extremely valuable to practicing scientists and their supporters, it's also true that that *same* research may be used, in the broader culture, to *undermine the authority* of the sciences themselves.[20]

Consider the differences between an historical account celebrating a scientific discovery – in which the brilliance and integrity of the discovers, as well as their ultimate scientific triumph, are emphasized – and one recounting the details of the ego battles, missteps, logical leaps, and massaging of data *involved* in that discovery. These two historical accounts will inevitably be very different in emphasis, and they may even draw different morals or lessons; nevertheless, because they operate on different levels, with the celebratory account emphasizing the long run, ultimate success, and the equally dramatic detailed one emphasizing the bumps and turns of the journey, they can both be accurate. The *primary* difference is that they serve different purposes, and are aimed at different audiences.

Once we focus on the variety of purposes that might be served by developing and disseminating descriptions and analyses of scientific discovery, struggle, and change, a deep misapprehension on the part of the opponents of science studies can be better understood. In brief, they tend not to distinguish between "demystifying" science and "discrediting" it. The publication of "inside" information about the workings, decision-making, and standard assumptions of particular scientific projects, people, or fields, *does* make them vulnerable; specifically, such information *could* be used to discredit these projects, people, and fields, and to challenge any social and political authority they might have. There is nothing in the information itself, however, which necessitates its use to discredit; what the information does do is *demystify* – it opens for examination and understanding the workings of social institutions that look quite mysterious from the outside, just as biochemical information and theories opened the living cell. Demystification is the shared purpose of all scientific investigation, including investigations into the operations *of* scientific investigation. Indeed, this is why science studies are important; if we are to be able to *compare* various systems and methods of investigating nature and producing scientific knowledge, we must understand them.[21] Even the critics we're considering here wish to discriminate among various instances of scientific processes, rejecting Nazi science, denouncing racist science, and science used toward the sole end of individual profit.[22] Nevertheless, they consistently attack others for attempting to "demystify" the sciences; this puts the critics in the untenable position of defending scientific authority by insisting on its "mystification."[23] We need to understand the conflicting interests and motivations which have led to this unstable position.

Internal Tensions

A more incisive analysis of the various uses of information would help us to see who is using inside information to further the sciences' objectives and authority, and who is using it with the genuine intention – or result – of undermining that essential authority. The critics' most profound error is that they have misidentified feminist and social scientists as being utterly *hostile* to science, rather than accepting them as allies who genuinely share their concerns and foundational political goals regarding science.[24] Those who advance the sciences are in a uniquely difficult position because of a double pressure – from scientific and democratic ideals – for openness and participation; this is why it is so damaging when their reactions are insufficiently refined, and in some cases border on paranoia – seeing enemies where there are pro-scientific allies.

In summary so far: historical and methodological accounts of science have long served in the roles of legitimating origins-myths, and this is fine, because such legitimation is necessary to public confidence and social order. These accounts can also serve a vital role in recruiting talented individuals to engage in scientific enterprises, and they are also naturally going to be the focus of those responsible for maintaining and funding scientific institutions. But studies of science can also be undertaken to provide more complete and accurate pictures of how the sciences actually work, how they succeed, how they fail, and what the most conducive conditions and standards are for attaining any scientific goals – this necessarily involves a "demystification" of science – as a science of science, it demystifies the workings of an otherwise magical-looking system.

An inevitable product of the investigations into science is, thus, *inside information* which, like military intelligence, has the capacity to be used in various ways, towards various social and political ends. *It has this capacity precisely because the sciences are so intimately tied to political authority.* Everyone involved in debates about the nature, authority and proper roles of the sciences understands this on some level – the creationists, social activists such as feminists and environmentalists, and the critics I discuss in this paper. It is precisely because of the fundamental, architectural role of the sciences in political, economic, and social life that social activists *target* the social authority of specific scientific programs and specific scientists.[25] No one in the US who has heard any discussion of *The Bell Curve* should doubt that contemporary science is as powerful a political player as any in action today. What remains at issue, then, is how this inside information should be analyzed, towards what ends it should be put to use, and how far and in what manner it should be disseminated.

There is thus an inevitable set of *tensions* facing scientists and those of us who wish to support the authority of the sciences vs., e.g., the authority of particular religious principles. The standards of openness in science, public education, and a democratic political system, all make it difficult for scientists to maintain control over information regarding the actual doing of science, and over anyone who chooses to analyze that information. The model of military intelligence – analysis for insiders, pep-talks for the troops, and secrecy and disinformation for the enemy – is therefore not one of the available options for the channeling of appropriate information to the right places.

Generally speaking, there are a variety of goals involved in human communication, and a central one of these goals is certainly to disseminate truth rather than falsehood; additional goals include cultivating broad rather than narrow perspectives, focusing on long-term or distant consequences as well as immediate ones, and updating and maintaining broadly-shared vocabularies for the purposes of business, civic life, and social stability. The critics who object to feminist and social analyses of the sciences are sympathetically read as focusing on the *potential harm* of such studies to social life, economic prosperity, and the civic responsibility which is essential to democratic government. It isn't at all clear how studying sciences in action will generate this harm, but the assumed mechanism seems to be something like the following: social studies, especially feminist studies, of the sciences sometimes challenge the accuracy and scope of specific scientific conclusions and inferences; by doing so, they're undermining the *authority* of scientific knowledge in the wider culture.

Hold it right there. How and why do challenges to specific instances or aspects of scientific activity amount to undermining the political and social authority of scientific knowledge itself? Doesn't that depend on the basis or standards which are used to generate the challenges? In other words, if appeals to empirical evidence, consistency, and other scientific standards are the substance of such challenges, then they are properly seen as operating *within* the sciences, and as such, legitimating them. The undermining and delegitimation of the authority of science is genuinely challenged, in contrast, when the *basis* of a challenge is, for example, a fundamentalist religious view, such as the belief that evolutionism teaches atheism.[26] Again, it is a mistake to equate demystification, and any potential improvements in the activities of science that may come to mind with that demystification, with delegitimation and the rejection of scientific authority *uberhaupt*.

This vital distinction between ameliorative reform and all-out rejection is being blurred by the critics' reactivity and defensiveness.[27] Holton, for example, advises, unselfconsciously: "for specific facets of anti-science, see the essays ... in ... *Counter-Movements in the Sciences*,"[28] apparently not noticing that the book is about cases *within* the sciences. Gross and Levitt threaten to eject humanities and social studies from the university unless they become more hospitable to being judged by natural scientists.[29] Regaining control over *how the sciences are understood* is the aim of Gross and Levitt's entire book, and it is an echo, in less sophisticated wrappings, of Holton's and Wolpert's conclusions, as well. So here is the tactic: mobilize pro-science constituencies to oppose the academic and socio-political legitimacy of those who demystify science. But this can't be right; *some* analysis of inside information is desirable for the effective functioning of the sciences themselves. In fact, Wolpert makes this very point, while not seeming to notice the resemblance between his own list of important issues to investigate, and the issues prominently addressed by feminists.[30]

So what is the plan? Is it that informed and intelligent analyses and criticism of sciences should remain behind closed doors? Not tolerated in the academy? Does such criticism paradoxically *encourage* superstition – even when one of its aims is to uncover or reveal unsupported beliefs and assumptions within the sciences which might be called superstitions themselves? Given the overriding value of openness of inquiry and information to both a scientific and a democratic world-view, there ought to be very good reasons – and no viable alternative – for limiting such openness.

The recent attacks on science studies have come out openly with their political and social worries, which must be read as their reasons for wanting to limit openness about inside information on the sciences. Their appeal to the importance of the authority of science has turned out to be very valuable, because they have simultaneously revealed their lack of insight into the overall situation involving friendly, democratic, and hostile uses of information. Furthermore, it makes inaccessible the option of taking the "high road": other people are contaminating discussions of the sciences with their political concerns, whereas real science is not appropriately dragged into the political mud.[31] If anyone is in a position to adopt this approach, it is *not* the critics we're discussing, because of the explicitness of the social and political goals of their own writings. Still, understanding the internal tensions inherent to this situation can help us to remain sensitive to the complications and possibilities for misunderstanding that will be inherent when two parties have differing perceptions of the security, dominance, and legitimacy of any particular sphere in which the exercise of scientific authority plays a role with substantive social consequences.

I conclude this section by sketching a few strategies, borrowed from the context of intelligence work, for managing "inside" information to reinforce the authority of the sciences.

Strategies

Control: The most direct way to keep potentially damaging information from being used against the sciences is to keep it secret or private, or, even better, not to gather or analyze it at all.

Discredit: Discrediting supposedly-inside information is also very effective. This can be done through: (1) claiming that it is not accurate; or (2) claiming that the purveyor is unqualified and incompetent; or preferably, both.

Exclude: The essence of exclusion is to define a potential source of information or analysis as *hostile* – as an enemy – and thereby to justify their exclusion from any legitimate access to or discussions of that information.

Each one of these strategies has its costs, because each one is potentially in conflict with the ideals and interests of both scientific inquiry and a democratic political system. Thomas Jefferson, quoted approvingly by Wolpert, makes clear the appropriate democratic solution: "I know no safe depository of the ultimate powers of the society but the people themselves, and if we think them not enlightened enough to exercise that control with a wholesome discretion, the remedy is not to take it from them, but to inform their discretion."[32]

In the following section, I will discuss how critics of social studies of science impose a condition of exclusivity on social and 'internal' scientific explanations of science, even though, *prima facie*, such explanations ought to be considered complementary. I think this move is usefully seen as a set-up for a struggle for control over information, especially when it is combined with a perception that any demystification and inside understanding of the workings of science are necessarily damaging to scientific authority and functioning. The problem is that, according to scientific standards themselves, a certain amount of internal analysis is necessary and desirable. As Wolpert concludes, "the Greek commitment to free and critical discussion was essential for science to flourish … once one rejects understanding and chooses dogma and ignorance, not only science but democracy itself is threatened."[33] Moreover, to read all demystification as the "enemy use of information" is to make a grave tactical error: it goes against scientists' own principles, and is hence hypocritical; it also underestimates the power of the sciences to withstand scrutiny and to absorb and reform any picture of the sciences' workings.

3. REASON VS. RELATIVISM

3.1. *The Exclusivity Doctrine*

One of the best ways to understand these critics is to examine their attachment to a certain dichotomy, a division of intellectual labor, in which social explanations and 'internal' ones are seen as exclusive and competing – rather than complementary – explanations. The basic move is to ascribe to authors in science studies a commitment to a *unitary* and *exclusive* model of explaining events in scientific practice: *all* such events are to be explained completely and exclusively in social terms, allowing no complementary and interlaced explanation in terms of the scientists' own ("internal") beliefs, theories, and evidence.

Let me tell a story. Imagine that we set out to study a diversity of religious rituals practiced in a variety of human societies. We can go in and watch and describe what people are doing, what they won't do, what they respect, what they say, what they refuse to do, etc. And we also may investigate by asking, by inference, by history, by interview, by any number of ways – what these people are doing and what it means – that is, *why* they're doing what they're doing. In other words, we need to understand both their various beliefs, *and* their beliefs about the relation between their beliefs and the world, in order to fully comprehend what they're doing, in order to make sense of – give an account of – these events, these religious rituals. Now, I take it that others would agree that these observations and investigations of these ritual behaviors are recognizable as part of a general area of study or approaches called Anthropology.[34]

It has been recognized as an important step in understanding the vast varieties of human ways of living, that we understand what human actions *mean* to the person doing them. Also, grasping such meanings must include the theories and assumptions about the ways the world works and the way the world is. Such detailed study of the comparison of actual beliefs and belief-systems – for example, in religious rituals – is probably most analytically and competently handled by theologians, who are familiar with the many aspects of comparing religious belief systems. Still, these studies of the belief systems are only part of the story; they are a necessary part, but only a part.

I think that there is an *exact* parallel to the practice of science, and if we're interested in as full and as complete an understanding of science as possible, I believe that (just as in the comparative religion case) some sorts of social and anthropological investigations are necessary, and that without them we do not have a full and complete account, on *anybody's* standards. This also means that the theoretical background and assumptions, and the beliefs about the relations between the theory, the theoretical backgrounds, and the aspects of the world being investigated, are important. Any claim that theoretical problems and their objects are essential factors in an account or explanation of scientific events sounds right to me, in exact parallel to the religious case. I take it though, that the objection we're considering is that social or anthropological studies of scientists *neglect* these central factors – the beliefs about theory and object, and the truth, accuracy, and evidence for these beliefs – in concentrating exclusively on the social, psychological, economic, or political contexts of scientific work. Potentially, *all* social analyses of scientific activity suffer from this fatal flaw.

The real question is whether the pursuit of social or anthropological descriptions of scientific activities and ways of life is *incompatible* with discussing the content of sets of beliefs and theories, and the reasons and standards that are used to evaluate them. An examination of the *contents* of beliefs would give us only an incomplete understanding of spiritual practices; for the same reasons, a summary of scientific beliefs would give us a very incomplete understanding of scientific practices. As far as scientific theories and their objects go, natural philosophers have spent centuries in detailed investigations of the theoretical problems, the objects, their relations to a set of scientific beliefs and systems, and so on. It's not as if nobody's looked at those problems. That's precisely where the action has been.[35] The concern seems to be, rather, that the social descriptions are intended as *exclusive*. Are social and anthropological accounts of science supposed to give a *complete account of science?* If such completeness is not being attributed, why would Gross and Levitt complain that: "To concentrate on the idea of empirical science as a manifestation of cultural and political imperatives is to *omit* important dimensions of the story, both human and philosophical."?[36] Similarly for Wolpert's objection that a "scientific discovery cannot be judged only in social terms but must also take into account the new understanding or knowledge it provides";[37] using a social approach to scientific knowledge, he complains, "*says nothing* about the belief's contribution to understanding, its correspondence with reality or its internal logical consistency."[38] The crucial difference here is between "saying nothing" and saying *there is nothing else to say*.

It thus seems possible that the critics do not understand the incomplete nature of the internally focused explanations they favor. Analyses of scientific theories and evidence and explanations are not made inaccessible or irrelevant by the existence of anthropological or sociological approaches; rather, they are complementary. Philosopher of science Sandra Harding makes the point very clearly: "Nature causes scientific hypotheses to gain good empirical confirmation, but so, too, does the 'fit' of these hypotheses' problematics, concepts and interpretations with prevailing cultural interests and values. It is a *maximally objective understanding* of science's location in the contemporary social order that is the goal here. This is *far from a call for relativism*. Instead it is a call for the *maximization of criticism of superstition, custom and received belief* for which the critical, skeptical attitude of science is supposed to be an important instrument."[39] Without a persuasive reason why social studies cannot contribute or add completeness to our available reasoned understandings, there is no argument – especially not one based on incompleteness – against pursuing them. Because any relevance or intellectual value of the critics' discussions rests on their providing *reasons* for rejecting the intellectual legitimacy of specific approaches to describing the sciences and scientific activities, the rest of this essay will be concerned with developing clues and hints regarding such reasons and arguments, and evaluating the results.[40]

In sum, an implausible and unpromising explanatory doctrine, in which social and scientific/evidential explanations are seen as strictly mutually exclusive and individually complete, is being used by the critics to cause much of the trouble here. (For the sake of brevity, I shall hereafter refer to this explanatory doctrine as "the exclusivity doctrine.") I shall suggest that many of the reactions against science studies, especially the accusations tying "relativism" to "irrationality," are best understood as arising from the exclusivity doctrine. The most important worries revolve around the vital roles of *truth* and of the *reasons* that scientists have for holding specific beliefs and theoretical commitments. The fundamental concerns seem to be that social analyses and critiques of science *neglect* the *content and truth of the scientific theories* they discuss: because content, truth, and evidence constitute the very foundations of science, any analysis which neglects them must fail to reveal anything significant about the sciences themselves. I argue that it is the critics of science studies who, paradoxically, tend to *insist* upon the exclusive nature of social and evidential explanations, rather than the authors to whom they attribute this position.[41] The paradox is illuminated by highlighting the strategic value of attributing the exclusivity doctrine; it can be an effective basis for both the "discredit" and "exclude" strategies introduced in section 2.

3.2. *Illustrations: The Burdens of Proof*

In this section, I interpret and analyze two standard objections to the "relativist" and "strong cultural constructivist" views, as they are perceived by our critics of science studies. While the objections differ in their details and targets, they share an underlying theme: they *insist* that social explanations and "internal", evidential explanations of scientific activity are mutually exclusive – in spite of indications by the authors they cite that they reject the exclusivity doctrine.[42]

The Control of Description

Gross and Levitt attack philosopher and anthropologist of science, Bruno Latour, by accusing him of holding the view that "laboratory politics *accounts for science as such* and is the *real story* behind the emergence of scientific theories";[43] "Latour's reports on the activities of scientists *are* to be accorded factual status," they exclaim, "while scientists' reports on nature are *not*."[44]

One predictable objection to pursuing the social study and critique of scientific practices is to challenge the supposed "privileged viewpoint" of such research. "Why should we believe," the usual argument goes, "so-and-so's claim that socio-economic factors are playing a significant role in the scientific community pursuing or believing theory or research program *A* rather than *B*?"

This is an important and interesting challenge. But note what the objection relies on setting up as the alternative: "why should we believe so-and-so's claim" ... *instead of the scientists' own claims* regarding why they're pursuing *A* rather than *B*. There are several serious problems here. Most obviously, it is a mistake, as I argued above, to *assume* that social explanations are posed – by Latour, in this case – as *replacements* for all the reasons adopted by any individual scientists. Second, the implicit alternative – that the scientists involved can simply reflect on and report their reasons for pursuing *A* rather than *B* – does seem to assume that scientists have super-human powers or insight. As philosopher of science Sandra Harding has pointed out, if "the 'science of the natural sciences' is best created by natural scientists ... [then] the sciences would be the only human acitivity where science recommended that the 'indigenous peoples' should be given the final word about what constitutes a maximally adequate causal explanation of their lives and works."[45]

It is important to remember that we need make no assumption that *others* necessarily have privileged information about scientists' motives, any more than we need grant the exclusivity doctrine. Note that the appeal of Gross and Levitt's objections, quoted above, relies utterly on attributing the exclusivity doctrine to Latour, even though he doesn't hold it.[46] This is a distraction from the fundamental issue, which is simply that giving *unquestioned and exclusive authority* to scientists' own descriptions of their actions and motives is – given common-sense acquaintance with the complexities of human self-understanding – an impoverished and silly way to understand scientific activities. The *relevant standard* here comes from social and anthropological studies *of other complex human activities*.[47] Once this is explicit, it is difficult to see how these critics could defend a blanket intellectual rejection of such studies. Should models of market forces in economics be applied to grains, gold, women, antibiotics, and petroleum but never to scientists and their grant money? Should admittedly ethnocentric studies and interpretations of social structures, values, and cultural practices be performed on non-European groups of people but never on groups of scientists? Should correlations between biological sex, wealth, and labor be investigated and detailed for communities around the world but never for scientific laboratories? The weaknesses – predictive, description, and explanatory – of these ways of studying our social lives are well-known, and I agree that we would do well to maintain the visibility of those weaknesses. The opponents of social studies of the sciences, however, cannot *legitimately exempt* the sciences *as subjects of study*, without some additional argument.

The Control of Judgement

In highlighting another example of their insistence on the "total opposition" between social and internal scientific explanations of scientific practices I appeal to certain passages quoted by or paraphrased by Gross and Levitt; even though these passages are offered as evidence that Steven Shapin and Simon Schaffer, in their *Leviathan and the Air-Pump*,[48] claim to give an exclusive and complete explanation for the science they investigate, they support no such claim. Rather, they appear to map out claims of *partial* explanation.

For example, Gross and Levitt quote Shapin and Schaffer's own claim: that "these political considerations *were constituents* of the evaluation of rival natural philosophical programmes."[49] Their paraphrases of Shapin and Schaffer's conclusions include the following: "the nascent Royal Society was, from the first, the creature and deputy of a political and social viewpoint",[50] "[Robert Boyle's and his co-workers'] supposedly empirical rules, it is said, constituted a specific social practice";[51] and finally, "The [Royal] society's supposedly objective science is thus to be read, *in large part*, as a construction of its ideological commitments."[52]

Note that claims to empirical adequacy, predictive power, or scientific explanatoriness of a scientific theory or approach are *nowhere here reduced to or exclusive of* its political commitments. That is, the position which Gross and Levitt attribute to Shapin and Schaffer *here* does not, in fact, include the exclusivity doctrine; rather, they attribute a more moderate view, in which both "objective scientific" and "ideological" explanations play roles.

In order to support their claim that Shapin and Schaffer adopt the exclusivity doctrine, Gross and Levitt must ignore not simply the immediate context, but the entire book surrounding the statement they chose to represent its most damning conclusion. Gross and Levitt claim that the authors think that this case study embodies a radical relativism; they write that the following passage "is Shapin and Schaffer's last word on the general epistemic principle that their particular case is supposed to illustrate." Quoting Shapin and Schaffer: "As we come to recognize the conventional and artificial status of our forms of knowing, we put ourselves in a position to realize that it is ourselves and not reality that is responsible for what we know."[53] Gross and Levitt then remark, "So, in the end, we come back to the dichotomy – fallacious in that it posits total opposition between 'reality' and 'convention' where there is, in fact, intense and continuing interaction – so favored by Latour and other constructivists."[54]

I shall say more about this attribution of the exclusivity doctrine to the "constructivists' in a moment. First I must admit, however, that of all the sentences in Shapin and Schaffer's 440-page book, this one almost begs to be plucked out and misinterpreted. Nevertheless, there are compelling reasons against construing it to mean: "*it is exclusively ourselves and not reality at all* that is responsible for what we know," as Gross and Levitt apparently do. One serious problem with such an interpretation is that it fails utterly to take account of the ongoing discussions and debates within the history of science of which this book is a part: in the wake of centuries of historical accounts of science in which the overpowering majesty of nature and the beauty and perfection of its rational order were portrayed as the engine driving scientific progress itself, recent historians have been developing more accurate and complete accounts, focusing on the specific decisions, activities, and desires of the investigators, and on the social and cultural contexts in which their interactions with nature came to yield scientific knowledge. There is, thus, a corrective force to Shapin and Schaffer's book regarding the history of science itself. More importantly, there is a plausible and reasonable interpretation – supported by abundant textual evidence – of the quoted passage, as follows.

In the context of the case study in the book – the development and use of the "air pump", or vacuum container – "artificial" signifies that we – members of a scientific community – crafted our apparatus (specifically, the air pump) through which we experiment and come to know things, while "conventional" signifies that the procedures and experiments that were worked out by Boyle had to be argued about and agreed upon through social processes of persuasion and interpretation – otherwise, the demonstrations wouldn't be understood *as* experiments. That this is a reasonable understanding of the situation is clear from Shapin and Schaffer's examination of the explicit debates about those methods, apparati, procedures, and interpretations; they document the arguments and the winning moves. In this context, it does *not* amount to any radical rejection of scientific standards, truth, or the existence of the real world, to conclude, as they did, that what we know *depends on* the existence and acceptance of specific equipment and procedures – that is, their existence and acceptance is *necessary to* our knowledge; hence *we*, and *not exclusively the reality independent of us*, are "responsible" for what we know. The point is both simple and incontestible: more than the *existence* of reality is necessary for us to know it; it has been there all along, and yet we have not known it until very recently; in order for us to know it, it must necessarily come within the purview of whatever *methods and experimental apparatus we do have*, which are, themselves, *our own creations*.

I find this to be a fairly typical example of Gross and Levitt insisting on the exclusivity doctrine. Interestingly, Gross and Levitt are aware that their criticisms appear designed to discredit and exclude, and they make an effort to display their good faith and openness regarding science studies by actually engaging Shapin and Schaffer on one issue: explaining why Thomas Hobbes' views were not accepted.

This little argument is extremely revealing because Gross and Levitt display their own ideas about the standards by which social studies of science should be judged. In brief, Gross and Levitt argue that Hobbes' repeated failure in mathematical arguments "provides a concrete and substantive reason, *in contrast to an ideological one*, for Hobbes's notoriety in scientific circles."[55] Note that they have introduced a dichotomy between scientific or mathematical explanations, and social or "ideological" ones. It is clear that *they think* the two types of explanation are exclusive; if, they say, Shapin and Schaffer had addressed Hobbes' failure as a mathematician, they "would have put themselves in the position of conceding the existence of sound, objective reasons for deciding at least some scientific controversies."[56] But Shapin and Schaffer did not deny the existence of any reasons that might be categorized today as defensible, scientific ones, nor is there a reason given to think that social explanations cannot be sound or objective. After Gross and Levitt legislate that *no social explanation is compatible with a [potential] 'internal' one*, they go on to condemn Shapin and Schaffer for "insisting that all such disputes are ideological."[57]

One could interpret all of these exchanges as a consequence of a benign lack of subtlety combined with defensiveness about maintaining the myth-serving aspect of the history of science. That would be a mistake, as Gross and Levitt appear to be very sensitive to exactly what is wrong with both a myth-preserving view and the view they foist on the "constructivists"; remember, the target of much of Gross and Levitt's criticism is the dichotomy posits *total opposition between 'reality' and 'convention' where there is, in fact, intense and continuing interaction*."[58] This vision of "intense and continuing interaction" between reality and convention, is the basis of the very texts they've attacked. While Gross and Levitt claim to reject the completeness and exclusivity of social and other explanations, their own arguments and conclusions rely extensively, and sometimes completely, upon just such exclusivity; furthermore, they seem so wed to their beliefs about the "constructivists" that they miss all of the textual evidence supporting a more moderate reading, even when it is given in their own paraphrasing (e.g., previously quoted uses of "constituents of" and "in part"). The climax comes near the end of their book, where they admit that "those who insist that science is driven by culture and by politics...are not for that reason alone to be dismissed as wrongheaded. On the contrary, these assertions, if 'driven' is replaced by 'influenced,' come near to being truisms."[59]

In sum, Gross and Levitt utilize the full battery of strategies against those they have branded as anti-science. In the text just considered, they misrepresent the social history of science they wish to reject; specifically, they ignore even *their own* references to partial causes, influences, and implications of shared social and empirical accounts. Such a pattern shows something far more interesting than its evident scholarly impairment: it clearly embodies the strategies of (1) attributing programmatic hostility; (2) discrediting interlocutors for being ignorant of the sciences; and (3) attempting to gain control over the terms of "legitimate" social studies of science. This last is never so clearly revealed as when Gross and Levitt propose their own standards and limits for studies of science: "inasmuch as the specific content of the [cultural-constructivist] thesis challenges the reliability of scientific conclusions ... and inasmuch as it does so, roughly speaking, on the basis of the same argumentative paradigm as scientists use in practice, the logic, evidence, and pertinacity of the thesis must be weighed against that of specific scientific arguments."[60] Gross and Levitt demand that, in order to make their case, "cultural constructivists must demonstrate that their arguments for unreliability outweigh those of conventional scientific papers for reliability *in the realm of phenomena addressed by the latter* [their emphasis]. They must show that their arguments are stronger than those put forth by Professor X in his paper on the role of transforming growth factor beta in the morphogenesis of the optic tectum, while simultaneously outweighing those of Dr. Y in his monograph on the classification of compact Lie group actions on real projective varieties! If they are to demonstrate that their arguments *contra* science are anything but sheer bluff, then clearly *they must play on the scientists' court*."[61] And Gross and Levitt are confident that such arguments will fail, because, "to put the matter brutally, science *works*."[62] Hence, in spite of their lip-service to "intense and continuing interaction," Gross and Levitt don't grasp or don't accept the essentially complementary nature of social, cultural, and evidential explanations.

Wolpert has similar difficulties, even though his is generally a much more subtle and penetrating understanding of scientific investigation than Gross and Levitt's. While Wolpert notes that Andrew Pickering's study of particle physics is self-consciously oriented towards the social aspects of scientific activities, Wolpert concludes, with some astonishment, "there is really nothing in his analysis that reflects such an approach or that is in conflict with an image of scientific advance *that scientists themselves would readily find acceptable*."[63] Thus, he seems to be suggesting that real constructivist analyses of science would, necessarily, *conflict* with scientists' self-understanding. This situation is illuminated by Wolpert's description of Pickering's account: "[it] shows just the sort of complex interactions between theoreticians and experimentalists that *one might expect*.... What Pickering does make clear is the symbiotic relationship between theoreticians and experimentalists: both are looking for new opportunities to advance their work."[64] Here, Wolpert is clearly speaking *as an insider*, these are insider's expectations, and an insider's understanding of the "symbiotic relationship" and the goals of advancing one's own work. In other words, because Pickering has "merely" made public "inside information," in an accurate fashion which is acceptable to (some) insiders,

Wolpert concludes that he has not, in fact, taken a genuinely "constructivist" approach; the obvious but abandoned alternative is that social and internal explanations of scientific exploration and change are *not* mutually exclusive or inherently conflicting.

Again, the overall tension is between a view of science used to inspire, promote, and legitimate, and a framework which can be used to inform scientists, philosophers of science, and others interested in the intimate details of science-in-action; a deep *conflict* can arise only out of insistence on the legitimacy of only *one* of these approaches to science.

3.3. *Strategic Intelligence*

The distinctions I've illustrated can be used to reveal a profound problem with the basic position which the critics wish to maintain. Consider the pictures of science that these authors wish to defend and disseminate: they explicitly claim that *their goals* for such public communications are social, economic, and political, just as they criticize their chosen opponents for *undermining* the accomplishment of these same goals. So far, so good; differences in strategies for accomplishing public, shared goals are the bread and butter of civic and policy debates, and this is their proper forum in democratic societies. *But these authors, in entering these debates, deny that they are even participating in them.* This surrealist move is accomplished by claiming that their views are the *only* views which are rational and true: they are not debating policy, they are announcing fact, and anyone disagreeing with them is necessarily advocating falsehood and perpetuating ignorance.[65]

If Wolpert, Perutz, Gross, and Levitt are making factual claims about the social, conceptual, and historical facets of the sciences, then the intellectual and scientific standards they champion demand that the truth or accuracy of factual claims must be evaluated with respect to evidence, as must accusations regarding the falsehood of the factual claims of their opponents. Instead, the critics tend to emphasize that their targets neglect scientific and evidential accounts of the sciences – we have seen that this complaint is damaging only in the context of the exclusivity doctrine, which they impose on the authors they criticize.

Through assuming and *imposing* an "exclusive and complete" picture of social vs. "internal" evidential explanations, the scope of social studies of science becomes indefensibly narrowed. Under such an explanatory doctrine, the legitimate topics of social studies of science encompass only whatever the scientists chose to regard as *unscientific*, i.e., as *not explainable* using the internal standards of that science – in which case they might better be called "studies of non-science." Fundamentally, this move amounts to setting the detailed decision-making, exploratory, and evaluative activities of scientists *off-limits* as subjects of analysis, which in turn eliminates the possibility of legitimately pursuing many social critiques of science.[66] Because any *overt* demand that scientists have a special, exempt status with regard to studies of human social systems and cultures is recognized as being in extremely bad form – as anti-scientific itself – the critics I examine offer disclaimers to the effect that no exemption is being demanded, while they simultaneously implement the strategies of secrecy, discrediting, and exclusion, all of which are justified by an appeal to the security of the democratic and social order.

Hence, the critics engage several of the available strategies for coping with the tensions regarding scientific and socio-political control of information are engaged in these criticisms; I find that their use has serious disadvantages, not the least of which is their dubious strategic effectiveness. In the contemporary US context, it is unclear whether insisting on the purity and apolitical sanctity of the sciences is a viable intellectual, political, or social option. What Gross, Wolpert, Perutz, and others, seem to be in denial about is that the cat's already out of the bag. Just as the rash of political assassinations, the Vietnam war, the Pentagon Papers, Watergate, Nixon's resignation, and changing journalistic standards contributed to a long-term, if not irreversible, public mistrust in elected officials and public institutions, so the spectacles of corruption and waste in the manufacture and design of our best military technology, lack of responsiveness within all parts of the health and medical technology professions during the first decade of the AIDS pandemic, and lying and cheating for money and prestige within the top universities and research institutions in the world, and even the waffling on diet guidelines regarding cholesterol, eggs, or oat bran – all have produced a public mistrust of both the disinterestedness and competence of scientists in general, and thereby of science itself.[67]

These events suggest that there might be an aspect of *scapegoating* to the fury that Holton and others direct towards anyone they perceive as "demystifying" science. The scientists involved in the incidents mentioned above needed no science studies sleuths to appear foolish, greedy, dishonest, biased, overreaching, or blinded by ambition; they did it all by themselves, and often in public view. The contributions of the science studies researchers whom Gross *et al.* target might best be described *not* as "revealing" and advertising the inside information that such unseemly events have happened, but rather as investigating the possibilities that such things are *built into* the social systems of the sciences as they stand: they might be structural, institutional, and predictable. In that case, they would *not* be freak accidents, any more than Oliver North was a rogue elephant single-handedly violating US policy.[68]

Conclusions

Scientific activities are unique ways of arriving at knowledge which depend on free exchange of information and criticism; they also *present unique problems*, precisely because of this necessity of openness of information and responsiveness.

I began this paper with an acknowledgement that there are legitimate and important issues which arise from the tensions between the essential social and political roles of the sciences and the standards internal to democracy and to science. Scientists and supporters of science must therefore reckon with the fact that we don't have control over all information about scientific acitivities which might be relevant to various parties, including genuinely hostile ones; moreover, this information *cannot* be controlled the way that military intelligence is controlled, because of the open critical exchange essential to the sciences themselves. Specifically, scientific standards make it difficult for scientific leaders, or those most interested in maintaining scientific authority, to use the strategies of propaganda. We could read the desire to "restore the old dominant view,"[69] as a *wish* that they could somehow shield the general public (or the university student) from these portraits of science at work, because *none of the usual mechanisms are available to make sure that propaganda goes out and secrets stay home*. They cannot – because of a fundamental, overriding interest in open scientific communication – classify very much as "top secret"; hence, they can't control access to the nitty-gritty details of scientific processes, successes, or failures. In other words, the first strategy listed in section 2, "control," is out; the burden therefore falls to heavy use of the strategies of discrediting and excluding. These are, I shall argue in the rest of the paper, the primary maneuvers that the critics have used in their reactions to the feminist authors they cite.

4. FEMINIST CONTRIBUTIONS TO SCIENCE AND SCIENCE STUDIES

4.1. *Objectivity, Truth, and the Standards of Scientific Communities*

With the foregoing analysis in hand, it becomes much easier to see that it is *not* a reasonable or adequate response to feminist contributors to science, to point out that they are politically motivated. Playing fair – that is, according to standards for scientific conduct – a *scientific* alternative, challenge, criticism, or commentary, must be evaluated and answered scientifically. There has been an enormous amount of confusion about this seemingly obvious point, but for now, you don't have to be a radical philosopher of science to see this; it is sufficient to buy John Herschel's distinction between the context of justification and the context of discovery.[70] Under this quite conservative view of science, the *source* of an alternative hypothesis or a criticism is seen as irrelevant to its scientific merit. This standard and ideal of scientific practice is important for two reasons. First, it highlights the fact that any refusal to consider and respond to feminist scientific contributions embodies a double standard: chemists wondering about the structure of hydrocarbons did *not* dismiss Kekule's benzene-ring structure because it came to him in a dream of a snake swallowing its own tail;[71] indeed, tolerance of a wide variety of explicitly political ideologies (among male scientists, anyway) has been one of the hallmarks and points of pride of the international communities of twentieth century scientists, and rightly so, because it fulfils a *standard of open-mindedness* essential to the practice of science itself. Second – and this is a closely related point – those who must respond to feminist ideas in the sciences do not have to accept *any* feminist views regarding the sources of the issues, problems, or omissions being debated: that is, they don't have to believe any feminist doctrines whatsoever, in order to *address* feminist scientists. Furthermore, they don't have to agree with feminist views to agree with some feminist scientific *claims and conclusions*; one would expect, in fact, to have many results of feminists doing science accepted *as* simply "good science."[72]

A quick glance at developmental geneticist and feminist Anne Fausto-Sterling's aims and methods makes it abundantly clear that she is *not* recommending that "political interests dictate scientific answers," either in the science she criticizes or in that which she promotes. In presenting her analyses of scientific claims regarding human sex differences, Fausto-Sterling advises her readers to apply perfectly ordinary scientific standards: "look at the data, think about the logic of the argument, figure out how the starting questions were framed, and consider alternate interpretations of the data."[73] Fausto-Sterling is signaling precisely that she is interested in enforcing adherence to usual standards of "good science": in her criticisms of specific claims regarding sex differences, she objects to their "gross procedural errors," their "striking errors in logic," and their "inaccurate understanding of biology's role in human development."[74] Similarly, neurobiologist and feminist Ruth Bleier argues that otherwise-good scientists "have shown serious suspensions of critical judgement in interpretations of their own and others' data," that they have ignored the known "complexity and malleability of human developments" and that they have made "unsubstantiated conjectures," not one of which "is known to be descriptive of scientifically verifiable reality as we know it today."[75]

Feminist contributions to and critiques of the sciences have also long been concerned with the structure and dynamics of the self-corrective processes of producing scientific knowledge: the now-standard feminist argument has been that it makes for *better science*, to encourage the training and full participation of informed researchers with a variety of background experiences, preconceptions, and viewpoints, precisely because such inclusion will encourage a wider variety of working hypotheses, as well as more thorough challenge and testing of any given scientific hypothesis or theory that is under consideration.[76] Philosopher of science and feminist Helen Longino, for example, has argued that "scientific method involves equally centrally the subjection of hypotheses and background assumptions to varieties of conceptual criticism and the subjection of data to varieties of evidential criticism."[77] In her explication and endorsement of standards internal to the community of science that lead to its objectivity, Longino writes: "Effective criticism produces change, and a community's practice of inquiry is *objective* to the extent that it facilitates such transformative criticism."[78]

Thinking about "objectivity" in scientific knowledge is very difficult due to the multiplicity of meanings and contexts of the term itself; I have delineated four basic meanings that are in wide current use, as follows.[79] When applied to knowers, "objective" means *detached*, disinterested, unbiased, impersonal, or invested in no particular point of view; in such cases, objectivity is not a property of whatever is known through these methods. Other uses of "objectivity" are more complicated, in that they involve relations between things: when "objective" means *public*, publicly available, observable, or accessible (at least in principle), some relation between reality and knowers is involved. Similarly, when "objective" means *existing independently* or separately from us, it directs us towards some relation between us as knowers and the reality we're trying to gain knowledge of. Finally, there is a current meaning of "objective" as really existing, "Really Real," or the way things really are. This last usage is supposed to apply no matter what the relations are between reality and knowers. I have found that untangling the various meanings of "objectivity" is absolutely essential to interpreting and evaluating the numerous claims involving feminism, objectivity, and knowledge.

Returning to Longino's discussion of the attainment of objectivity in scientific thought, she emphasizes the mechanics of self-correction in her descriptions of how individual variation in scientific opinion "is dampened through critical interactions whose aim is to eliminate the idiosyncratic and transform individual opinion and belief into reliable knowledge."[80] Hence, her emphasis is on a certain sort of detachment, and on public critical agreement. Longino lists four key features necessary to the knowledge-productive capacity of scientific communities: "avenues for the expression and dissemination of criticism; uptake of, or response to, criticism; public standards by reference to which theories, etc. are assessed; and equality of intellectual authority" among qualified practitioners.[81] On this analysis, the objective and self-correcting nature of scientific inquiry, which is counted among its most profound strengths, is actively *reinforced* by feminist participation.[82]

4.2. *Good Science*

We need to investigate briefly an interesting dynamic that is at work when feminist contributions to the sciences which *are* compelling and accepted, are written off as "good science at work" and as having nothing whatsoever to do with feminism.[83] It might be thought, especially in light of the discussion in section 4.1 concerning scientific standards of evaluation and their divorce from the origins of any candidate scientific contribution, that feminist scientists must abandon any claims regarding their own ideological commitments. This may be true with regard to the ultimate evaluation and acceptance of any particular feminist scientific claim; that is, feminist scientists have not been in the business of demanding that their scientific claims be accepted or rejected on purely ideological bases. Quite the contrary, as I reviewed above.

One crucial point is easily lost, though, in the context of feminist insistence on applying rigorous scientific standards, and that is the *vital role* of the *participation* of feminist thinkers in the celebrated self-corrective processes of science. As Fausto-Sterling argues, in her discussion of feminist corrective contributions to medical and behavioral sciences: "These ideas, although they may represent good science, arose in the context of a vast and multiply branched political–cultural movement, that of modern Western feminism. [To apply a purely] good versus bad science analysis is to ignore the important role feminism has played in *forcing* the re-evaluation of inadequate and often oppressive models of women's health and behaviour."[84] Fausto-Sterling then elaborates on the crucial role feminism has, *in fact*, played in the dynamics of supporting corrective scientific challenges: "In the past, legions of highly trained doctors and scientists have failed to see and criticize what is wrong with the biomedical and behavioral models of female behavior. Why? Because ... they had no alternate framework within which to develop new sight. Feminism provided that new vision, allowing many scientists – even those who do not consider themselves political feminists – to move in a new direction."[85]

The feminist scientists and philosophers of science we've considered have argued that the *self-critical and participatory* aspects of scientific inquiry, when combined with empirical evidence and its analyses, are able to produce the *most reliable and objective scientific practices*; furthermore, the full participation of qualified feminist scientists and critics constitutes an essential part of this self-corrective dynamic.

4.3. *Illustration*

Contrary to the accusation that feminist views of science involve sweeping anti-scientific and anti-reason biases, actual feminists' contributions to and critiques of the sciences are projects which address particular assumptions in particular versions of some people's ways of doing science; they frequently argue that, on the standards accepted *by the scientists in question*, the goals of a particular pursuit in science would be better served by rejecting or changing certain assumptions and practices. Consider the following example of feminist science at work.

Primatology

The changes wrought in scientific understanding of primate behavior through feminist interventions would be difficult to overstate.[86] I focus here on a single article written by statistician and primatologist Jeanne Altmann, and published in 1974. In "Observational Study of Behavior: Sampling Methods," Altmann surveyed sampling methods that could be used by researchers in the field, who typically have little or no control over the movements or conditions of the animals they are studying.[87] For decades, primate studies had relied heavily on sampling *ad libitum*, which tended to result in reports of rare or dramatic events, or in emphases on events of particular interest to the observer; such observational results could not, Altmann pointed out, provide the evidence necessary to answering a host of crucial questions, including those involving differences in behavior patterns among individuals and across sub-groups, such as male and female, adult and adolescent. Altmann also provided a devastating methodological criticism of a widely used and supposedly sophisticated sampling method, in which the occurrence (or non-occurrence) of a particular type of event was recorded; one consequence of her statistical critique was an immediate and irrevocable reduction in the scientific import of the work of some of the field's leading researchers.

But Altmann's primary contribution was positive: she articulated the procedures and advantages of a method that she dubbed "focal animal sampling," in which a focal animal or group of animals is followed for a pre-set time period, and all occurrences of a specified action or interaction are recorded. As Donna Haraway points out, "The embarrassing truth was that many of the regularly cited field studies ... both gathered and analyzed data in a way that did not justify the conclusions reached."[88] In essence, *Altmann raised the standards of evidence accepted within the entire community of field primatologists*, and her paper became one of the most cited in the entire modern literature on animal behavior.[89]

Nothing in the foregoing reveals the fact that Altmann herself was acutely dissatisfied with the skewed visions of primate social-structure that arose out of pervasive biases towards focusing on male animals and on dominance interactions, and that her active participation in the modern U.S. feminist movement contributed to her awareness of the significant theoretical implications of these biases.[90] Nevertheless, her revision of sampling practices cleared the way for pursuing a central problem in available theories of primate evolution; specifically, Altmann thought that differential reproductive success – the "motor" of natural selection – was much more significant among females (and much less so among males) than had ever been acknowledged. As leading primatologist Sarah Blaffer Hrdy described the situation twelve years later: "changes in methodology (e.g., focal animal sampling of all individuals in a group) and the emergence of long-term studies played critical roles in revising male-centered models of primate social organization."[91]

Note also that Altmann's feminist contribution does not fall along the lines that some might expect. Altmann's methods enforce – by any account – *an increase in objectivity and precision*, and she makes no mention of her beliefs regarding the sexism in the prevailing sampling methods she so effectively replaced. This illustrates that feminist contributions cannot be identified by their content; rather, feminist scientists have changed those fields in which they have participated through active engagement with particular experimental, analytical, and theoretical problems.

4.4. *Making Enemies*

In stark contrast to their portrayal by the critics, feminists contributing to the sciences and science studies insist – repeatedly, consistently, in detailed and appropriate contexts – on the centrality of truth, evidence, and objectivity; furthermore, some have effected an increase in intellectual and evidential standards within their own scientific or philosophic communities.

I would like to address briefly the nature of "the feminist menace"; *why* have these feminists provoked the grotesqueries of misinterpretation which I document in the next section? Given that the feminist scientists and analysts in question operate within, and appeal to, the explicit goals and standards of their own scientific and intellectual disciplines, they cannot be read as introducing *conflicting bases* on which knowledge claims should be evaluated – in contrast, for example, to the creationists. Some feminist concerns do center around social abuses of scientific authority, but even these do not challenge the fundamental desirability of a scientific basis for knowledge claims, i.e., scientific authority itself. The authoritative mantel of science is supposed to be hard-earned; the science involved must be good science, if appeals to its authority are an important part of the sociopolitical landscape.[92] Hence, social abuse of scientific authority arises when: (1) specific scientific theories and conclusions which are serving social roles do not fulfil standards of good science (i.e., the "scientific authority" is not earned); (2) efforts to defend these "scientific" results utilize tactics intended to exclude and control qualified critics, rather than relying on the standards of scientific inquiry (i.e., the scientific standing is not defended scientifically); and finally, (3) the scientific results in question reinforce and legitimate the social status quo (i.e., the science is used socially to protect those in power). Therefore, the feminist criticism of the social abuse of scientific authority can *only* be seen as threatening to *scientific* authority by *identifying* scientific interests with specific social ones.

It is also essential to understand that there are clear remedies to suspected cases of social abuse; the accusation of abuse of authority can be blunted effectively by *responsive* scientific communities, with an emphasis on self-correction, fair-mindedness, and intellectual responsibility. The canonical "Mistakes were made; scientists are human. We've responded rapidly and effectively to well-grounded objections," *does* involve some demystification and admissions of fallibility, and such "demystification" may reduce, temporarily, the untouchability of scientific authority concerning the matter at hand. The alternatives, however – corruption, dishonesty, and strong-arming – are unlikely to do the authority of science any good in the long run.

Thus, it seems that the threatening aspect of feminist and social analyses of the sciences is the revealing of the interdependence of social and cultural context, scientific practices and products, and the actions of individuals; this is, in fact, clear from the critics' confusion about the exclusivity doctrine. They seem willing to acknowledge that, in *general*, scientists are social, human, beings, but unwilling to admit that the social, cultural context played any significant role in any *particular* case. In spite of the resulting aggravation, I would argue that this resistance serves a vital function: it is a method of maintaining scientific standards within science. In fact, many of the debates in which feminist scientists have engaged have involved efforts to apply critical scientific standards to scientific claims; in other words, they were ordinary contributions to scientific debates. There is, hence, no *necessary* "abuse of authority" accusation involved in feminist participation in ongoing scientific debates. The actual abuse of scientific authority arises only when the social-authority functions of science *overrule* the knowledge functions; this is why demonstrations of responsiveness are the most effective response to accusations of abuse.

Feminist contributions to the sciences, then, are best described, not as challenges to the legitimacy of science, or even as accusations of the abuse of scientific authority, but rather as developments of the *information and analyses necessary for temporary, local, scientific self-correction*. The self-corrective capacities of the sciences serve as the *foundations* of scientific legitimacy and authority. Short-term fallibility and the willingness to admit error are the price to be paid for maintaining both individual authority within the sciences, and the social authority of the sciences in a democratic state. Such willingness to admit error is everywhere a matter of judgement, of course, a careful balancing of whether to keep on, give in, or investigate alternatives. For scientists to admit they were incorrect in the past, other members of the scientific community who share investigative interests must provide the evidence, interpretations, and alternatives, in addition to exerting pressure to meet the relevant scientific standards. Let us now consider some responses to feminist efforts to do just this.

5. IDEOLOGY VS. TRUTH

5.1. *Critics*

Margarita Levin: "*The real threat to feminist ideology*, it turns out, is the scientific method itself, with its *promise of objectivity* no matter who the scientist is"; Levin describes "the first of three fundamental errors that form the basis of the feminist account of science," which is that feminists commit the genetic fallacy, i.e., they "confuse something with its origin and reject it on that basis." She then outlines the feminists' "second basic error, which is their failure to take seriously the fact that so-called masculine science *works*. Science makes predictions that can be and are verified every day."[93] What feminists just don't understand, she claims, is that "the *self-correcting character of the scientific method*, with its emphasis on observation, the replication of experiments, and open discussion, insures that ['"deviations from the ideal of objectivity'"] will eventually be seen as such."[94] Ostensibly against feminist views of science, Levin explains that masculinist metaphors that might have helped formulate scientific theories are "completely irrelevant to the verifiability and accuracy of scientific theories inspired by those metaphors."[95]

Clifford Geertz: "The worry is … that the *autonomy of science, its freedom, vigor, authority, and effectiveness, will be undermined* by the subjection of it to a moral and political program – the social empowerment of women – *external to its purposes*."[96] In considering the long-range prospects for feminist science, Geertz concludes that its development "depends most critically on how the *tension gets resolved between the moral impulses* of feminism, the determination to correct gender-based injustice and secure for women the direction of their lives, *and the knowledge-seeking ones of science*, the no-less-impassioned effort to understand the world as it, free of wishing, 'really is.'"[97]

Paul Gross and Norman Levitt: "Recent feminist theorizing about the sciences … contains heavy doses of dogma," and like "other dogmatisms," it is "beyond the reach of rational argument."[98] Feminists aim for "the dethronement of Western modes of knowledge and their claims to objectivity … [and] champion 'women's ways of knowing' … [and think] that Western paradigms have been effectively demolished."[99] More specifically, "the relativism of cultural constructivist doctrine is the perfect tool for *discounting science as biased or corrupt if and when it inconvenience's one's political program*."[100] "It is a commonplace among relativists of all kinds [including feminists] *to ignore or dismiss the self-correction process by* *which good science survives and bad science* – that which is not verifiable by others of different tastes and tendencies – *vanishes in due course*."[101]

One persistent theme elaborated by critics of feminist contributions to science, such as those sampled above, is that feminists pursue ideology instead of truth; their activities are, therefore, in conflict with the aims of science. There are two readily-available versions of this claim: Feminists formulate and evaluate specific scientific claims by having their ideological commitments *override* standard scientific goals of discovering truth. That is, feminists use ideological criteria *in place of* scientific criteria. A more radical interpretation is that feminist ideology itself involves a *wholesale rejection* of scientific standards such as objectivity, scientific methods, and scientific criteria of evaluation. On this view, feminists are *anti-science*, and are wholly incapable of contributing to or furthering scientific goals or discovering truth.

Another, partly orthogonal, set of claims centers around the theme of science's *self-corrective* capacities and their role in the effectiveness and genuine objectivity of scientific inquiry. One version of this complaint is that feminists simply don't understand that scientific inquiry is self-correcting. A more damaging accusation is that feminists reject or devalue the scientific process of self-correction. Because their critics recognize the essential links between the self-correction of science and its *objectivity*, feminists are seen as either *neglecting or violating* the very core of scientific inquiry – scientific objectivity.

The critics' claims are striking for two reasons: first, their attributions so vividly conflict with what the feminists in question have actually said and done; and second, the misattributions are so clearly aimed at discrediting and excluding these feminists. I am *not* suggesting a conspiracy of any kind; rather, I have argued that the strategies of discrediting and excluding are fairly general ones for dealing with the publication of inside information which is perceived as threatening. In the following sub-section, I detail several of the specific forms these strategies take in the hands of our critics, including: discrediting by claiming that feminists place ideological goals before scientific ones; excluding through omitting or obscuring *scientific* challenges offered by feminist scientists; discrediting by using a definition of "scientific objectivity" intended to disqualify feminists; excluding by painting feminist scientific contributions as essentially and radically hostile to scientific aims; and finally, excluding by obscuring the feminist sources of scientific self-correction, i.e., "good science." I conclude, in section 5.3, with some of the most overt attempts at exclusion.

5.2. *Strategies: The Discrediting and Exclusion of Feminists*

Pursuing Ideology and Pursuing Truth

Feminism might be seen as an endorsement of a specific set of doctrines or dogmas. One fear is that feminists will give top priority to pursuing their political goals – and to protecting the truth of any dogmas that they deem necessary to those political goals – and will reject genuinely *open* inquiry into the scientific strengths and weaknesses of the dogmas themselves. If they do this, they separate themselves from the goals of *scientific* inquiry, which puts open investigation into the truth of any and all empirical claims as its top priority. Because feminists cling to specific dogmatic views – involving the eliminability of certain gender roles, the social aspects of the development of sex differences – they disqualify themselves as participants in open, scientific inquiry.

The portrayal of feminists as dogmatic can be found in Gross and Levitt's discussion of Longino's analysis of the interpenetration of "contextual values" (i.e., norms and values of the social and cultural context), and "constitutive values" (norms and values internal to the sciences), within the scientific investigations of hormonal influences in sex differences. Having quoted Longino's claim that the studies in question "are vulnerable to criticism of their data and their observation methodologies," they launch the following complaints:

> nowhere in the body of Longino's work do we find identified specific, recognizable flaws in the data and the methodologies.... Indeed, the criticisms are not directed toward those at all. Instead, they are either banal (e.g., the argument that data from rodents should not be used to infer processes in people), or indictments of the investigators for making value judgements about departures from sex-stereotypical behaviors.... Led to expect serious criticism of data or methodologies, we find, not cooked data, uncontrolled experiments, or statistical gaffes, but implicit attitudes claimed to have been detected – by a hypersensitive anti-essentialist.[102]

At this point, Gross and Levitt deliver their verdict: "By and large, the logic here is that, since the conclusions are unacceptable by feminist lights, the science must be flawed."[103]

Gross and Levitt accuse Longino of rejecting specific scientific results *because* "What Longino is really after is a way of doing science that will negate *any* possibility of biological determination,"[104] i.e., that her overriding commitment is to a specific scientific conclusion for ideological reasons. Gross and Levitt's conclusion, that "science as-it-is becomes, for such critics, an intolerable constraint, terrible danger,"[105] provides the real leverage used to *discredit* feminist scientific criticisms *as scientific*. The accusation is so common that it has a standard form:[106]

(1) feminists *reject* particular scientific findings and explanations *exclusively because of their political content or implications*;
(2) this is a wrongheaded basis on which to *evaluate* scientific claims;
(3) its use demonstrates that the feminists are *not being scientific* – i.e., are not using scientific standards – either because they don't know how, or because they are motivated to "twist the facts" or "reject the truth" in order to attain their political ends (which is precisely what they criticize others for doing);
(4) Therefore, because the feminist contributors are not being scientific, no scientific *evaluation of or response to* their claims is warranted or merited.

A brief diversion into Gross and Levitt's discussion will lead us to the primary point that is being so insistently buried here. They claim to have sought – in vain – for any substantive scientific content in Longino's discussion of studies on hormonal influences on human sex differences; therefore, they were (reluctantly) forced to conclude that the feminists were being unscientific. Under these circumstances, it becomes rather suspicious, if not deceptive, that they ignore Longino's repeated references to the detailed scientific critiques offered by Bleier and by Fausto-Sterling.[107]

The fact is that Bleier and Fausto-Sterling offer detailed analyses of "the cooked data, uncontrolled experiments, or statistical gaffes" which Gross and Levitt claimed were nowhere to be found.[108] In fact, Gross and Levitt do cite Fausto-Sterling's relevant work once – calling it "her polemical book" – when they accuse her of firmly denying the existence of "significant biological differences between men and women."[109] Their mischaracterization of Fausto-Sterling's book – and their *non-engagement* with her detailed *scientific* objections – play a pivotal role in the soundness of their entire analysis, as does the invisibility of neurobiologist Bleier's work.[110] Evidently, Gross and Levitt feel that they need this genuinely scientific work not to exist; in the presence of undeniably scientific *and feminist* contributions, their blanket dismissal – as I've outlined it in this section – of any possible relevance of feminist contributions will fail.

The use of the standard "purely ideological rejection of science" accusation is quite risky: even the *appearance* of possible scientific objections must be eliminated; otherwise, the authors would be expected *to engage their colleagues in ordinary scientific debate*. This brings us to the fundamental point: feminists have everything to gain from being included in such "ordinary" scientific debate. This has been, in fact, a primary feminist goal. And Gross and Levitt have – inadvertently – shown some awareness of the importance of this goal to feminists, in their determination to resist its satisfaction. Far from advocating an abandonment of the principles of open, critical inquiry, feminist scientists have consistently attempted *to enforce* them; far from being "anti-scientific," respected scientists have aimed to *improve* their sciences through demonstrating and insisting upon the highest standards of scientific evidence and justification.

Contrary to the caricature of dogmatism discussed above, feminist commitments to openmindedness and the value of fair and responsible scientific inquiry have served as the linchpins for feminist contributions to the sciences.

Feminism as Anti-objectivity

As reviewed in section 4.1, appeals to the strengths of *objective methods*, and to the centrality of *objectivity* in our searches for knowledge about reality play pivotal roles in feminist scientific self-understanding. The accusation of dogmatism – with its lack of fidelity to genuinely open and objective inquiry – is sometimes used within the sciences to *disqualify* feminist participants from full membership in the scientific community, in the following way:

(1) Our scientific projects are concerned with objective knowledge and objective reality; hence, by definition, they must involve objective inquiry;
(2) feminists are promoting specific ideologies, or social values; this is what it *means* to be a feminist;
(3) but being "objective" just means being *free* of values or biases, or commitments or ideologies; therefore:

(4) the pursuit of objective scientific inquiry is *incompatible* with simultaneously pursuing a specific ideology or any particular social values.

(In brief: the type of science we are pursuing is objective. Ideology is not. Therefore it is impossible to pursue both at the same time; they're incompatible.)

The most important thing to notice about the standard argument I outlined above is that it utilizes only part of one of the meanings of "objectivity" which has currency today. If any of the other meanings listed in section 4.1 are substituted into the argument, it becomes invalid. Consider a plausible alternate premise: (3*) Being "objective" just means pursuing truths about things that exist completely independently from us.[111] *It simply doesn't follow* that "(4) The pursuit of objective scientific inquiry is *incompatible* with simultaneously pursuing a specific ideology or social values." In order to sustain this conclusion, additional arguments must be provided that pursuing truths about independently-existing parts of the universe is logically, psychologically, or at least statistically incompatible with performing behaviors aimed at achieving a particular social organization or dynamic (i.e., acting as a feminist). One of the challenges facing this further claim is that feminists have already provided arguments and examples in which using feminist analytic tools has *contributed* to the long-term empirical adequacy and objectivity of scientific research (see section 4.3). Consequently, those who wish to exclude feminist analyses *a priori*, because of some apparent conflict between feminism and "objective" scientific method, ought not base their arguments on simplistic, misleading, and inaccurate views about the nature of "objectivity."

Feminism as Anti-scientific

Gross and Levitt have claimed that "strong cultural constructivism," in which science is viewed as "a wholly social product, a mere set of *conventions* generated by social practice,"[112] is a "chief ideological mainstay" of feminist views of science.[113] Remember, Gross and Levitt say that on such an approach: "Scientific questions are decided and scientific controversies resolved in accord with the ideology that controls the society wherein the science is done. Social and political interests dictate scientific 'answers.'"[114]

I have already noted that Gross and Levitt's position has required them to neglect the plain scientific content in the feminist views they dismiss. Before it can be taken seriously, any characterization of feminist contributors to science which postulates some deeply anti-scientific agenda which is accompanied by rejection of the most basic scientific standards, will have to reckon with the scientific competence and scientific attitudes displayed by those feminists.

Feminists vs. Self-correction: Good Science at Work

In a bizarre twist in the plot, the very dynamic of participation and self-correction emphasized by the feminists we've been considering has been used *to dismiss* the significance of feminist participation in the sciences;[115] the issues raised deserve scrutiny.

One claim is that feminist scientists just don't comprehend that the processes, methods, and standards of scientific inquiry make it self-correcting. This accusation is uninteresting because it is patently false, although some might find it amusing that Gross and Levitt, in their untouchable faith that feminists are enemies of science, apparently believe themselves to have trapped in a contradiction a group of feminist scientists who appeal to the importance of "controlling for [gender] bias" in order to improve biological science.[116] Gross and Levitt gleefully lecture: "to 'control for bias' is an ancient house rule of empirical science. ... It is one of the hallmarks of the 'good science' that the postmodernist critics of science [among whom Gross and Levitt include even those feminists who claim to be empiricist] disparage."[117]

But we could imagine a more substantive claim along the following lines: feminists wish to have a free rein to scrutinize the sciences, but wish themselves to be exempted from the critical scrutiny which enables science's self-correction. In fact, Gross and Levitt seem to make this charge when they complain that feminist theorizing has "an unprecedented immunity to the scrutiny and skepticism that are standard for other fields of inquiry."[118] As I have shown earlier, it was Gross and Levitt who *refused to scrutinize* the feminist contributions to science; this is precisely the opposite result the feminist scientists in question hoped for and deserved.

Nevertheless, opponents of feminist approaches to the sciences rely heavily on an appeal to the invisible hand of the marketplace of scientific ideas: when they cannot get away with dismissing feminist scientific thought as "unscientific," or when the authors cannot be discredited as operating outside the sciences, feminist work must be characterized as an inevitable product of science-at-work. This appeal to the properly-functioning scientific community does not sustain the conclusion drawn by opponents of feminist scientists, according to which the feminist origins of accepted scientific contributions are rendered irrelevant.

Gross and Levitt acknowledge that the central issue at stake with feminist approaches to science is "the extent to which the prevalent feminist critique, as agent of methodological or conceptual change, is relevant to [science's] advance."[119] Now, consider the fact that Gross and Levitt help themselves to the following conclusions: "At times, baseless paradigms in medicine and the behavioral sciences have been pretexts for subordinating women. Pseudoscientific doctrines of innate inferiority and moral frailty have been used to discount female capacity for achievement and to confine women to subservient roles. All this is beyond dispute and generally recognized in intellectual circles."[120] We may expect a modest capitulation to the uses of feminist thought – at least in helping to unmask the baselessness of these paradigms and the "pseudo-" nature of these scientific doctrines – but no. Even though Gross and Levitt accept "that in scientific debate and in the process by which a preference for one paradigm over another emerges, attitudes of mind come into play that are in some measure dictated by social, political, ideological, and religious preconceptions," they maintain that it is the self-correcting dynamic of scientific method that does the real work: "Our reading of the history of science suggests ... that theories leaning heavily on such props tend to be fragile and ephemeral [their emphasis], and that part of the increasing *power of scientific methodology* derives from always-increasing *awareness* of the danger that reasoning can be corrupted in this way if one is not careful [my emphasis]."[121] And finally, "We are *not* trying to deny that social interests and nonscientific belief systems often enter into the very human business of doing creative science," but "in the long run logic, empirical evidence, and explanatory parsimony are the masters ... in the house of science."[122] This echoes Levin's claim that "the self-correcting character of the scientific method, with its emphasis on observation, the replication of experiments, and open discussion, insures that ['"deviations from the ideal of objectivity'"] will eventually be seen as such."[123]

The only gesture towards explaining how this process of self-correction actually functions, lies in the claim that "If [scientific results] survive, they do so because they work, for a large number of people of hugely varied backgrounds and interests."[124] Gross, Levitt, and Levin are unequipped to reckon with the details of *how* that self-correction – among people with "varied backgrounds and interests" – *works*: who offered the corrections and "increased the awareness," and why did they do so? How could such inadequate science have had such a long and influential run in the hands of the top scientists at the time?[125]

The *heart* of feminist analyses answers just these questions: including feminists among those with "varied backgrounds and interests" has and will contribute to the self-correction and advance of the sciences. Given that Gross and Levitt have already acknowledged the scientific importance of including different types of scientists, and that they admit that "expanding the pool of scientists will produce more and perhaps better science,"[126] they bear the burden of proof for concluding that *feminist* scientists are a type that ought either to be excluded, or to not count among the *scientifically significant* varied backgrounds and interests.

There is, thus, a striking *logical* problem for Gross, Levitt, and Levin's blanket dismissal of feminist contributions to the sciences, namely, they must launch and defend a rather demanding counter-factual: even if feminist scientists had not been the "correcting" force confronting this science, someone, sooner or later, would have provided such correction. In some cases, this is probably true; there is much common ground between feminist and other scientists' critiques of certain programs and explanations anyway (e.g., anti-reductionism, favoring models with higher complexity of interactions). But what is the significance of the claim that – even without the feminist contributions – the rest of the scientists would have eventually realized that something was wrong? I would remind readers that the critics' portrayal of the dynamics of self-correcting and objective sciences closely resembles the views articulated and advocated earlier by Longino, Fausto-Sterling, Harding, and other feminists. The only point of contention, it seems, lies in the counterfactual life of Gross, Levitt, and Levin, who appear to be making the petulant claim that even if feminists hadn't been there as participants in science's self-correction, everyone would have (eventually) gotten along fine without them. Whether or not this is so is completely irrelevant. Unless the critics can provide *reasons* – other than those that failed, above – that feminists *cannot* contribute to the usual processes of scientific inquiry, then their attempts to exclude feminists from the sciences amount to unvarnished dogmatism.

5.3. *Exclusion at Any Price*

Having evaluated the contradictions between some of the attacks on feminist approaches to science and their targets, we are faced with a mystery: if the feminists attacked by the critics for being anti-science or ignorant of science are actually neither, then what is going on here?

Many of the attacks involve the demonization that is essential to the identification of any group as essentially *hostile*, i.e., they embody the strategy of exclusion. Consider Gross and Levitt's description of feminists' "acute and apocalyptic oppositionism": "The announced goal, upon which feminists of the most disparate schools agree, is a science transformed ... [this is an enormously] ambitious project: to refashion the epistemology of science from the roots up."[127] Similarly, we have Holton's improbable casting of a scientist and a philosopher of science as having an "alliance to a 'science' very different from conventional science", and as working towards "the end of science as we know it."[128] This is all in the face of: Longino's insistence that what she is urging as "feminist science" "*does not demand a radical break* with the science one has learned and practiced. The development of a 'new' science involves a more dialectical evolution with established science than the familiar language of scientific revolution implies;"[129] as well as Sandra Harding's insistence, in the very article quoted by Holton, that she aims to restore "the ability of the sciences to provide objective empirically defensible descriptions and explanations of the regularities and underlying causal tendencies in nature and social relations."[130] Note that the critics have misrepresented the feminist views in a particular fashion, one which attempts to distance the goals of feminists from recognizable and accepted goals of the sciences.

Under these circumstances, one might think it prudent for the critics to address themselves only to feminist analysts of science whose writings don't *directly* contradict their picture of feminist views. Instead, Gross and Levitt declare that "cultural constructivism is the underpinning of *all* these attacks, *even when they are made by self-styled empiricists*."[131] Now recall how Gross and Levitt attempted to show that Longino – in spite of her self-proclaimed attachment of empiricist goals – was *really* a card-carrying anti-scientist: they sifted out the substantive scientific criticism and pretended that it didn't exist; the remaining appeals to empirical standards were brushed off as ordinary or scientific common sense.[132]

Why would intelligent critics like Gross and Levitt, who appear to have made efforts to read the relevant feminist works, place themselves in such a vulnerable position – one in which their frequent declarations of their "honest" evaluation of this literature[133] are so blatantly violated – by their studied avoidance combined with uncredited appropriation, of some of the strongest arguments offered by the feminists they claim to investigate? Gross and Levitt spare us from having to speculate about what they're really after: they want to *exclude* feminists from university and intellectual life, period.[134] Take a close look at this part of Gross and Levitt's description of "feminist criticism of science": they complain that "the best-known critics *are accepted as legitimate historians and philosophers of science, in circles far wider than their feminist peers*."[135] Here, we have the real story. Two of the researchers Gross and Levitt are most concerned to examine are Sandra Harding and Helen Longino, who were both trained as philosophers of science in highly-ranked philosophy departments. Nevertheless, according to Gross and Levitt, we ought not consider historians and philosophers of science who have feminist views to be *peers* of *other* historians and philosophers of science; rather, the peer-group – the colleagues – of these scholars *is to be defined by their political views*. There could hardly be a clearer attempt to exclude participation by fully-qualified professionals on the basis of ideology.

Gross and Levitt also repeatedly characterize feminist contributors to science as "non-scientists," and claim that what they have to say about science is naive and misguided, at best.[136] In addition to the problem, discussed above, that they bury, neglect, or misrepresent the *scientific* nature of feminist contributions, Gross and Levitt put themselves in the awkward position of explicitly dismissing as scientifically incompetent their other two prime feminist targets: Evelyn Fox Keller, who has a PhD in physics from Harvard and has produced laboratory results in molecular biology, and Donna Haraway, whose PhD is in biology from Johns Hopkins.[137] How do the critics justify their apparent belief that holding feminist views actually, or effectively, *subtracts or neutralizes* even very high status scientific credentials?[138] I mention these cases to illustrate the stunningly arbitrary assignments of scientific competence made by Gross, Levitt, and Holton, depending upon political or ideological affiliation. Such re-assignments of scientific stature cannot be defended by any simplistic and unsophisticated appeal to an apparent conflict between holding political goals and the "objectivity" of science.

Finally, Gross and Levitt's book is peppered with phrases which suggest that feminist participation in scientific inquiry involves a category mistake. "Feminist science criticism" is represented as taking place in "women's studies programs," and they consistently refer to feminist writings on the science as "cultural" – rather than "scientific" – criticism. Both of these moves make it much easier for them to reach their conclusion that "the academic left's" *cultural* criticism doesn't make any real difference to practicing scientists.[139] To see the work being done by this misrepresentation, one only needs to consider their claim that: "most scientists are made aware of the academic left's critique only by fragmentary and sporadic contact"; how does this sound, when the truth is that feminists are themselves scientific colleagues in the same departments?

6. THE REAL ENEMIES OF OBJECTIVITY

I think that what motivates these books is fear – most significantly, fear of the loss of the sciences as authoritative resources for a host of social, political, and economic aims. In section 2, I sketched some of the tensions and conflicts that arise organically when the standards and ideals of scientific inquiry must be balanced with the interests of social and political control regarding the dissemination and publicity of information from and about the sciences. I have analyzed several of the most influential recent discussions and critiques that address these issues. While these critiques may enhance awareness of the existence and shape of the tensions which I think are inherent in our democratic-cum-scientific society, I have found that their strategies for dealing with these conflicts are profoundly unsatisfactory on numerous counts.

On the more empirical side, it would seem that appropriate public consideration of their criticisms of social analyses of the sciences would involve careful examination of the relative evidential merits of each side. This would involve the critics in two difficult tasks: doing the work; and then engaging with the scholars they are attacking on their own turf. In some cases, the critics attempt to circumvent this necessity by transforming the appropriate questions and analytic frameworks of the fields of history, philosophy, and social sciences of science, in such a manner as to render them more tractable to the control of select natural scientists. Their other primary strategy involves equating feminist and social discussions of science with the rejection of objectivity, critical thought, and rationality. The fundamental problem with this strategy is that it involves *denying* that feminist contributors to science *share the pursuit of objective scientific knowledge and its rational applications in our world.*

The misrepresentations of feminist analyses, some of which I have documented in this paper, produce a double bind. Feminists have described some of the ways that aspects of the broader culture which involve beliefs about sex and gender have historically shaped, and continue to have uncritical acceptance in, various aspects of scientific activities and products. While they clearly mark their analyses and contributions as *adhering to and advancing the basic principles and standards guiding scientific thought*, they are nevertheless *treated as enemies* of the processes of scientific investigation. For example, one of the consistent themes of feminist discussions of the sciences is the importance of the full inclusion of qualified women – and others perceived as "feminine," such as racial minorities – in the development of scientific knowledge. The mass of arguments and evidence center around the value *to the sciences* of such inclusion. The cruel irony is that when these analyses of exclusion and arguments for the desirability – to the sciences – of inclusion are presented, they are not read as arguments for inclusion in social activities and processes that are *valued by feminists*. Instead, they are seen as *attacks by the enemies* of those pursuing scientific knowledge.

The significance of being labeled as "enemies" can become more visible through the analogy to military intelligence; such a label delineates a person's or group's *relation* to "inside" information. On the military model, the "sides" are clear, and one of the defining characters of the enemy is precisely "one who will use information against us" – this is why the controlling of information (including selective release of disinformation) is one of the cores of military strategy. Perhaps it will be easier to understand the level of alienation experienced by women and by feminists through recognizing that being denied full participation in producing analyses of scientific information – which they have earned the credentials to do – *amounts to being treated like "the enemy."* The urgency and vehemence which are sometimes seen in feminist critiques and analyses of the sciences ought to be understood in just this light; they *reflect* the depth of the hostility which is embodied by exclusion.

In other words, feminists object to being treated like "the enemy" when they are not. They object to having their scientific and scholarly credentials effectively *taken away* solely on account of their ideological interests, especially when they demonstrate, again and again, that they have *not* abandoned the fundamental standards and values of the scientific or philosophical communities in which they have earned full participation. Yet these very objections are also seen as further proof that *they are the enemy*. This is a vicious double bind.

There is an obvious conclusion to be drawn: those promoting the misrepresentations of feminist contributions do not want to see the feminists as friendly rather than hostile, perhaps because then they would have to include them; the goal *is* exclusion. The problem is that this goal is in direct conflict with the explicit ideologies of both scientific investigation and democratic society, each of which counts responsible participation as essential. The inherent conflicts that arise concerning the control and dissemination of information about the sciences can be resolved, pragmatically speaking, in a variety of ways, all of which include some degree of violation of the ideals of fairness, participation, or openness. The necessity of some level of pragmatic compromise, however, cannot *justify the level and scope* of misrepresentation and exclusion that are present in the critiques I have examined. And the reason is not simplemindedly moralistic, as the critics have contended. When social structures that rely for their functioning and success on ideologies of openness, fairness, and responsible participation rely instead – any more than they absolutely must – on secrecy, exclusion, and deception, *they cannot survive, much less thrive*.

The critics I have considered here are right to worry about defending the authority of the sciences against religious fundamentalism and superstition. Against science studies and feminism, however, they themselves stand as the true enemies of rationality, objectivity, and the long-term success of the sciences.

ACKNOWLEDGEMENTS

I am especially grateful to Mathias Frisch, Jim Griesemer, David Hull, Matthew Nichter, Ina Roy, Michael Selgelid, and David Smith, for their attention and criticism. Deepest thanks to Lynn Hankinson Nelson and Jack Nelson for their advice, encouragement, and patience.

Department of Philosophy, UC Berkeley

NOTES AND REFERENCES

[1] *Amazon Odyssey* (New York, NY: Links Books, 1974), p. 131.

[2] Gerald Holton, *Science and Anti-Science* (Cambridge, MA: Harvard University Press, 1993), p. 143.

[3] Holton, pp. 152–154; my emphasis.
[4] Holton, p. 181; my emphasis.
[5] Lewis Wolpert, *The Unnatural Nature of Science: Why science does not make (common) sense* (Cambridge, MA: Harvard University Press, 1992), p. 101; my emphasis.
[6] Wolpert, p. 103; my emphasis.
[7] Wolpert, p. 115; my emphasis.
[8] Wolpert, p. 117; my emphasis.
[9] M. F. Perutz, "The Pioneer Defended"; Review of Gerald L. Geison's *The Private Science of Louis Pasteur*'(Princeton, NJ: Princeton University Press), *New York Review of Books*, **XLII** (20) (December 21, 1995), p. 54; my emphasis.
[10] Perutz, p. 54; my emphasis.
[11] Paul Gross and Norman Levitt, *Higher Superstition: The Academic Left and its Quarrels with Science* (Baltimore, MD: Johns Hopkins UP, 1994), p. 9.
[12] Gross and Levitt, p. 2; my emphasis.
[13] Gross and Levitt, pp. 45–46. That is, strong cultural constructivists "view science as a wholly social product, a mere set of *conventions* generated by social practice" (1994, p. 11, their emphasis).
[14] Gross and Levitt, p. 45; my emphasis. Or, it "affords *no special leverage* among competing versions of the story of the world" (1994, p. 38; my emphasis).
[15] Gross and Levitt, p. 47.
[16] Gross and Levitt, p. 4. Cf. p. 15, on the potential for these authors having a "great and pernicious social effect."
[17] Gross and Levitt, p. 7.
[18] Perhaps precipitated by Darwinism, Freudianism, the atheism of the dialectical materialism of Marx and Engels, or by the rise of industrial capitalism and the urgent need to replace the masses' preoccupation with the afterlife with a preoccupation for worldly goods that would provide the expanding markets essential to capitalism, and so on.
[19] *See* David Hull's important analysis and documentation of a variety of dynamics in scientific inquiry, in David L. Hull, *Science as a Process* (Chicago, IL: University of Chicago Press, 1988).
[20] A particularly slick appropriation of such research can be found in the creationist arguments of law professor Phillip Johnson, who is fond of twisting Stephen Jay Gould's evolutionist, in-house criticisms of parts of evolutionary theory to discredit evolutionary biology as a science. Karl Popper's philosophy of science is also cunningly misrepresented and recruited for the creationist cause.
[21] In describing the *value* of social analyses of science to scientific success, Sandra Harding writes: "we can hold that certain social conditions make it possible for humans to produce more reliable explanations of patterns in nature just as other social conditions make it more difficult to do so," 'Why "Physics" is a Bad Model for Physics,' in *The End of Science? Attack and Defense* (25th Nobel Conference, 1989) (Lanham, MD: University Press of America, 1992), p. 7. Cf. Gross and Levitt's claim: "scientists welcome the sort of 'social' explanation that examines minutely and honestly the intellectual, attitudinal, and … the moral preconditions of culture that encourage and sustain the practice of science" (1994, p. 128).
[22] E.g., Holton, 1993, pp. 114–123, 155–156, 181–184; Gross and Levitt, 1994, p. 110; Wolpert, 1992, Ch. 8.
[23] For example, Gross and Levitt claim that the aim of the "oppositional social critic" is "to demystify science and topple it from its position of reliability and objectivity" (1994, p. 50; see also p. 234). Later, they write of "the emerging dogmas of the left concerning the innate fallibility of Western science" (1994, p. 243). Surely, Gross and Levitt do not want to defend a position wherein the sciences are *infallible*; they elsewhere claim that success in science requires "unremitting self-scrutiny and attention to the possibility of error" (1994, p. 27).
[24] See their section heading, "The Face of the Enemy" (1994, p. 34).
[25] For instance, Barbara Ehrenreich, Stephen Jay Gould, Donna Haraway, Ruth Hubbard, Evelyn Fox Keller, and Richard Lewontin, to mention a few prominent American scientist-activists from the natural sciences.
[26] E.g., a popular bumper-sticker in San Diego: "Creation: God said it, I believe it, and that settles it."

[27] "The central appeal of [science studies] is the pretext is provides to *disparage the natural sciences* – to dismiss their astounding achievements as so much legerdemain on the part of a ruling elite" (Gross and Levitt, 1994, p. 240; my emphasis).

[28] Helga Nowotny and Hilary Rose, ed. (Dordrecht, The Netherlands: Reidel, 1979).

[29] 1994, pp. 245–257.

[30] "What is required [of science studies] is an analysis of, for example, what institutional structures most favour scientific advance, what determines choice of science as a career, how science should best be funded, how interdisciplinary studies can be encouraged" (1992, p. 122). *See* section 4.

[31] "Modern science is seen, by virtually all of its critics, to be both a *powerful instrument of the reigning order* and an ideological guarantor of its *legitimacy*" (Gross and Levitt, 1994, p. 12; my emphasis). Do they think the sciences play important, legitimating, social roles, or not?

[32] Quoted in Wolpert, 1992, p. 170.

[33] Wolpert, p. 178.

[34] Or 'cultural anthropology', 'comparative religion,' etc.; it makes no difference to this discussion what we call these endeavors.

[35] In other words, the natural philosophers have served as the theologians here.

[36] Gross and Levitt, p. 68; my emphasis.

[37] Wolpert, pp. 113–114.

[38] Wolpert, p. 110; my emphasis.

[39] Harding, 1992, p. 19; my emphasis.

[40] Again, these authors *claim* they're not against pursuing the questions asked in history, philosophy, anthropology, or sociology of science; they are only against how these studies are actually done (Gross and Levitt, 1994, p. 69; but see nn. 21 and 23). Their proposals for a proper or more appropriate standard of practice for these studies will be discussed in later sections; at this point, however, the burden of proof is on them to reject the present standards in science studies.

[41] Although it is possible to locate passages in which social explanations do seem to be presented as complete (or as replacement) explanations, even those science researchers most closely linked to this thesis in the early-eighties have since published views which make no such claim. James Griesemer has pointed out that, construed within their context, the more sweeping claims are most obviously read as efforts to carve out disciplinary turf; Wolpert's casual defense of the necessary excesses of discipline-establishment is therefore of great interest: "Of course there was some resistance to the new ideas and the molecular biologists were evangelical in trying to persuade others. They undoubtedly also used rhetoric" (1992, p. 104; James R. Griesemer, comments, public lecture by Paul Gross, November 1995, UC Davis).

[42] One point to note immediately is that the critics' insistence on the exclusivity doctrine raises the stakes for both of the supposed completely-social and completely 'internal' or evidential explanations; in fact, it virtually demands an over-extension of explanatory tools to areas in which they are inappropriate.

[43] Gross and Levitt, p. 58; my emphasis.

[44] Gross and Levitt, p. 58; their emphasis.

[45] Harding, 1992, p. 14. Harding also emphasizes the unsuitability of the training of natural scientists for the task at hand: "Natural scientists are trained in context stripping, while the science of science, like other social sciences, requires training in context seeking" (1992, p. 16).

[46] They conveniently deny the fact that *things* – parts of the world that scientists interact with during their investigations – play essential roles in Latour's 'networks' (1994, p. 59).

[47] Remember that we are considering the *legitimate scope* of pursuing social and/or anthropological analyses of the processes of scientific research and explanation. The decisive problem with the question concerning the 'privileging' of scientists' own explanations is that it changes the subject; *it is irrelevant.* The specific challenge we're considering asks, in its strictest form, whether *any* anthropological or social explanation of human activity ought to be *believed.* The problems are legion: the epistemic nightmare of attributing intentions (and to what?); the metaphysical issues concerning the relative reality of subatomic particles, organic chemicals, organisms, and their psychological states (and the relations among these); and the ontological status of groups and their parts, systems and their components, functions and roles and goals. These are, indeed, some of the central issues in the philosophy of social sciences. But the

actual question being addressed has been lost in this methodological swamp; it is, as posed by its enemies: are anthropological studies and social critiques of science legitimate topics of university study? (See Chapter 9 of Gross and Levitt, 1994, in which they threaten humanists and social scientists with expulsion from universities because the physical scientists won't tolerate them anymore.)

[48] *Leviathan and the Air-Pump: Hobbes, Boyle, and the Experimental Life* (Princeton, NJ: Princeton University Press, 1985).

[49] Shapin and Schaffer, p. 283, quoted in Gross and Levitt, p. 63; my emphasis.

[50] Gross and Levitt, p. 64.

[51] Gross and Levitt, p. 63; surely Gross and Levitt would not want to deny this.

[52] Gross and Levitt, p. 64; my emphasis.

[53] Shapin and Schaffer, p. 344, quoted in Gross and Levitt, p. 65.

[54] Gross and Levitt, p. 65.

[55] Gross and Levitt, p. 67; their emphasis. One hotly debated question at the time concerned the proper *role* of mathematics, scientifically; the issue was especially pressing, given the prominent place that Descartes had given mathematics in the definition of knowledge itself, contrasted with the deficiencies of his physics. Gross and Levitt ignore this.

[56] Gross and Levitt, p. 67.

[57] Gross and Levitt, p. 68.

[58] Gross and Levitt, p. 65; my emphasis.

[59] Gross and Levitt, p. 234.

[60] Gross and Levitt, p. 49.

[61] Gross and Levitt, p. 49; my emphasis. Given *this* view of appropriate explanation and evidence, we must wonder about the ingenuousness of Gross and Levitt's disclaimer that "working scientists are not entitled to special immunity from the scrutiny of social science" (1994, p. 42).

[62] Gross and Levitt, p. 49; their emphasis.

[63] Wolpert, p. 116; my emphasis. Andrew Pickering, *Constructing Quarks: a sociological history of particle physics* (Edinburgh, UK: Edinburgh University Press, 1984).

[64] Wolpert, p. 116; my emphasis.

[65] "We are accusing a powerful faction in modern academic life of intellectual dereliction. This accusation has nothing to do with political correctness or 'subversion'; it has to do, rather, with the craft of scholarship" (Gross and Levitt, p. 239).

[66] See Gross and Levitt's appeal to *scientists' right to judge, as experts*, all work concerning "scientific methodology, history of science, or the very legitimacy of science" (1994, p. 255).

[67] The majority of adult Americans receive their information about the world from TV news, with radio news running second. Among the science scandals aired on national network news within the past 24 months, I would mention: the manufacture of data for the Pittsburgh Breast Cancer study; the reinstatements of eggs into the recommended anti-cholesterol diet; the well-publicized omission of women from nearly all of the most extensive and expensive heart disease studies, which led to a special initia tive by Congress; and earlier, the Dalkon Shield devastation; the fanciful claim by President Ronald Reagan that there is no evidence that radiation causes cancer (see Philip Fradkin, *Fallout* (Tuscon, AZ: University of Arizona Press, 1989)); or the revelations of the horror of radiation experiments done on unsuspecting civilians from the 1940s through the 1970s. The public perceptions of some of the scientists involved in these events is far from the genius with special insight into nature, and closer to Drs Frankenstein or Mengele.

[68] The fact that 'executive deniability' has been an essential part of CIA operations *policy* since its inception is well-documented; see John Ranelagh's sympathetic history, *The Agency: The Rise and Decline of the CIA*, 1986.

[69] Perutz, 1995, p. 54.

[70] John F. W. Herschel, *A Preliminary Discourse on the Study of Natural Philosophy* (London: Longman, Rees, Orme, Brown, and Green, 1831). I must note that this distinction has come under sustained criticism within philosophy of science, especially by feminists. My focus here, however, is on the most conservative views of science held by working scientists. The point is that even under these views, objections to the feminist *source* of specific scientific contributions violates the canons of scientific conduct.

[71] Friedrich A. Kekule, 'Origin of the Benzene and Structural Theory,' *Chemistry*, **38** (1965): 9.

[72] See sections 4.2 and 5.2 for elaboration. For the most recent work on why sexist science is not properly characterized as 'bad' science, see *Synthese*, **104** (September 1995).

[73] Anne Fausto-Sterling, *Myths of Gender: Biological Theories about Women and Men*, (New York: Basic Books, 1985), p. 10.

[74] Fausto-Sterling, pp. 8, 60.

[75] Ruth Bleier, 'Sex Differences Research: Science or Belief?' Ruth Bleier, ed., *Feminist Approaches to Science*, (New York, NY: Pergamon, 1986), p. 149. Also, Ruth Bleier, *Science and Gender: A Critique of Biology and Its Theories on Women* (New York: Pergamon Press, 1984).

[76] See esp. Helen Longino, 'The Essential Tensions – Phase Two: Feminist, Philosophical, and Social Studies of Science,' *A Mind of One's Own: Feminist Essays on Reason and Objectivity*, ed. Louise Antony and Charlotte Witt (Boulder, CO: Westview Press, 1993), 257–272; Longino, *Science as Social Knowledge* (Princeton, NJ: Princeton University Press, 1990); John Dupre, *The Disorder of Things* (Cambridge, MA: Harvard University Press, 1993); Sandra Harding, *Whose Science? Whose Knowledge?* (Ithaca, NY: Cornell University Press, 1991); Harding, 'Rethinking Standpoint Epistemology: What is Strong Objectivity?,' *Feminist Epistemologies*, ed. Linda Alcoff and Elizabeth Potter (New York, NY: Routledge, 1993), 49–82; Harding, '"Strong Objectivity": A Response to the New Objectivity Question,' *Synthese*, **104.3** (1995), 331–349; Longino, 'Gender, Politics, and the Theoretical Virtues,' *Synthese*, **104.3** (1995), 383–397; Lynn Hankinson Nelson, *Who Knows: From Quine to a Feminist Empiricism* (Philadelphia, PA: Temple University Press, 1990); 'Epistemological Communities,' *Feminist Epistemologies*, ed. Linda Alcoff and Elizabeth Potter (New York, NY: Routledge, 1993), 121–159; Nancy Tuana, 'The Values of Science: Empiricism from a Feminist Perspective,' *Synthese*, **104.3** (1995): 441–461; Tuana, ed., *Feminism and Science* (Bloomington, IN: Indiana University Press, 1989); Alison Wylie, 'Methodological Essentialism: Comments on philosophy, sex and feminism,' *Atlantis*, **13.2** (1988), 11–14. Cf. Paul Feyerabend, *Against Method* (London, UK: New Left Bookstore, 1975); John Stuart Mill, *On Liberty*.

[77] Longino, 1993, p. 266.

[78] 1993, p. 266; my emphasis.

[79] Treated at length in E. Lloyd, 'Objectivity and the Double Standard for Feminist Epistemologies,' *Synthese*, **104** (September 1995), 351–381.

[80] 1993, p. 265. *See also* Longino, *Science as Social Knowledge*, 1990, esp. Chapters 4 and 9.

[81] 1995, p. 384. Cf. Wolpert, on science's "rigorous set of unstated norms for acceptable behaviour": "Included in these norms are the ideas that science is public knowledge, freely available to all; that there are no privileged sources of scientific knowledge – ideas in science must be judged on their intrinsic merits; and that scientists should take nothing on trust, in the sense that scientific knowledge should be constantly scrutinized" (1992, p. 88). Like Longino, Wolpert emphasizes the community-level process over the individual traits of researchers: "leaving aside the question of whether scientists are more objective, rational, logical and so forth, scientists have developed a procedure in which there are free discussion, accepted standards of behaviour and a means of ensuring that truth will, in the long run, win. Truth will win in the sense that open discussion and observing nature *constitute the best way of making progress*" (1992, pp. 122–123; my emphasis).

[82] The critics' evasion of this core component of feminist views is discussed in section 5.2.

[83] Gross and Levitt assert: "there are as yet no examples...of scientific knowledge informed, reformed, enhanced by feminism" (1994, p. 112). Their strategies for dealing with the numerous feminist contributions to the sciences they subsequently cite are instructive: briefly put, if feminist work is persuasive and is accepted as correct, it's simply good science; if not, it's bad science tainted by ideology. In other words, the feminist contributions to science are either *not feminist* or *not contributions*.

[84] Fausto-Sterling, p. 213; her emphasis.

[85] Fausto-Sterling, p. 213.

[86] See Donna Haraway, *Primate Visions: Gender, Race, and Nature in the World of Modern Science*, (New York, NY: Routledge, 1989), for a comprehensive bibliography and analysis. I have borrowed from Haraway's discussion of J. Altmann in presenting this case.

[87] Jeanne Altmann, 'Observational Study of Behavior: Sampling Methods,' *Behaviour*, **49**, 227–267.

[88] Haraway, p. 307.

[89] Cf. Wolpert, in discussing the *scientific importance* of a contribution: "As the mathematician David Hilbert once expressed it, the importance of a scientific work can be measured by the number of previous publications it makes it superfluous to read" (1992, p. 86).

[90] As Haraway documented and analyzed, *Primate Visions*, 304–310; also, personal communication with Jeanne Altmann.

[91] 'Empathy, Polyandry, and the Myth of the Coy Female,' *Feminist Approaches to Science*, Ruth Bleier, ed. (New York, NY: Pergamon), 135–136; cf: Linda Fedigan, *Primate Paradigms* (Montreal, Can: Eden, 1982); Shirely Strum, *Almost Human: A Journey into the World of Baboons* (New York, NY: Random House, 1987).

[92] Wolpert makes this very point, when he identifies the sole social responsibilities that scientists must take on: "Scientists have an obligation to make the reliability of their views in ... sensitive social areas clear to the point of overcautiousness. And the public should, wherever possible, demand the evidence and critically evaluate it" (1992, p. 163).

[93] Margarita Levin, 'Caring New Science: Feminism and Science,' *American Scholar*, **57** (Winter 1988), 100; my emphasis.

[94] Levin, p. 100; her emphasis.

[95] Levin, p. 104; my emphasis.

[96] Clifford Geertz, 'A Lab of One's Own,' *NY Review of Books*, **37** (8 November 1990), 19; my emphasis.

[97] Geertz, p. 23; my emphasis.

[98] Gross and Levitt, p. 111.

[99] Gross and Levitt, p. 38.

[100] Gross and Levitt, p. 162; my emphasis.

[101] Gross and Levitt, p. 123; my emphasis.

[102] Gross and Levitt, pp. 145–146.

[103] Gross and Levitt, p. 146.

[104] Gross and Levitt, p. 147; their emphasis.

[105] Gross and Levitt, p. 147.

[106] Other instances can be found in: Levin, 1988; Michael Ruse, *Is Science Sexist? And Other Problems in the Biomedical Sciences*, (Dordrecht: Reidel, 1981); and John R. Searle, 'Rationality and Realism: What is at Stake?,' *Daedalus*, **122.4** (1993), 55–84.

[107] Longino, 1990, pp. 119, 127, 131, 134; Helen Longino and Ruth Doell, 'Body, Bias and Behavior: A Comparative Analysis of Reasoning in Two Areas of Biological Science,' *Signs*, **9** (1983), 206–227.

[108] (See block quote, above). Bleier, 1984, *Science and Gender*; Fausto-Sterling, 1985, 133–141. The work of Harvard biologist and feminist Ruth Hubbard is also very important, especially: Ruth Hubbard, Mary Sue Henifen, and Barbara Fried, eds, *Biological Woman, the Convenient Myth: A Collection of Feminist Essays and a Comprehensive Bibliography* (Cambridge, MA: Schenkman, 1982); Hubbard, *The Politics of Women's Biology* (New Brunswick, NJ: Rutgers University Press, 1990).

[109] Gross and Levitt, p. 125. Since Fausto-Sterling acknowledges the existence of biological differences between males and females throughout her book, it remains mysterious how Gross and Levitt could defend this statement, unless they put all the weight for its truth on whatever they mean by "significant".

[110] In spite of Bleier's high scientific status as a research scientist in the specialty in question at a major research institution (the University of Wisconsin, Madison), the book in which her sceintific objections to this research are summarized and defended, *Science and Gender*, is not listed in Gross and Levitt's bibliography. Nor do they ever mention, in parallel to their treatment of Ruth Doell (Longino's early co-author and a biologist), and Fausto-Sterling (a senior research scientist at Brown), that Bleier is even a scientist.

[111] Gross and Levitt appeal to *this* meaning of 'objective' when discussing "the objective nature of the phenomena" of physics, while depending on the 'detached' view when criticizing Keller (1994, pp. 128, 141). Wolpert, on the other hand, *rejects* the 'detached' view as desirable for science (1992, p. 92).

[112] Gross and Levitt, p. 11, their emphasis.

[113] Gross and Levitt, p. 47.

[114] Gross and Levitt, p. 46.

[115] See, e.g., the passages by Levin, and by Gross and Levitt, quoted in section 5.1.

[116] Gross and Levitt, p. 122, quoting the Biology and Gender Study Group, 'The Importance of Feminist Critique for Contemporary Cell Biology,' *Hypatia*, **3** (1): 61–76 (Spring 1988). Reprinted in Nancy Tuana, ed., *Feminism and Science* (Bloomington, IN: Indiana University Press, 1989).
[117] Gross and Levitt, p. 274; also Levin, p. 100.
[118] Gross and Levitt, p. 110.
[119] Gross and Levitt, p. 112.
[120] Gross and Levitt, p. 110.
[121] Gross and Levitt, p. 44.
[122] Gross and Levitt, p. 56; their emphasis.
[123] Levin, p. 100.
[124] Gross and Levitt, p. 112.
[125] My analysis of the inconsistencies and grave evidential problems in recent evolutionary theorizing about women's orgasm has been met repeatedly with the response that it is 'simply good science'; this reaction fails to engage the problem I address, namely, *why* it took decades for these able scientists to become *aware* that the evidence they cited undermined their own explanations. Elisabeth A. Lloyd, 'Pre-theoretical Assumptions in Evolutionary Explanations of Female Sexuality,' *Philosophical Studies*, **69** (1993), 139–153.
[126] Gross and Levitt, p. 131. They also admit that women's scientific "contributions have often in the past been undervalued" (1994, p. 123). How do they account for the fact that this happened in the first place? How do they account for the fact that it has, according to them, changed?
[127] Gross and Levitt, pp. 32, 108.
[128] Holton, pp. 152, 143. The targets are Evelyn Fox Keller and Sandra Harding, respectively.
[129] 1990, p. 193, my emphasis.
[130] Harding, 1992, p. 1. She also states, "it's a very conservative notion of objectivity that I'm ... proposing here ... there are important aspects of the traditional notion of [scientific] objectivity which need not be challenged in order to accomplish the goals that I have in mind" (p. 20).
[131] Gross and Levitt, p. 109; my emphasis.
[132] They make parallel moves vs. Harding (1994, p. 249).
[133] 1994, pp. 113, 128, 241, 256. Their editors should have advised Gross and Levitt that their avowal, "we are not dishonest," cannot help but recall Nixon's, "I am not a crook" (1994, p. 257).
[134] They argue, for example, that the high academic standing of these feminist researches *itself* "raises serious questions about the presumed intellectual meritocracy of the academy" (Gross and Levitt, p. 235).
[135] Gross and Levitt, p. 108, my emphasis.
[136] Gross and Levitt, pp. 108, 122, 127, 159, 235, 236, 251.
[137] While Holton clearly is referring to Keller's work, he perversely refuses to name her or to cite any of her books or articles (1993, p. 154).
[138] Gross and Levitt's inclusion of Haraway and Keller among the four chief representatives for feminist views of science belies their earlier aside that "a handful of figures with scientific credentials, as well as the occasional refugee from an unsatisfactory scientific career, can be found on the movement's *fringes*" (p. 14; p. 6, my emphasis).
[139] "Science will not, in any serious way, be influenced, deflected, restricted, or even inconvenienced by these critics and those they influence" (1994, p. 236; see pp. 3, 11, 112, 253–256).

PART IV

VIEWS FROM MULTICULTURAL AND GLOBAL FEMINISMS, AND FROM FEMINIST PHENOMENOLOGY

SANDRA HARDING

MULTICULTURAL AND GLOBAL FEMINIST PHILOSOPHIES OF SCIENCE: RESOURCES AND CHALLENGES

1. INTRODUCTION: THE CONTENT OF MODERN SCIENCE IS HISTORICAL

Many of the issues still centered in much contemporary Anglo-American philosophy of science have already appeared problematic from the perspective of the last three decades of the social studies of science. These histories, sociologies and ethnographies of modern sciences have revealed how surprisingly deeply immersed in their historical eras have been the cognitive, technical cores of modern sciences. Feminist science studies have played an important role in this history, showing how in the midst of more general social struggles commitments to symbolic and structural gender hierarchies have helped to shape the content of modern science – its models, metaphors and narratives, its choices of topics, its favored ways of producing knowledge, its standards for what does and does not count as legitimated methods and knowledge claims. Such accounts do not support the usefulness of preoccupations with purportedly universally valid, transcultural concepts and principles, transcultural "logics" of inquiry and of explanation that have remained centered in much mainstream Anglo-American philosophy of science.

Of course everyone always knew that much of the "production process" of science was thoroughly social, historical and local. The motivations of scientists, their funders and sponsors, the distinctive historical trajectories of scientific theories, the historically changing structures and meanings of the institutions of science, the metaphors used to explain scientific claims to non-scientists, the technologies, applications, uses and abuses of the sciences – these have already been well understood as thoroughly social, historically local phenomena.[1] But post-Kuhnian science studies (to use a not-quite-accurate but convenient short-hand) show that modern sciences are local in far deeper ways that undermine the very possibility of identifying which elements of a science are social and which are not. Apparently even the most abstract levels of the content of modern sciences are part of the distinctively local social formations. It turns out that scientific instrumentation has epistemological consequences; sciences' socially distinctive metaphors, models and narratives direct the growth of scientific knowledge and infuse its most abstract claims; historically-local interests make some aspects of nature objects of knowledge and distort our understandings of others, and so forth. Of course this is not to deny that nature plays a crucial role in generating scientific claims. While no scientific claims may be trans-historically, uniquely *congruent* with nature's regularities, only certain ranges of them are *consistent* with such regularities, and thus can generate reliable predictions (cf., e.g., Hayles, 1993).

L. H. Nelson and J. Nelson (eds), Feminism, Science, and the Philosophy of Science, 263–287.

When even the purportedly distinctively trans-historical methods of modern science, their mathematical tools, and their standards of truth have been shown by some of the most distinguished scholars in these fields to be historically local to their very cores (e.g., Restivo, 1992; Schuster and Yeo, 1986; Shapin, 1994; Shapin and Schaffer, 1995), philosophers have had to begin to rethink the conventional assumptions of philosophies of science. A number have made important contributions to the post-positivist philosophies and social studies of science (e.g., Cartwright, 1983, 1989; Dupre, 1993; Hacking, 1983; Rouse, 1987).

Feminist histories, sociologies and ethnographies of modern sciences have contributed to such analyses and shared the emerging understanding that there appears to be little left of philosophers' of science's conventional preoccupations that can today, in light of these historical and social studies, safely be assumed to be universally valid, trans-historically objective, rational in the sense of unlocatable on maps of the emotions, and the like. These researchers and scholars have added gender relations, both structural and symbolic, to the dimensions of the social formations within which sciences, too, are lodged. Perhaps in the future it will be possible again to justify some kind of philosophic focus on trans-historical features of science; but for those familiar with these three decades of research, it is hard to have confidence about such a possibility. Thus feminist philosophers of science have had to face the problem of generating standards for "good science" that do not depend upon ahistorical assumptions about sciences' histories, cultures and practices. They have begun to develop notions of such conventional philosophic concerns as good evidence, maximizing objectivity, rational argumentation, the possible relations "between science and society", for example, that do not depend on a-historical assumptions about modern sciences' universality (e.g., Barad, this volume; Gilbert, 1995; Haraway, 1991; Harding 1986, 1991, 1996; Jaggar, 1989; Keller, 1984; Longino, 1990).

In the sections that follow, I want to expand these concerns of the last three decades of mainstream and feminist concerns with the social embeddedness of sciences by asking what role the newly emerging multicultural and global feminisms might play in developing feminist philosophies of science. Even readers ready to think about the specifically feminist contributions to philosophy of science might wonder what the relevance to them could be of these multicultural and global feminisms, whose contributions have heretofore been more visible in such fields as literature, history and cultural studies. However, there exists an important literature that can lead us to answers to such questions – provide a bridge to them – as well as being of interest to philosophers of science in its own right. That is the emerging field of postcolonial science studies. Post-Kuhnian philosophies and social studies of science, including most feminist science studies, have not yet turned to make use of the kinds of maps of global histories that are beginning to become the standard histories in K-12 history classrooms and in new popular understandings of, for example, what I learned as "Columbus' Discovery of America". To sharply pose one issue here: what did the two great events of modernity – European expansion and the emergence of modern science in Europe – have to do with each other? How do the answers to this question emerging from postcolonial science studies expand and reinforce the arguments of post-Kuhnian social studies and post-positivist philosophies of science, and how do they challenge them? And what are the locations of feminist issues in such new accounts? In short, I want here to set the scene for asking what would be central issues for multicultural and global feminist philosophies of science.

"Set the scene" for asking such a question is about all I can do in the space available here. Obviously these questions are very large ones. My goal is a much more modest one, namely to outline briefly the concerns of postcolonial science studies, and then of multicultural and global feminisms and the issues they have already begun to raise about the natural sciences. Finally, in conclusion, I point to what might be fruitful philosophic issues for multicultural and global feminisms in which we all, men and women, from Europe, the US and elsewhere, play significant roles – whether or not we are aware of them.

2. POSTCOLONIAL SCIENCE STUDIES

Postcolonial science studies have arisen at the convergence of several literatures that in the US only occasionally have appeared aware of each others' concerns and resources. Historians and ethnographers of non-Western cultures have been providing studies of the science and technology traditions of Chinese, Arabic, Indian, Pacific islander, Andean and other cultures, including those referred to today as indigenous or local knowledge systems.[2] Another group, consisting of historians, geographers and political economists looking at European cultures, has been writing anti-Eurocentric accounts of European expansion and its effects on the societies into which Europe expanded from roughly the fifteenth through nineteenth centuries. A third group of science and technology policy analysts, largely separate from the first two, has examined the nature and effects of post World War II Northern development policies and practices on and in the so-called "under-developed" countries.[3]

As these fields begin to converge, the concerns of each are increasingly recognized to have implications for the concerns of the others. Scholars from the developing societies have had an easier time seeing the relations between the three, initially conceptually separated, kinds of studies, and have been more motivated to do so. Their societies have borne a disproportionate share of the costs of European expansion, including its scientific and technological components, from the origins of modern European sciences through today's militarism and science and technology "development" practices. Scholars from at least some European societies have also had easier entrance into such studies because of the distinctive patterns of their colonial pasts. Colonialism insured that their histories of national sciences (French science, Dutch science, British science) are also histories of European expansion, "Third World development", and the interactions of these processes with the empirical knowledge traditions of the cultures in their colonies. This colonial history of British, French, Dutch and other such national sciences has left rich "paper trails" in national archives, providing just the resources to tempt today's history of science graduate students in these countries.

Three concerns of Multicultural and Global Feminisms can gain resources from central focuses in such postcolonial science literatures. First, these accounts explore the strengths and limitations of different non-Western cultures' scientific and technological traditions. Second, they look at the historically changing relations between European and non-European scientific and technological traditions in their historical contexts. Third, they chart the distinctively European features of European sciences and technologies.

Non-European science and technology traditions have been far more accomplished both before and after 1492, to use one important historical marker, than standard Eurocentric histories reveal. Indeed, modern European/American scientific and technological development has borrowed extensively from these traditions; often these other scientific traditions have been way ahead of European ones. Obviously this is so with respect to the regularities and causal determinants of parts of nature with which Europeans were unfamiliar – non-European climates, land and sea formations, flora, fauna, diseases, peoples, their languages and cultures. Moreover, postcolonial analysts note that Europeans were exposed to, but saw no need to borrow, such subsequently important elements of modern science as the sophisticated mathematical concepts developed in India and Islamic cultures (Goonatilake, 1984; Joseph, 1991).

Such contingencies of cultures' locations in heterogeneous nature combined with their culturally distinctive interests, discursive traditions, and ways of organizing the production of knowledge to produce robust and still valuable sciences, the practical and discursive strengths of which have been made invisible in conventional Eurocentric accounts. These four categories of "localness" of knowledge – locations in nature, interests, discursive traditions, and ways of organizing knowledge- production – are just what make any science and technology tradition able to produce both systematic knowledge and, necessarily systemic ignorance about nature and social relations. These culturally distinctive features make sciences able to "see" nature in ever new ways, a theme that appears in recent Northern science studies, also (cf. Van Fraassen and Sigman, 1993; Harding, forthcoming b).

The claim that these sciences have been far more accomplished may seem strange, since for many Northerners the phrase "non-Western sciences" appears as a misnomer, conjuring up images of witchcraft and prayers before such events as Spring planting and travels into unexplored territories. Nevertheless, the sciences and technologies reported in the accounts indicated have in many respects proven highly effective at prediction and control of the parts of nature with which these societies regularly interacted. In many respects they were more advanced than European sciences until the seventeenth century and, in some cases, until well into the nineteenth century. Moreover, postcolonial theorists point to ways in which they are still more advanced. Some know how to maintain kinds of fragile environments with which Northern sciences have little experience. Others have developed local pharmacologies that Northern pharmaceutical companies are now busy appropriating. Other kinds of health practices that have not fit into Northern biomedical assumptions are only recently becoming appreciated in the North – for example, pain management through acupuncture, and the concept of bodies as in or out of balance rather than as healthy or ill (cf. Kaptchuk, 1983). Many have learned to interact effectively with their natural environments by negotiating between conflicting knowledge systems, such as their local one and European biomedicine, pharmacology, or environmental sciences. In some ways they are no more local than are Northern sciences, even though the latter are immensely more powerful and effective in some respects. It is such considerations that make it worthwhile simply to bracket for the moment the familiar contrast between universal modern sciences and these local knowledge systems in order to see what the postcolonial authors can illuminate for us by ignoring the way this distinction is usually drawn.

A second theme of postcolonial science studies provides resources for the Multicultural and Global Feminist concern with gender relations in changing political economies. It traces the relationship between these science traditions of other societies, the emergence of modern sciences and technologies in Europe, and European expansion – or "Science and Empires" as the title of one recent collection names the topic (Petitjean *et al.*, 1992). Most of us learned isolated "multi-stream" histories of Europe, the Americas, and possibly something about Africa and Asia. The history of science was usually contained within the history of Europe – at least until the US achieved membership in the European "international" scientific community post-World War II. However, postcolonial science studies locates its accounts in the now over 30-year tradition of "single-stream histories" that tell the history of Europeans as part of the history of the peoples of the Americas, Africa, Asia and the rest of the world, and vice versa. Such accounts are anti-Eurocentric in a second way, for they do not restrict their perspective to the way such processes tend to appear in familiar European histories. They start off from the lives of the peoples Europeans encountered, from their histories prior to the arrival of the Europeans on their shores, and in doing so are able to provide a more balanced, less Eurocentric account of the encounters and interminglings of peoples throughout human history (cf., e.g., Blaut, 1993; Brockway, 1979).

According to this kind of account, just which aspects of nature European sciences describe and explain, and how they are described and explained, have been selected in part by the purposes of European expansion. Of course these are not the only purposes shaping these sciences, but they are significant ones. The problems that have gotten to count as scientific ones in the modern North are disproportionately ones that expansionist Europe needed solved. One historian points out that during the British occupation of India, in effect "India was added as a laboratory to the edifice of modern science" (Kochhar, 1992–93, 694). The point is generalized throughout these writings: the world was added as a laboratory to modern science through European expansion. This process continues today through the science and technology components of "development" that are controlled by the World Bank, the International Monetary Fund, elites within the cultures of the North, and the elite groups in the South allied with them. Most people directing such processes certainly have not intended to benefit the already over-advantaged at the expense of the already most economically vulnerable. Moreover, it is certainly not the case that everything done by these institutions and cultures has had bad effects on the peoples Europeans encountered, though clearly this has happened in many cases. Rather the point is that the great majority of peoples who bear most of the costs of the science and technology decisions made in these institutions have not had a proportionate share in making them.[4] With respect to Northern expansion, the picture of nature produced by solving the expansionist North's problems ignores or hides those aspects of nature that are assumed to be irrelevant to success at expansion. Thus culturally distinctive patterns of both systematic knowledge and systematic ignorance are easily detected from the perspective of cultures with different purposes.

Postcolonial science studies point out, for example, how modern scientific and technological development answered questions about how to improve European land and sea travel; mine newly needed ores; identify the economically useful minerals, plants and animals of the other parts of the world; manufacture and farm for the benefit of Europeans living in Europe, the Americas, Africa and India; improve their health and occasionally that of the workers who produced profit for them; protect settlers in the colonies from settlers of other nationalities; gain access to the labor of the indigenous residents, and do all this to benefit only local European citizens – for instance, the Spanish vs. the Portuguese, French or British. As one historian puts the point, modern sciences and technologies have been crucial in enabling Europeans to monopolize "energy", in the form of precious metals, plants, animals, fertile land and human labor, for the benefit of elites in Europe (Brockway, 1979). These sciences have not been concerned to explain how the consequences of interventions in nature for the benefit of Europeans would change the natural resources available to the cultures they encountered, or what the other social, cultural, psychic, environmental, economic and political costs of such interventions might be. They have not been concerned to explain how to eradicate diseases or other causes of poor health that do not much affect peoples of European descent, and, especially, the already advantaged within this group, or how to sustain fragile environments and resources.

"Development" has from its beginnings been defined as transferring Northern models of industrialization, with their central reliance on modern sciences and their technologies to the South (see earlier citations). Already by the 1960s the perception had spread in the development agencies that while transferring to Southern societies Northern scientific applications, technologies, and forms of industrialization certainly has had some significant successes, often this has not occurred. When it has occurred, the poor – who constitute the vast majority of members of these societies – have not benefited. Moreover, in either case, such processes have continued to accumulate control of the world's resources and peoples in the North. Just one obvious example is immediately apparent on university campuses in the North: consider the scientific "brain drains" so visible in the faculty and student composition of Northern mathematics, science and engineering departments.

Consequently, critical questions multiply about who bears the costs and who benefits from scientific and technological change. After all, even physics, supposedly the "purest" and most value-neutral of sciences, evidently has been far more shaped by its pursuit of militarily useful knowledge than is generally recognized (Forman, 1987).

A third theme in postcolonial science studies has been to explore the distinctively Northern characteristics of modern sciences. This work provides resources for the Multicultural and Global Feminist focus on androcentric aspects of Eurocentricism. Conventional Northern accounts assume that modern science is trans-cultural, that it bears no distinctive cultural fingerprints, and that it is precisely the lack of such cultural presuppositions, values and interests that is responsible for its ability to predict and control nature at places far removed in time and space from the sites of original observation. But the postcolonial analyses point out feature after feature of the constitutive cognitive cores of modern sciences that are culturally specific – culturally "local". Here I simply mention three that have been discussed elsewhere (cf. Harding, 1994, and forthcoming b).

First, as argued above, the pattern of systematic knowledge and systematic ignorance characteristic of modern sciences represents what the dominant Northern groups have and have not wanted to know about nature's regularities and their underlying causal determinants. Even the choice of "basic research" topics is by no means immune from cultural presuppositions about on which aspects of nature it is most important to have basic research. Moreover, many models of nature and method that constitute and direct modern scientific research are culturally distinctive – they would not have appeared reasonable or illuminating in other cultures. Mechanistic models, the notion of "laws of nature", and experimental method are just three of such models that have been examined by historians (see Adas, 1989; Merchant, 1980; Needham, 1969; Shapin and Schaffer, 1985). This local "European" feature of Northern sciences was prefigured in the discussion above about how non-European cultures have come to know aspects of nature to which they were led by their specific locations in nature, cultural interests, discursive resources and ways of organizing the production of knowledge. Both European and non-European science and technology traditions make use of these kinds of local resources that advance some and retard other aspects of the growth of knowledge.

Again, the way peoples of European descent both distribute and account for the consequences of modern sciences appears distinctively Northern, as indicated above. The benefits are distributed disproportionately to already-overadvantaged groups in Europe and their allies elsewhere, and the costs disproportionately to everyone else. Northerners account for this distribution by "internalizing" the benefits and "externalizing" the costs, a practice not free of local values, interests and presuppositions. Thus the social benefits sciences bring are supposed to be a consequence of scientific method, but the costs are not the consequence of anything scientific at all; they are the result of politics, or science policy, but not of any purportedly internal features of science itself. The postcolonial critics point out that this is a culturally distinctive way to distribute and account for sciences' benefits and costs – one that cultures that disproportionately bear the costs do not find plausible. It is inevitable, they argue, when nature is represented as consisting of isolated parts.

Finally, modern sciences' purported value-neutrality would itself mark them as culturally distinctive, which is, of course, paradoxical. However, most cultures do not value neutrality; they value Confucian, or Maori or Islamic values. So a culture that values neutrality is itself culturally distinctive. Moreover, as postcolonial theorists see the matter, claims for modern sciences' objectivity and value-neutrality are themselves unsupportable, and very much "a politics of disvaluing local concerns and knowledge and legitimating 'outside experts'" (Bandyopadhyay and Shiva, 1988).

My point here is that the three larger themes reviewed in this section – the high value of local scientific and technological traditions, the effects of European expansion on them and on European sciences and technologies, and the paradoxically "local" character of purportedly culturally value-free European sciences and technologies – echo the distinctive conceptual frameworks of Multicultural and Postcolonial Feminisms, to which we next turn. Postcolonial science studies already has formulated these themes in their science and technology analyses, though these studies only rarely explore how gender relations have shaped their objects of study.

3. MULTICULTURAL AND GLOBAL FEMINIST SCIENCE STUDIES

Feminist theory is not now and has never been a monolithic whole. Indeed, a powerful stimulant to its growth and increasing usefulness has been the continual expansion of the historical, political and cultural locations from which it has spoken. Different political philosophies have guided the analyses and forms of activism characteristic of Liberal, Marxist, Radical and Socialist Feminisms – to name four of the approaches that have most shaped public agenda feminist theory and political projects since the eighteenth century. Multicultural and Global Feminisms are two of the most recently appearing new feminisms, though each has far older roots. They are providing resources for everyone to use to think about gender relations within the multicultural communities and global political economies that constitute our worlds today (cf., e.g., Anzaldua, 1987, 1990; Enloe, 1989; Jaggar, 1983; Jaggar and Rothenberg, 1993; Mies, 1986). Such concerns have not been central in the conceptual frameworks of the Liberal and Socialist Feminisms within which contemporary European–American feminist philosophies of science have been developed. So, the question for this paper is how the concerns of Multicultural and Global Feminisms call for distinctive focuses in the philosophy of science.

Now is none too soon to acknowledge that this may seem an odd question to many readers since most philosophers (and others) think that we should have just one philosophy of science. We should not have many of them, each developed within a distinctive political theory, let alone distinctive to historical, cultural locations. Multiplicity is a sign of error from this conventional perspective; or, at least, acceptance or appreciation of it is taken to reflect a damaging seduction by soft-minded relativists.

However, it is easy to see why the social theories of these diverse feminisms lead to different concerns about scientific and technological change, and thus to different and sometimes conflicting focuses of their philosophies of science. For example, Liberal Feminism and its contemporary empiricist philosophies of science have retained important elements of their partial origin within the conceptual framework of eighteenth-century European educated classes' concern with reason, direct observation ("experience"), the formation of distinctively qualified, authoritative professional communities, and separation between public and private worlds (cf. e.g., Shapin, 1994; Shapin and Schaffer, 1985). And philosophies of science indebted to Socialist Feminism retain elements of their partial origins in nineteenth-century Marxian analyses of the causal relations between what people actually do in everyday life ("labor") and the kinds of conceptual schemes that they tend to find most plausible (what they can know), and thus a focus on how hierarchical social relations shape conceptual frameworks for producing both systematic knowledge and systematic ignorance (e.g., Hartsock, 1983; Rose, 1983; Smith, 1988, 1990). The feminist empiricism and feminist standpoint theory that are contemporary heirs to these two philosophies of science have incisively challenged and usefully transformed key elements of each of their conceptual legacies (cf. Chapter 5 in Harding, 1993; Chapter 11 in Jaggar, 1983).

Strains of the earlier feminist analyses appear in both Multicultural and Global Feminisms. Multicultural Feminisms focus on the culturally distinctive histories and practices shaping women's and men's conditions, interests and desires in different local, national and trans-national cultures, and the necessity for thinking from women's distinctive histories and practices in theorizing and designing social change. They also focus on the epistemically powerful resources of bicultural and multicultural communities in creating "outsiders within" (Collins, 1991), thinkers socially positioned to see the world "from margin to center" (hooks, 1984), and from the "borderlands" (Anzaldua, 1987). In even more complex contexts than standpoint theorists have analyzed, Multicultural and Global Feminists show how much communities and individuals have had to learn to negotiate between unequally powerful, conflicting cultures. They have not been permitted the dangerous luxury of assuming that one and only one conceptual framework can provide all the answers to every important question one needs answered in order to survive and flourish. Cognitive dissonance is for them an uncomfortable but necessary and valuable resource for negotiating daily life.

Global Feminist approaches are concerned to explain the role of gendered institutions, practices and cultures in the global political economy and, to put the point another way, the role of global political economies in gender relations. They want to show how institutions of gender relations are crucial to the global political economy. The latter depends upon the maintenance of androcentric cultures in militaries, governments and multinational corporations such that the continued appropriation of women's labors in different ways in different parts of the global economy can be organized and legitimated. These two feminist theories have challenged the adequacy of Liberal and Marxian analyses of both gender relations and of world systems, showing the (often-unintended) commitments to Eurocentrism and male supremacy that have shaped such accounts and the policies directed by them.

The rich and complex analyses of Multicultural and Global Feminisms contribute to and challenge earlier feminist theories in at least the following three ways. They focus on the resources and limitations to be found in the different cultures within which women struggle against male supremacy, cultural imperialism and economic immiseration. They focus on analyzing the mutually causal effects of changing gender relations and changes in global political economies. They focus on how the Eurocentrism of Northern assumptions exacerbates the effects of each culture's existing androcentrism, leading to further deterioration of women's resources; Eurocentrism is also androcentric. While they focus on such themes without especially taking up science and technology issues, the postcolonial science studies we examined in the previous section focuses on science and technology issues in such multicultural and global contexts, but without especially taking up gender issues. So the project is to explore the missing terrain between these two kinds of studies, and the philosophic issues they generate. Of course a third participant in such a project is the existing Northern feminist science studies.

Contemporary Northern feminist science theories have only begun here and there to think about gender issues in scientific and technological change from the perspective of such concerns, even though these concerns have so visibly begun to benefit research and scholarship in other disciplines. A major obstacle to centering such issues in feminist science studies is the Eurocentrism of conventional philosophies, histories and social studies of science and technology within which Northern feminist analyses have been perched, if ever so fragilely, on the margins. Multicultural and global analyses beckon feminist science studies with the powerful resources they obviously have provided to research and scholarship in other disciplines such as history, sociology, literary and cultural studies. Many of us teach these writings in our feminist social theory courses. These writings are represented in the introductory women's studies and feminist philosophy texts that we assign to freshmen (cf. e.g., Jaggar and Rothenberg, 1993). And our students at most US universities increasingly are multicultured. Moreover, students are learning through their other courses and the media about the intellectual and cultural traditions of non-Western cultures, and of the ways Eurocentrism has devalued and suppressed them.[5]

However, conventional philosophies and social studies of science and technology have been highly resistant not only to feminist analyses, but also to locating their accounts of the development of modern science on the postcolonial maps of "single-stream", global and multicultural history that are now becoming widely available. Even the more recently developed ethnographies of European sciences, case studies in the cultural history of European sciences, and diverse forms of philosophic "post-positivisms" have given their readers little reason even to imagine such postcolonial maps as relevant to understanding scientific and technological change.

One way to understand this situation is to note that while most of the leading thinkers within the new philosophies, sociologies, ethnographies and histories of science insist that they are "post-positivist", from the perspective of feminist and postcolonial studies, they are not post-positive enough. They have usefully focused on how scientific facts are always shaped by presuppositions, values and interests. But they have neglected to analyze the systematic ways in which a culture's presuppositions, values and interests in nature's regularities and their underlying causal tendencies are, for better *and* worse, shaped by "the facts" of their social relations as well as of nature's order. They have only half-way dismantled the fact/value dichotomy, for values and interests are left much as the positivists understood them as only the a-rational and idiosyncratic expression of the emotions and preferences of individuals and "special interest groups". They have over-stated the weakness of influence from historical "is's" to cultural "oughts" (cf. Rouse, forthcoming). "Semi-post-positivism" would be a more accurate designation for such a conceptual framework. The emerging postcolonial science studies examined in the last section move past such a position to ask how "the facts" of historical processes have tended to make appear desirable to Europeans the constitution of modern sciences by certain presuppositions, interests, discursive resources, and ways of producing knowledge that are not universal, but only local to those historical processes.

Important strains of feminist analyses, namely standpoint epistemologies, have called for identifying and using the distinctive resources of our location in the social relations that structure societies in order to provide maximally objective understandings of the presuppositions shaping dominant conceptual frameworks (Collins, 1991; Haraway, 1991; Harding, 1986, 1991; Hartsock, 1983; Smith, 1988, 1990). Yet our feminist science studies, including philosophies of science, have largely remained blithely grounded in world histories and political philosophies that are now widely understood to be Eurocentric. Such accounts may very shortly look to our students, colleagues in other disciplines, and to others much like those beautiful ancient maps with all their threatening sea dragons and their cartographical silences marked with "terra incognita"s.

The task of locating what some of the important philosophical issues are for Multicultural and Global Feminist science studies is not as difficult as might first appear, since we now have available not only the background feminist and postcolonial literatures mentioned, but also a small though rapidly increasing number of specifically feminist works located on the map of postcolonial science studies in particular. Whether or not they are by official philosophers, philosophical issues are raised in the writings by Donna Haraway (1989, 1991), Joni Seager (1993) and Sharon Traweek (1988, forthcoming), for instance. Moreover, the gender and (Third World) development analyses of the last few years, in contrast to the earlier "Women in Development" accounts, also provide resources for such a project (e.g., Agarwal, 1993; Dankelman and Davidson, 1988, Harcourt, 1994; Mies, 1986;

Shiva, 1989; Sparr, 1994). We shall turn in the final section to identify some specifically philosophical concerns that Multicultural and Global Feminisms raise for science studies. But first we must outline that missing terrain that lies between postcolonial science studies, Multicultural and Global Feminisms, and Northern feminist science studies. The writings in such existing fields suggest at least the following four important focuses for multicultural and global feminist science studies.

What is Gender?

First, we must begin by noting that these two feminist theories have made crucial contributions to the development of a far richer conception of gender than that assumed by earlier theories. This account developed for analyses in history, economics, sociology, literary criticism and other fields must be centered also in science studies (cf., e.g., Agarwal, 1993; Anzaldua, 1987, 1990; Collins, 1991; Dankelman and Davidson, 1988; Enloe, 1989; Haraway, 1989; Harding, 1996b, hooks, 1983; Mies, 1986; Parker *et al.*, 1992; Shiva, 1989; Trinh, 1989).

At risk of belaboring the now familiar for some readers, let me briefly review the recent shift in this notion. Until the last few decades, most people thought that the observable differences between men and women – their physical appearances, emotional and mental tendencies, and their social conditions – were a consequence of biological differences. The "discovery" of gender – of the systematic cultural construction of many of such differences – is one of the notable achievements of late twentieth-century feminisms. However, much contemporary social science research and popular thought takes gender to be only a property of individuals; gender is treated as only a variable of individuals. In an earlier work I argued that this was a mistake. More fundamentally, gender is a relation between groups, and thus a property of material social structures (of the labor force, for example). Gender appears as social structure through the assignment of different human activities to different groups, so some societies have more gender than others insofar as their activities are more gender-segregated. Equally importantly, it is also a totemic relation that is a property of symbolic systems (Harding, 1986). Here gender appears as a way of giving meaning to things and processes that have little or nothing to do with biological differences. Thus nations, ships and, until recently, hurricanes were coded feminine, and objectivity, rationality and value-neutrality coded masculine. It is the differences imputed between masculine and feminine, rather than any properties of either which are purportedly identifiable apart from the relation between them, that make gender a useful concept for symbolizing other things. Moreover, in all cases such gender relations are hierarchically organized, although to different degrees and in different ways from one culture to another. Men usually have more control over the distribution of social, political and economic resources, and manliness is usually taken to designate whatever counts as the distinctively or ideally human in a culture. Historical, sociological and anthropological studies of cultural differences have made gender relations a much more useful tool for understanding differing social relations between men and women.

Furthermore, gender relations are interlocked with class, "race", ethnicity, sexuality and whatever other hierarchical social relations organize a culture's discursive frameworks, institutions and practices.[6] That is, masculinity and femininity as individual variables, structural locations and discursive resources appear only in class, ethnic/"race", and other culturally specific forms. Gender, class, sexuality and ethnicity/"race" are mutually constructing and maintaining. They form a social matrix in which each of us has a determinate location (individual, structural and symbolic) at the juncture of these different hierarchical social relations. Men and women bear different costs and receive different benefits in different classes, ethnicities/"races", etc. This feature of gender is sometimes referred to as its "intersectionality" with other social formations. Last, gender relations are dynamic, historically-changing ways of obtaining and distributing scarce social resources. They are sites of political contestation during every kind of social change, including scientific and technological ones (Haraway, 1989; Merchant, 1980; Mies, 1986; Shiva, 1989; Wacjman, 1991).[7]

These developments of the concept of gender are a major reason why the concerns of Multicultural and Global Feminisms cannot simply be added to the conceptual frameworks of the earlier feminist theories. It is not that nothing of value can be learned by trying to do so, but rather that conceptual frameworks chosen to accommodate simpler notions of gender (in fact, ones that are elitist, since they assume that one can explain everything worth explaining about gender relations without looking at how they are affected by ethnicity/race, global political economies, etc.) will systematically block even raising some of the issues most central to these more comprehensive feminist theories (cf. Harding, 1991, Chapters 8 and 9). Since all of the feminist theories have criticized the similar "add women/gender and stir" approach to dealing with androcentric thought, the limitation of attempts to "add multiculturalism and globalism" to Eurocentric thought should not surprise.

We are now in a position to use the themes of postcolonial science studies and Multicultural and Global Feminisms to sketch out the terrains lying between them that can generate Multicultural and Global Feminist philosophies of science.

Gender Relations in Local Knowledge Systems (LKS)

One concern of Multicultural and Global Feminisms is with culturally distinctive histories and practices shaping women's conditions, interests and desires in different cultures. Postcolonial science studies is interested in the strengths and limitations of non-Western science and technology traditions. The "gender and development" debates are one place where these concerns conjoin in analyses of how LKS are gendered, and of women's roles in them. The "gender and development" discussions are the latest stage in attempts to evaluate the effects of development policies on women. We shall return in the next section to look at some additional resources that they offer for thinking about feminist philosophical issues.

What kinds of distinctive knowledge do women in developing societies have, according to this "gender and development" analysis? One UN report points out that women in particular are the repositories of generations of knowledge about the local environment and agri-forestry practices, plant and animal biodiversity, breeding, healing, harvesting and processing techniques and integrated resource management. The gender dimension of these knowledge systems may manifest itself in several ways. Women and men may have "different knowledge about similar things; or knowledge of different things; or have different ways of organizing knowledge; or have different ways of preserving and transmitting knowledge" (Hill and Appleton, 1994, 3).

Here is one place where the more complex understanding of gender relations is valuable. Insofar as a culture is structured by the distribution of activities by gender (e.g., childcare, agriculture and other daily subsistence activities to women, fishing for export, negotiating with other cultures, and waging war to men), and insofar as it uses gender difference to give meaning to aspects of daily life – insofar as gender exists in any culture – women and men will have different knowledge. One can see the development theorists arguing that through a culture's gender relations, women and men at least partially tend to interact with different parts of nature's regularities, have different interests in nature's resources and dangers, stand in different relations to their culture's discursive traditions (that include gendered meanings), and organize differently the production of knowledge. Gender functions like other cultural differences to create, in effect, gender cultures within any larger society.

Attention to gender relations in Southern LKS draws attention also to women's similar situations in the metropolitan centers. There are also LKS in the North in the sense, first, of systematic alternatives to modern scientific knowledge, many of which increasingly are becoming legitimated by scientific institutions as valuable complements to Northern sciences. Women have distinctive roles in these, too. Many, many Americans and Europeans regularly draw on alternatives to the knowledge provided by bio-medical sciences. Acupuncture, chiropracty, massage, nutrition, vitamin and exercise therapies are just a few of the many LKS widely in use – and among highly-educated citizens, even scientists! – in the North. Women are thought to be high users of this knowledge not because they are ignorant of or don't value bio-medical knowledge, but because they are assigned responsibility for monitoring and maintaining the health of themselves, children, household members, and elderly kin, and they gain valuable knowledge in such contexts that is often ignored or devalued by physicians and bio-medical institutions. Evidently Northern women by no means abandon bio-medical knowledge systems when the latter do not meet their needs but, instead, learn effectively to negotiate between these systems and various kinds of LKS that complement what bio-medical knowledge can provide. One study reports on how this process works when the LKS is women's first-person knowledge of their pregnant bodies, their "bodily knowledge", and, in a second case, when the LKS is their "emotional knowledge" of frail elderly kin for whom they are caretakers. In both cases they develop distinctive skills in negotiating between the different resources provided by bio-medical knowledge and LKS, and the legitimation that each system can sometimes give for the other (Abel and Browner, forthcoming; cf. also Martin, 1987). These are precisely the skills at assessing the resources and limitations of partially conflicting local knowledge systems that the postcolonial critics point out have been developed by members of colonized non-Western cultures. Developing such skills at everyday evaluation of competing knowledge claims moves crucial knowledge from systems to the individuals who must negotiate between them.

Let me be clear here. LKS are often just plain wrong, as everyone knows, and in certain ranges – e.g., processes and events that are very small, very large, far distant – to state one highly probable error range – they are clearly more likely to be wrong than in others. But so, too, is modern scientific knowledge just plain wrong in those ranges where it does not, in fact, have knowledge – where its characteristic limitations are not adequately appreciated. Indeed, as the last thirty years of post-Kuhnian philosophy and social studies of science have argued, modern science also is a LKS, constituted by its culturally local interests in the nature with which it does or wants to interact, its discursive traditions, and its particular ways of organizing the production of knowledge (cf. e.g., Dupre, 1993; Hess, 1995; Kuhn, 1962; Pickering, 1992; Rouse, 1987, 1993; Shapin, 1994; Shapin and Schaffer, 1995; Schuster and Yeo, 1986; Watson-Verran and Turnbull, 1994). It is a LKS that is immensely powerful at many projects, but often not at ones for which it was not designed – such as sustaining fragile environments in the South (or North), understanding complex interrelations between regularities of nature and culture. Colonized cultures have had to learn to distinguish the limitations and the strengths of both their own knowledge tradition and modern European sciences. This is a respect in which their scientific practices can provide a model for the rest of us. Insofar as women's repositories of and skills at generating knowledge have been devalued, their heightened alertness to the strengths and limitations of working with two or more knowledge systems that often are grounded in conflicting assumptions and conceptual frameworks leads toward a different kind of epistemological framework than the prevailing Northern ones – a point to which we shortly return.

Gendered European Expansion and the Growth of Modern Science in Europe

What issues lie in the terrain between the postcolonial science studies discussion of the resources that European expansion and the growth of modern science in Europe provided for each other, on the one hand, and the Multicultural and Global Feminist interests in gender in the global political economy?

The gender and development debates are the feminist part of the latest stage in a three-decade process of reevaluating the effects of post-World War II, Northern-directed transfers to the South of those models of "modernization" that were productive in Europe and the US. Thus development processes imported Northern industrialization models, in which Northern scientific and technological knowledge figured centrally. However, already by the 1960s significant failures in such a conception of development had become visible. By the 1990s, even most of the international and Northern development agencies recognized that though their policies certainly could claim some successes, they had also played a significant role in creating widespread environmental degradation, cultural destruction, and "dedevelopment" of the majority of the South's populations that were already the most economically and politically vulnerable (cf., e.g., Sachs, 1992; Sardar, 1988). The European models are not the right ones for these other cultures, or in the current global political economy that is largely controlled by Northern interests. This kind of reevaluation of development policy can be regarded as bringing the concerns of postcolonial science studies to bear on post-World War II "European expansion": they see Northern-controlled development policies as, in significant respects, "colonialism by other means".

Within this three-decade reevaluation of Northern development policies has been a continuing analysis of the especially bad effects that such policies have had on women, who in the South, as in the North, are disproportionately represented among the economically and politically most vulnerable. Such analyses have moved from demands to include women in development (WID) – that is, in modernization through Northern-style industrialization – to recent arguments that such models of development are androcentric as well as Eurocentric. This is so in that a narrow, economic model of "development" itself is androcentric, devaluing in ways familiar to Northern feminists both natural environments and just social relations in order to favor "rational economic man". Such models fail to recognize that the environments upon which depends women's ability to provide daily subsistence to their households and communities (food, water, fuel, food for animals), and the social relations through which such provisioning is made justly, must also be sustained and improved if development is to deliver benefits rather than primarily losses to its recipients. If no attention is paid specifically to providing needed resources to women, the import of Northern sciences and technologies will invariably further dedevelop women. The topic must be "women, environments and alternatives to development", or "gender and development" (which emphasizes the focus on gender relations rather than only on women's conditions) instead of only a narrowly conceived economic model of development. Androcentric models of what counts as development distribute disproportionate costs of development to women, thereby further dedeveloping not only women, but all those whose well-being depends upon their labor (cf. Agarwal, 1993; Dankelman and Davidson, 1988; Harcourt, 1994; Mies, 1986; Shiva, 1989).

In addition to the above-cited studies of gendered development, other kinds of feminist accounts have begun to map the centrality of gender relations to Northern expansion in the late twentieth century through scientific and technological changes. For example, Donna Haraway has studied how a distinctive US focus on primatology as "about" how to control race and gender relations was central to evolving nationalist and imperialist political agendas during the twentieth century. Historian/ethnographer Sharon Traweek has studied how the culturally distinctive gender relations in and between European/American and Japanese high-energy physics have effects on that field. Geographer Joni Seager shows how mainstream and much of the existing Northern feminist environmental approaches fail to focus on how the policies of militaries, governments and multi-national corporations are constructed through what are very much the masculine cultures of such institutions. These policies are the major causes of global environmental degradation and the worsening conditions of women in particular (Haraway, 1989; Traweek, 1988, forthcoming; Seager, 1993).

Modern Sciences' Androcentric Eurocentrism

Shared by all of these accounts are new ways of understanding how the meanings and practices of Northern scientific and technological institutions contributing to and shaped by colonial projects – in the past and today – that on the face of it have little to do with gender relations are, in fact, deeply permeated by them. This terrain also lies between postcolonial science studies, Multicultural and Global Feminisms, and Northern feminist science studies. Such meanings and practices are centrally shaped by the need to maintain socially desirable models of masculinity and femininity. To put the point another way, processes of scientific and technological change are always sites of gender struggles, as they are also of class, ethnic, imperial, and other forms of struggle over scarce social resources (cf., e.g., Wacjman, 1991). Success at retaining or eradicating androcentric meanings and practices of scientific and technological institutions determines how the benefits and costs of scientific and technological change – and European expansion – will be distributed to men and to women.

It is easy to see in scientific and technological cultures and practices how models of the scientist and the engineer, the national hero, the military hero and the explorer have all been constructed in terms of ideals of manliness. Northern feminist analyses showed how the figure of the scientist was invariably an ideal of masculinity, and that of nature a feminine one. The scientist was to "marry nature", or torture her to make her reveal her secrets, or "strip away her veils" so that her true nature could be observed. Similarly, the national hero, the military hero and the explorer have been linked to important models of masculinity, and to each other. The national hero has been the freedom fighter, the revolutionary, the colonist adventurer and explorer who plants the British, Dutch or Spanish flag on Roanoke Island, New Amsterdam, the Cape of Good Hope, Darwin Bay, or the North Pole. European expansion, with its adventure, hardship and militarism was a playing field upon which models of national masculinity could be tested and established. The land, the nature that was discovered, explored, conquered, explained and named was imagined "native female" (cf., e.g., Enloe, 1989; Haraway, 1989; Hayles, 1992; Keller, 1984; Merchant, 1980; Parker *et al.*, 1992; Schiebinger, 1989).

My point is that the language of adventure, discovery, conquest, taming nature, and achieving national glory is also that of the public and professional discourse of scientific and technological growth. So the gendered discourse of European expansion has from its beginnings through late twentieth-century models of development helped to construct and maintain the gendered distribution of the benefits and costs of scientific and technological growth, and vice versa (Adas, 1989; Haraway, 1989; Kochhar, 1992–93; Stepan, 1993; Trinh, 1989).

Finally, it is time to identify some philosophical issues raised by these new images of scientific and technological change in history.

4. TOWARD MULTICULTURAL AND GLOBAL FEMINIST PHILOSOPHIES OF SCIENCE

Let us take Northern feminist philosophy of science as our reference point here in thinking about philosophical issues raised by Multicultural and Global Feminist interests in scientific and technological change. I take this reference point to include not only work by the handful of feminists with degrees in philosophy who work in philosophy departments, but also by scholars in the history, sociology and ethnography of modern science, and researchers in biology and other sciences, who have also raised philosophical issues. And I take this field as a reference point not to criticize it, and certainly not to criticize it alone among Northern science studies, but because I take it to be the most familiar reference point for this collection's readers. The terrain sketched out in the preceding section as Multicultural and Global Feminist science studies suggests further and more comprehensive evidence for existing positions in Northern feminist philosophies of science, on the one hand. On the other hand, it also challenges Northern feminist philosophies of science in other respects.

In a supportive relation, we have further evidence for the usefulness of standpoint epistemologies (Collins, 1991; Haraway, 1991; Harding, 1986, 1991; Hartsock, 1983; Jaggar, 1983; Rose, 1983; Smith, 1988, 1990). By starting off analyses from the standpoint of those women's lives that are excluded and devalued by the prevailing Northern scientific and technological conceptual frameworks, we can gain more accurate, comprehensive, and objective descriptions and explanations of scientific and technological change in both the North and the South. Sciences (and science studies) are not and could not be culturally neutral; they could not succeed in providing transparent windows onto the worlds that they would describe and/or explain. Knowledge is always socially situated. Standpoint epistemologies provide maps for more useful ways to conceptualize what counts as knowledge and how it gets produced than are available from the older "internalist", transcultural epistemologies. And conjoining the discourses of postcolonial science studies with the beginnings and possibilities of Multicultural and Global Feminist science studies points to yet more accurate and comprehensive representations of scientific and technological change within global social history.

Multicultural and Global Feminist science studies also support Northern philosophers' related concerns to locate standards for "strong objectivity" that delink maximizing objectivity from the requirement of value-neutrality and, instead, direct us to figure out how to start out thinking and research from outside the dominant conceptual frameworks in order to identify the culturally local features of those frameworks. In identifying the distinctively local features of purportedly value-free

Northern sciences and science theories, such approaches do/would not abandon the goal of maximizing objectivity; instead they generate more objective accounts than can be gained by thinking only within Eurocentric and androcentric conceptual frameworks.

Some have worried that standpoint theories are essentialist, assuming that there is some model woman or women from whose lives these standpoint approaches direct one to start thinking. Others have claimed that they are damagingly relativist; that they are committed to holding that there are no rational grounds for making judgements about the relative cognitive value of competing accounts. Some have thought that they are even ethnocentric "identity philosophies", claiming that only the oppressed can generate legitimate knowledge claims about or from the perspective of their own lives. These worries are understandable but ungrounded, as the projects indicated in Multicultural and Global Feminist science studies clarify.

Does this mean, then, that Multicultural and Global Feminist science studies will not need the resources of feminist empiricism or any other feminist science/knowledge theory? Those still gripped by the idea that one should hold one's favored theory of knowledge (or any other kind of theory) as absolutely, universally right for all times, places and purposes, will conclude that they must choose one or another of such partially-competing theories of scientific knowledge. However, for those who think, as I do, that epistemologies are justificatory and methodological strategies that are designed for historically specific purposes, a different conclusion follows. Here, theories about scientific knowledge are usefully thought of as distinctive maps, just as are scientific theories. We are used to thinking about how different scientific theories are useful for different projects for interacting with natural and social worlds. So, too, are different epistemologies useful for different justificatory and methodological contexts. A practice of strategically deploying different epistemologies, of constantly negotiating between partially incompatible theories of what should count as knowledge and how it gets produced, will be congenial for these feminisms for other reasons, also. It matches their experience with negotiating between partially incompatible LKS of different local cultures, including modern science.

Thus, one can foresee that Multicultural and Global Feminisms are at times going to need some of the resources of feminist empiricist and, no doubt, other feminist epistemologies already and/or not yet developed. For one thing, the scientific and policy worlds in which anyone trying to improve sciences must function are largely still contained by conventional Liberal/empiricist assumptions about the nature of science and of society. Feminist empiricist arguments about how to avoid "bad science", how to eliminate sexist bias, how to add women's concerns, and women in developing countries' concerns, to the existing foci of scientific and technological projects, and about the need to make professional scientific communities more inclusive to counteract biases that otherwise occur within them – these and other feminist empiricist arguments for improving existing sciences will also be useful to Multicultural and Global Feminisms.

Let us turn now to the ways that Multicultural and Global Feminist science studies can challenge Northern feminist philosophies of science. First, there is obviously the problem of the Eurocentric conceptual frameworks that have limited all Northern philosophies of science, including feminist ones. Northern feminist science studies have started off thinking about their topics from the standpoint of largely European women's lives in the North (and, often, primarily educated women's lives), while the Multicultural/Global frameworks start off from the realities of non-European women's lives in the South and the North. This difference in the standpoints of Northern and Southern feminist science studies produces conceptual mismatches in their accounts: they talk past each other in significant ways. We already noted earlier the more complex concept of gender needed for the Multicultural and Global projects. Northern feminists already identified a philosophic issue here in their insistence on understanding scientific concepts, as well as the philosophic concepts, metaphors, models and narratives used to explain scientific progress, as gendered, not trans-culturally "human". The more complex concept of gender proposes a more accurate and useful approach to this philosophic issue.

Additionally, the Northern studies' focuses on "nature", on sciences that are abstracted from any of their colonial or development history, and on "high sciences" mark three related sites of problematic difference from Multicultural and Global Feminist science studies. Take the gap between the focus on "nature" and, in Southern feminist science studies, on "environments". "Nature", with all its meanings and their histories, arrives from deployment in the distinctively European and US discourses of Mother Earth, and of wild and unruly Nature that man must tame lest he lose control of his fate. It is the Nature that speaks the language of mathematics that must be learned by those who would converse with her. It is modest Nature whose veils must be stripped so that science can discover her secrets. It is the nineteenth-century US invention of nature as "the wilds" to be revered in an imagined pristine state unsullied by human presences ... and more.

In contrast, "environment" in the Southern feminist science and technology discussions arrives from deployments in discourses about the surroundings with which humans regularly interact in daily subsistence struggles; that can be appropriated for the purposes of conquistadors, colonists and multinational corporations; that can be turned into uninhabitable wastelands by arrogance, carelessness, and immorality; that must be sustained if species, human and others, and the cultures dependent on them are to survive. Of course some of these notions also appear in Northern environmentalism, including environmental ethics. But they do not appear in philosophies of science because they are not present in the standard Northern histories of science on which philosophers reflect. Evidently ethics is not the only area of philosophy to which environmental issues are relevant. However, when ethics is the main area of philosophy concerned with the environment, and when sciences are conceptualized as requiring isolation from social "values", Northern philosophies of science are going to have to struggle to get environmental issues in focus.

The concept of "nature" in Northern thought is an abstraction of selected features of European environments from the cultural histories that have made these, and not others, the features of interest to Northerners. As postcolonial science studies point out, the dominant Northern philosophies then argue that these abstracted features have no cultural history, and are all that one needs to think about in order to predict and explain the regularities of "empirical reality". Possibly the concept of nature is a class concept, emerging as part of "class warfare" against peasants and farmers.[8] Obviously such philosophies have served well European expansion and its colonists and multinational corporations, but they do not serve Multicultural and Global Feminisms. The links between androcentrism and Eurocentrism in such a concept of "nature" deserve charting.

Relatedly, Northern feminist philosophies of science have been concerned primarily with the androcentric conceptual frameworks of Northern "high sciences" – physics, chemistry, and biology – as these are purportedly distinct from applied sciences, technologies, and social sciences; yet in the Southern studies it is neither easy nor, usually, useful to isolate the former from the latter. Multicultural and Global Feminisms want to see the gendered dimensions of environmental, health, military, and industrial technosciences that do and could impact their lives – both those of their cultural traditions and those from the North.

Finally, the contrast between universally-valid Northern sciences and the merely LKS of other cultures needs reexamination from the perspective of Multicultural and Global Feminist standpoints. The latter can draw on a "different cultures" model of the history of sciences developed in postcolonial science studies that sees modern science as different in important respects from other LKS; however, like other LKS, it is itself also very much a LKS. All systems of knowledge about nature/environments are constituted by the local interests, discursive traditions and ways of organizing the production of knowledge that they mobilize to explain their observations of their particular "corners of nature".

I have been arguing that Multicultural and Global Feminisms do raise distinctive philosophy of science issues. They link and expand the concerns of postcolonial and Northern feminist philosophies of science in their support of stronger standards for more adequate epistemologies and ontologies that are capable of detecting the androcentrism and Eurocentrism of the dominant philosophical frameworks not only of self-conscious philosophies of science of the North, but also of the less self-conscious ones of Northern sciences and science studies.[9]

ACKNOWLEDGEMENTS

Lynn Hankinson Nelson provided useful comments on an earlier version of this paper. Many individuals and audiences over a number of years have greatly enriched my understanding of postcolonial science issues in the larger project on which this paper draws. For the larger project, see, e.g., Chapters 8, 9 and 11 of Harding, 1991; Harding, 1993, 1994, 1996, forthcoming a, b, and c; Harding and McGregor, 1996.

NOTES

[1] Even this list gives too much away, since most engineers would argue that their technologies are not social at all in any meaningful sense of the term. They have social applications and meanings, but the technologies themselves, i.e., hardware, are "universally valid" in that they work in any and every

culture for what they were intended to do. By excluding from their definition of a "technology" not only its social applications and meanings, but also the knowledge of how to make it, use it, and maintain it, they can perpetuate the illusion that technologies are not cultural at all (cf. Wacjman, 1991).

[2] I shall use the term "science" here in a more expansive way than is characteristic of familiar philosophy of science discourses. Following the practices of many of the postcolonial science theorists, I shall refer to any knowledge about the natural world as scientific knowledge. For example, David Hess says "I will understand science to be knowledge about the natural world, just as social science is knowledge about the social world and technology is materially embedded knowledge about how to create effects via artifacts. ..." (1995, p. 1). Of course there are important differences of many sorts between high-energy physics and local botanical knowledge – major ones of which will be indicated as we go along. The point here and in others' writings of initially ignoring such differences is to come up with more accurate and useful representations of both the differences and similarities between modern sciences and the conventionally-labelled local knowledge systems than conventional conceptual frameworks of Northern science studies have permitted. (Of course there are other problems, also, with extending the term "science" to other cultures' knowledge of the natural world; many other cultures do not use such a term to refer to their own repositories of such knowledge, etc. It is worth committing this form of Eurocentrism for the moment, in my opinion, in order to explore possible ways to counter the profound form of Eurocentrism that sees Northern knowledge of the natural world as universally and uniquely valuable.)

[3] In addition to the examples of feminist science studies cited above, see Adas, 1989; Bass, 1992; Bernal, 1987; Blaut, 1993; Brockway, 1979; Crosby, 1987; Goonatilake, 1984, 1992; Hess, 1995; Joseph, 1991; Kaptchuk, 1983; Kochhar, 1992–93; Lach, 1977; McCallum, 1992; Moraze, 1979; Nandy, 1990; Needham, 1954ff., 1969; Petitjean *et al.*, 1992; Rodney, 1982; Sabra, 1976; Sardar, 1988; Shinn *et al.*, 1996; Todorov, 1984; Van Sertima, 1986; Watson-Verran and Turnbull, 1994; Weatherford, 1988.

In the parlance of this essay, "Northern" and "Southern" will refer to discourses, not to ethnicities, citizenship, or places of residence. (Though "North" and "South" will sometimes simply refer more familiarly to what used to be called the West and the Third World. Hopefully, the context will make the references of these terms clear.) Some of the Northern theorists, more or less uncritical of Eurocentrism, are nationals from Brazil, Nigeria, Japan and India who live in their homelands. Some of the Southern theorists, intensely critical of Eurocentrism in science and technology cultures and practices, as elsewhere, are French, Brits and Americans living in their homelands. Moreover, it should be recognized that such discourses are not themselves homogeneous; different presuppositions, resources, interests, institutions and practices create diversity in their analyses (cf. Williams and Chrisman, 1994).

[4] Much of the postcolonial science writings cited throughout this paper regard post-World War II "development" as a continuation of "colonialism by other means". For particularly pointed criticisms, see, e.g., Bass, 1992; Mies, 1986; Nandy, 1990; Harcourt, 1994; Petitjean *et al.*, 1992; Sardar, 1988; Sachs, 1992; Sparr, 1994.

[5] Note that, contrary to some media representations, these courses and discussions have by no means all been "on the Left", since, for example, many business schools and international relations programs have been major supporters of multicultural and global studies – though not, usually, of "postcolonial" focuses on them – because they want to train their students to learn to function effectively in the global political economy. The political uses of focusing on multiculturalism and global relations are diverse.

[6] "Race" is in scare quotes to signal that it is socially constructed, not a feature of nature (cf. Livingstone, 1993).

[7] "Last" for our purposes here. Poststructuralist analyses have shifted this concept yet further away from its origins in feminist thinking of two or three decades ago. They show how it is gender relations which direct us to identify sex differences; how gender discourses generate distinctive kinds of persons, etc. But these discussions lie beyond the terrain of this paper.

[8] Anthropologist Tom Patterson has made this argument in conversation and in a paper at this point still in draft.

REFERENCES

Abel, Emily K. and C. H. Browner: forthcoming, 'Selective Compliance with Biomedical Authority and the Uses of Experiential Knowledge', in *Women, Biopower and Resistance*, ed. Margaret Lock and Patricia Kaufert. Cambridge: Cambridge University Press.

Adas, Michael: 1989, *Machines as the Measure of Man*. Ithaca: Cornell University Press.

Agarwal, Bina: 1993, 'The Gender and Environment Debate: Lessons from India', *Feminist Studies*, **18**, no. 1.

Anzaldua, Gloria: 1987, *Borderlands/La Frontera*. San Francisco: Spinsters/Aunt Lute.

Anzaldua, Gloria, ed.: 1990, *Making Face/Making Soul*. San Francisco: Aunt Lute Foundation Press.

Bandyopadhyay, J. and V. Shiva: 1988, 'Science and Control: Natural Resources and their Exploitation', in Z. Sardar, ed., *The Revenge of Athena*. London: Mansell.

Barad, Karen: 1996, 'Meeting the Universe Halfway', this volume.

Bernal, Martin: 1987, *Black Athena: The Afroasiatic Roots of Classical Civilization*, Vol. I. New Brunswick: Rutgers University Press.

Blaut, J. M.: 1993, *The Colonizer's Model of the World: Geographical Diffusionism and Eurocentric History*. New York: Guilford Press.

Brockway, Lucile H.: 1979, *Science and Colonial Expansion: The Role of the British Royal Botanic Gardens*. New York: Academic Press.

Cartwright, Nancy: 1983, *How the Laws of Physics Lie*. New York: Oxford University Press.

Cartwright, Nancy: 1989, *Nature's Capacities and Their Measurement*. Oxford: Clarendon Press.

Collins, Patricia Hill: 1991, *Black Feminist Thought: Knowledge Consciousness, and the Politics of Empowerment*. New York: Routledge.

Crosby, Alfred: 1987, *Ecological Imperialism: The Biological Expansion of Europe*. Cambridge: Cambridge University Press.

Dankelman, Irene and Joan Davidson: 1988, *Women and Environment in the Third World*. London: Earthscan Publications.

Dupre, John: 1993, *The Disorder of Things: Metaphysical Foundations for the Disunity of Science*. Cambridge: Harvard University Press.

Enloe, Cynthia: 1989, *Bananas, Beaches and Bases: Making Feminist Sense of International Politics*. Berkeley: University of California Press.

Forman, Paul: 1987, 'Behind Quantum Electronics: National Security as Bases for Physical Research in the U.S., 1940–1960', *Historical Studies in Physical and Biological Sciences*, **18**.

Gilbert, Michael: 1995, 'Feminism, Argumentation and Coalescence', *Informal Logic*, **16.2**, 95–113.

Goonatilake, Susantha: 1984, *Aborted Discovery: Science and Creativity in the Third World*. London: Zed.

Goonatilake, Susantha: 1992, 'The Voyages of Discovery and the Loss and Rediscovery of the "Other's" Knowledge', *Impact of Science on Society*, no. 167.

Hacking, Ian: 1983, *Representing and Intervening*. Cambridge: Cambridge University Press.

Haraway, Donna: 1989, *Primate Visions: Gender, Race and Nature in the World of Modern Science*. New York: Routledge.

Haraway, Donna: 1991, 'Situated Knowledges: The Science Question in Feminism and the Privilege of Partial Perspective', in her *Simians, Cyborgs and Women: The Reinvention of Nature*. New York: Routledge.

Haraway, Donna: forthcoming, 'Universal Donors in a Vampire Culture: It's All in the Family. Biological Kinship Categories in the Twentieth-Century United States', in William Cronon, ed., *Reinventing Nature*. New York: Norton.

Harcourt, Wendy: 1994, *Feminist Perspectives on Sustainable Development*. London: Zed Press.

Harding, Sandra: 1986, *The Science Question in Feminism*. Ithaca: Cornell University Press.

Harding, Sandra: 1991, *Whose Science? Whose Knowledge?* Ithaca: Cornell University Press.

Harding, Sandra, ed.: 1993, *The "Racial" Economy of Science: Toward a Democratic Future*. Bloomington: Indiana University Press.

Harding, Sandra: 1994, 'Is Science Multicultural? Challenges, Resources, Opportunities, Uncertainties', in *Configurations* **2:2**, and in *Multiculturalism: A Reader*, ed. David Theo Goldberg. London: Blackwell.

Harding, Sandra: 1995, '"Strong Objectivity": A Response to the New Objectivity Question', *Synthese*, **104: 3**, 1–19.

Harding, Sandra: forthcoming a. 'What Makes Possible Women's Standpoints on Nature?', *Osiris*.

Harding, Sandra: forthcoming b. 'Is Modern Science an Ethnoscience?', *Sociology of the Sciences Yearbook 1996*, ed. Jack Spaapen and Terry Shinn. Dordrecht: Kluwer.

Harding, Sandra: forthcoming c. *Is Science Multicultural? Feminist and Postcolonial Perspectives*. Bloomington: Indiana University Press.

Harding, Sandra and Elizabeth McGregor: 1996, 'Science and Technology: The Gender Dimension', in Howard Moore, ed., *World Science Report*. Paris: UNESCO (preprinted by UNESCO as separate monograph in 1995).

Hartsock, Nancy: 1983, 'The Feminist Standpoint', in S. Harding and M. Hintikka, eds, *Discovering Reality*. Dordrecht: Reidel/Kluwer.

Hayles, N. Katharine: 1992, 'Gender Encoding in Fluid Mechanics: Masculine Channels and Feminine Flows', in *Differences*, **4.2**, 17–44.

Hayles, N. Katharine: 1993, 'Constrained Constructivism: Locating Scientific Inquiry in the Theater of Representation', in *Realism and Representation*, ed. George Levine. Madison: University of Wisconsin Press.

Hess, David J.: 1995, *Science and Technology in a Multicultural World: The Cultural Politics of Facts and Artifacts*. New York: Columbia University Press, 1994.

Hill, Catherine and Helen Appleton: 1994, in *Indigenous Knowledge and Development Monitor* (December).

hooks, bell: 1984, *Feminist Theory From Margin to Center*. Boston: South End Press.

Jaggar, Alison: 1983, *Feminist Politics and Human Nature*. Totowa: Rowman & Allenheld.

Jaggar, Alison: 1989, 'Love and Knowledge: Emotion in Feminist Epistemology', in *Gender/Body/Knowledge*, ed. Alison M. Jaggar and Susan R. Bordo. New Brunswick: Rutgers University Press.

Jaggar, Alison, and Paula Rothenberg, eds: 1993, *Feminist Frameworks*, 3rd edn. New York: McGraw-Hill.

Joseph, George Gheverghese: 1991, *The Crest of the Peacock: Non-European Roots of Mathematics*. New York: I. B. Tauris and Co.

Kaptchuk, Ted J.: 1983, *The Web That Has No Weaver: Understanding Chinese Medicine*. New York: Congdon and Weed.

Keller, Evelyn Fox: 1984, *Reflections on Gender and Science*. New Haven: Yale University Press.

Kochhar, R. K.: 1992–93, 'Science in British India', Parts I and II, *Current Science*, 63.11; **64.1** (India).

Kuhn, Thomas S.: 1962, *The Structure of Scientific Revolutions*. Chicago: University of Chicago Press.

Lach, Donald F.: 1977, *Asia in the Making of Europe*, Vol. 2. Chicago: University of Chicago Press.

Livingstone, Frank B.: 1993, 'On the Nonexistence of Human Races', in *The "Racial" Economy of Science*, ed. S. Harding. Bloomington: Indiana University Press.

Longino, Helen: 1990, *Science as Social Knowledge*. Princeton: Princeton University Press.

Martin, Emily: 1987, *The Woman in the Body*. Boston: Beacon.

McCallum, James E.: 1992, *Colonialism and Science: Saint Domingue in the Old Regime*. Baltimore: Johns Hopkins University Press.

Merchant, Carolyn: 1980, *The Death of Nature: Women, Ecology and the Scientific Revolution*. New York: Harper and Row.

Mies, Maria: 1986, *Patriarchy and Accumulation on a World Scale: Women in the International Division of Labor*. Atlantic Highlands: Zed Press.

Moraze, Charles (ed.): 1979, *Science and the Factors of Inequality*. Paris: UNESCO.

Nandy, Ashis (ed.): 1990, *Science, Hegemony and Violence: A Requiem for Modernity*. Delhi: Oxford University Press.

Needham, Joseph: 1954ff, *Science and Civilization in China*. Cambridge: Cambridge University Press.

Needham, Joseph: 1969, *The Grand Titration: Science and Society in East and West*. Toronto: University of Toronto Press.

Parker, A. *et al.*, eds: 1992, *Nationalisms and Sexualities*. New York: Routledge.
Petitjean, Patrick, *et al.*: 1992, *Science and Empires: Historical Studies About Scientific Development and European Expansion*. Dordrecht: Kluwer.
Pickering, Andrew, ed.: 1992, *Science as Practice and Culture*. Chicago: University of Chicago Press.
Restivo, Sal: 1992, *Mathematics in Society and History: Sociological Inquiries*. Dordrecht: Kluwer.
Rodney, Walter: 1982, *How Europe Underdeveloped Africa*. Washington: Howard University Press.
Rose, Hilary: 1983, 'Hand, Brain, and Heart: A Feminist Epistemology for the Natural Sciences', in *Sex and Scientific Inquiry*, ed. S. Harding and J. F. O'Barr. Chicago: University of Chicago Press.
Rouse, Joseph: 1987, *Knowledge and Power: Toward a Political Philosophy of Science*. Ithaca: Cornell University Press.
Rouse, Joseph: 1993, 'What are Cultural Studies of Scientific Knowledge?', *Configurations*, **1:1**, 1–22.
Sabra, I. A.: 1976, 'The Scientific Enterprise', in B. Lewis, ed., *The World of Islam*. London: Thames and Hudson.
Sachs, Wolfgang, ed.: 1992, *The Development Dictionary: A Guide to Knowledge as Power*. Atlantic Highlands: Zed Books.
Sardar, Ziauddin, ed.: 1988, *The Revenge of Athena: Science, Exploitation and the Third World*. London: Mansell.
Schiebinger, Londa: 1989, *The Mind Has No Sex: Women in the Origins of Modern Science*. Cambridge: Harvard University Press.
Schuster, John A. and Richard R. Yeo, eds: 1986, *The Politics and Rhetoric of Scientific Method: Historical Studies*. Dordrecht: Reidel.
Seager, Joni: 1993, *Earth Follies: Coming to Feminist Terms with the Global Environmental Crisis*. New York: Routledge.
Shapin, Steven: 1994, *A Social History of Truth: Civility and Science in Seventeenth-Century England*. Chicago: University of Chicago Press.
Shapin, Steven, and Simon Schaffer: 1985, *Leviathan and the Air Pump*. Princeton: Princeton University Press.
Shinn, Terry *et al.*: 1996, 'Science and Technology for the South', *Sociology of the Sciences Yearbook*. Dordrecht: Kluwer.
Shiva, Vandana: 1989, *Staying Alive: Women, Ecology and Development*. London: Zed.
Smith, Dorothy E.: 1988, *The Everyday World as Problematic: A Feminist Sociology*. Boston: Northeastern University Press.
Smith, Dorothy E.: 1990, *The Conceptual Practices of Power: A Feminist Sociology of Knowledge*. Boston: Northeastern University Press.
Sparr, Pamela, ed.: 1994, *Mortgaging Women's Lives: Feminist Critiques of Structural Adjustment*. London: Zed Books.
Stepan, Nancy Leys: 1993, 'Race and Gender: The Role of Analogy in Science', in *The "Racial" Economy of Science*, ed. S. Harding. Bloomington: Indiana University Press.
Todorov, Tzvetan: 1984, *The Conquest of America: The Question of the Other*, trans. Richard Howard. New York: Harper and Row.
Traweek, Sharon: 1988, *Beamtimes and Life Times*. Cambridge: MIT Press.
Traweek, Sharon: forthcoming, *Big Science in Japan*.
Trinh, Minh ha: 1989, *Woman/Native/Other: Writing Postcoloniality and Feminism*. Bloomington: Indiana University Press.
Van Fraassen, Bas and Jill Sigman: 1993, 'Interpretation in Science and in the Arts', in *Realism and Representation*, ed. G. Levine. Madison: University of Wisconsin Press.
Van Sertima, Ivan: 1986, *Blacks in Science*. New Brunswick: Transaction.
Wacjman, Judy: 1991, *Feminism Confronts Technology*. University Park: Penn State Press.
Watson-Verran, Helen and Robert Turnbull, eds: 1994, 'Science and Other Indigenous Knowledge Systems', in *Science and Technology Handbook*. Berkeley: Sage Publications.
Weatherford, Jack: 1988, *Indian Givers: What the Native Americans Gave to the World*. New York: Crown Publishing Co.
Williams, Patrick and Laura Chrisman, eds: 1994, *Colonial Discourse and Post-Colonial Theory*. New York: Columbia University Press.

SARA HEINÄMAA

WOMAN – NATURE, PRODUCT, STYLE? RETHINKING THE FOUNDATIONS OF FEMINIST PHILOSOPHY OF SCIENCE

INTRODUCTION

Phenomenology questions the foundations of knowledge and science in a specific sense. It studies the *meanings* of the phenomena measured and generalized about in the empirical sciences: What do we mean, for example, by "natural processes", by "nature", "fact", "culture", "human being", etc.? And it studies the *meanings* of concepts basic to science: knowledge, objectivity, mistake, false belief, etc. Understanding the philosophy of science broadly, we can say that phenomenology is a specific viewpoint or approach within it; it critically attends to the basic concepts at work in both the natural and human sciences.

A feminist approach in phenomenology also asks: What are the meanings of the basic and fundamental ideas/concepts/terms of *Feminist Studies* and *Women's Studies*? That is, what do we mean – what can we mean – by "woman", "sex", and "gender", and how did/do these categories get their meanings? The study of these questions *precedes*, for example, Sandra Harding's questions of "Is there a specific feminist method?", "Is science gender neutral?", and "Is Bacon's idea of science sexist?" That is, feminist phenomenology asks about the constitution of experiences and phenomena presupposed in feminist philosophy of science in the analytic tradition. It asks *what do we mean by* "feminist", "method", "woman", "oppression", "freedom", "science", "gender", etc. It is an attempt to critically study the foundations of feminist studies and women's studies, not to answer the skeptic by providing more knowledge, but to understand (to describe and to analyze) the meanings on which feminist and women's studies have been based.

The most important of these phenomena is sexual difference. What do we *mean* when we say that women and men are different? Do we mean that women are different *from* men, i.e., that men are the standard case and women the deviant (this was de Beauvior's analysis)? *And* can the meaning of this difference be conceived in a new way (this is Irigaray's question)?

Conceptions of sexual difference have been central both to specific research in the sciences and to feminist critiques of such research. In what follows, I analyze the *meanings* of sexual difference in both areas and, building on phenomenology of the body, suggest alternatives to these conceptions.

L. H. Nelson and J. Nelson (eds), Feminism, Science, and the Philosophy of Science, 289–308.

THE PROBLEM OF GENDER

For a long time, the concept of woman has been regarded as relatively unproblematic in Anglo-American feminist theory. Arguments and theories have been based on the conceptual distinction between biological *sex* and social *gender* and this distinction has not seemed, to many feminists or nonfeminists, to have required any philosophical analysis. The separation of sex and gender – female and feminine, male and masculine – has constituted a basic conceptual framework which has been used to acquire and structure information, and which has enabled scholars to formulate new detailed questions about the social and cultural relations between the sexes.[1] Not many theorists have felt the need to look behind or beneath the framework. Philosophical questions about the being and nature of women have been passed over as either solved, trivial or wrongly formulated.

Since the 1980s the concept of gender has been subjected to fundamental criticism. Its validity has been questioned both by poststructuralists who are concerned with the fragmentation of the subject, and by lesbians and African–American women who have pointed to totalizing tendencies within the women's movement and women's studies.[2] Poststructuralists have criticized, for example, feminist theories about "women's experiences" on the grounds that these are often based on an uncritical acceptance of the empiricist concept of experience. African–American and lesbian scholars have argued that feminist theory has generalized the experiences of white, middle-class, and heterosexual women to all women. Suddenly the unity and the being of the woman subject seems problematic. Simone de Beauvoir's questions have again become topical: "Are there women, really? – What is a woman?" (1949/93: 11, 1949/87: 13), and "How does one become one?" (1949/91: 13, 1949/87: 245).

The aim of this article is to present a philosophical analysis of the concept of woman. I shall begin by studying how the distinction between sex as a biological category and gender as a social category has structured attempts within Anglo-American feminism to answer de Beauvoir's questions. The basic limitation of the sex/gender approach proves to be that it is based on the (bio)scientific understanding which conceptualizes the human body merely as an *object* of knowledge and ignores the body's constitutive role in subjectivity and knowledge. Although feminists who have taken the sex/gender distinction as a basic organizing principle have criticized biological explanations for male bias, they have taken for granted the concept of body as a bio-mechanism. This has two problematic consequences.

First, the relations of the body to its environment are understood as external relations of cause or function. That is, it is supposed that the states of the body and its environment can be described and verified independently of each other.[3] So, if sex is an internal attribute of the body and gender is a socially-constructed interpretation superimposed on the body, then the relation between sex and gender is logically contingent. Ultimately, as Judith Butler has pointed out, this line of thinking ends up in the claim that "*man* and *masculine* might just as well signify a female body as a male one, and *woman* and *feminine* a male body as easily as a female one" (Butler, 1990a: 6).[4] The problem for feminist practice and theory, as I see it, is that such ease seems to be an illusion. Man and masculine seem to be intertwined with maleness, and woman and feminine with femaleness more intimately than causal descriptions allow. Masculine females and feminine males are conceived as abnormalities. The task is to understand the nature of these "departures" and the norms they embody.

The second problem is that the (bio)scientific concept of the human body limits the possibilities to theorize about the relations between knowledge and embodiment. This is because the sexed body is taken merely as an object of knowledge production, not understood as its starting point. Thus, although the (bio)scientific conception of the body allows for studying how the body conditions *some* knowledge claims, it also makes it plausible to assume that these relations of conditioning can be adequately described by empirical sciences: biological, psychological, social and historical. Hence, this conception also makes it plausible to think that at least some knowledge claims made within these sciences can be accepted and used as an unproblematic starting point for feminist epistemology, an assumption which is made, for example, by some naturalized epistemologies (e.g., Duran, 1991).

The main objective of my article is to suggest alternatives to the sex/gender distinction and the (bio)scientific concept of the human body implicit in the distinction. I shall argue that the phenomenological tradition, especially what has come to be called *phenomenology of the body*, offers promising possibilities for feminist theorists who want to talk and theorize about women. Phenomenology of the body introduces several concepts – for example the concepts of *sediment* and *style* – which refuse to abide by the familiar dichotomies: biological or cultural, corporeal or intentional, object or subject. Here the human body is ultimately not an object of knowledge, but a precondition for all objects and all knowledge claims. The body is our basic framework of meaning and truth, both the source and the "sedimentation" of significations and values, which constantly reproduces itself in practices and actions. Its relations to the world are not external relations of cause and function but internal relations of dependence. From this conception, I will argue, it becomes possible to understand the intimate connection between feminine and female, gender and sex, not by reducing one to the other, but by studying both as units or aspects of the same system of meanings and values.

Specifically, I shall argue that the phenomenology of the body as founded by Maurice Merleau-Ponty and Simone de Beauvoir – and critically developed by Luce Irigaray – offers a serious alternative both to the sex/gender distinction and theories which rely on it and to the sceptical views emerging within post-feminist thought that would reject the concepts *female*, *feminine* and *women* as relics of essentialist thinking (e.g. Fraser and Nicholson, 1990: 34–35, Haraway, 1991: 155, 170; Flax, 1993: 26–28). Anglo-American feminists do not usually connect de Beauvoir's work with Merleau-Ponty's theory. My claim, however, is that, despite differences in focus and contraries in details, de Beauvoir and Merleau-Ponty have a common phenomenological approach to body and sexuality, and that if this is realized, then de Beauvoir's statements about "the second sex" open up in fruitful new ways.

THE AMBIGUITY OF THE SEX/GENDER DISTINCTION

As said, the bulk of feminist criticism and theory produced in English is largely based on the conceptual distinction between sex and gender. The distinction is assumed to be self-evident and has been cited as the starting point of *all* feminist theory by some sources (e.g., Tuttle, 1986: 123; Humm, 1990: 84).[5]

It is somewhat surprising that such a widely accepted distinction is not completely unambiguous: the terms "sex" and "gender" are used in two different ways. In the following, I will distinguish between the *substantive* and the *criterial* senses of "the sex/gender distinction".[6] This section will make the two senses explicit and show that, although historically connected, they do not imply each other.

In feminist theory the sex/gender distinction was first made *substantively* by listing the attributes which belong to the two categories. The feminists of the 1970s defined "gender" as the *mental* and *behavioral* differences between women and men; "sex" was restricted to *genetic*, *hormonal*, and *organic* features (e.g., Millett 1969/70; Greer 1970/78; Oakley 1972). So, for example, the claim that women are more emotional than men would, according to this definition, be a claim about gender, rather than about sex.[7] Similarly, Carol Gilligan's (1982) feminist theory about the differences in women's and men's moral thinking would, on this understanding, concern gender differences.

The substantial use and definition has since given way to a *criterial* one which does not give the content of the category directly but only a criterion for belonging to it. "Gender" has come to refer to any differences between women and men – be they mental, behavioral or anatomic – which have their origin in society and culture. Correspondingly, "sex" refers to those properties which are biologically given. Thus, on the criterial understanding, it is the causal origin of a characteristic that determines whether it is to be classified as a sex or a gender feature.

It is important to notice that the criterial definition, unlike the substantive definition, leaves the content of each category a matter for empirical research to decide. In other words, the definition does not state explicitly which attributes belong to the category of sex and which to the category of gender. Maggie Humm's definition in her *Dictionary of Feminist Theory* is a good example of this openness:

> Gender: a culturally-shaped group of attributes and behaviors given to the female or the male (Humm, 1990: 84).

Currently, there are many different, competing hypotheses about the content of the category of gender. *Biological determinism* and extreme *social constructivism* represent extreme positions here. According to the first, most differences between women and men – perceptual, intellectual, emotional and social – are determined by their different biological structures, independent of social environment (e.g. Moir and Jessell, 1989).[8] According to the second, only certain attributes directly related to reproduction, if any, are biologically determined.[9]

Although dictionaries and encyclopedias today define the sex/gender distinction often criterially, the substantive use is still found in feminist theory and women's studies. For example, Lisa Tuttle's definition in her *Encyclopedia of Feminism* combines the criterial and the substantive understanding. On the one hand, Tuttle presents the idea of cultural influence, and, on the other hand, she restricts "gender" to the mental:

Gender: Whereas sex refers to the biological, anatomical differences between male and female, gender refers to the *emotional* and *psychological* attributes which a given *culture expects* to coincide with physical maleness and femaleness (Tuttle, 1986: 123; italics mine).[10]

This combining of the criterial and substantive senses of a sex/gender distinction is not uncommon in recent feminist discussions. "Gender" is often used to refer to: (1) mental and behavioral attributes (2) which are determined culturally and socially. The implicit assumption here is that *only* mental and behavioral differences between women and men have their origin in culture and society. In other words, bodily differences are supposed to be largely independent of socio-cultural factors. Hence, it is not just bio-determinist hypothesis about natural behavioral differences which are rejected; also the possibility that the bodily differences between the sexes might have their origin in culture and society is ignored.

Recent criticisms of the sex/gender distinction have called attention precisely to this limitation. Moira Gatens argues, for example, that the distinction is problematic because it reproduces the "Cartesian" distinction between mind and body (1983/91: 144, 156, n. 23).[11]

The problem Gatens raises is real, but it would be misleading to claim that it is present in all gender theory. To be more precise, the mind/body distinction is taken for granted only in those theoretical frameworks in which the sex/gender distinction is used substantively. This is certainly a common way of using the words, but not dominant nor necessary; "sex" and "gender" are often understood in a purely criterial way which does not *presuppose* the mind/body distinction. Rather the problem with the criterial use is the separation of nature and culture.[12] The mind/body and nature/culture distinctions should not be confused, although both might be dichotomies in the sense feminist criticism has proved problematic (Cixous, 1975/86; Jay, 1991). They coincide only in those contexts where the substantive and criterial uses of a sex/gender distinction are conflated.

So, in summary, there are several layers to the vagueness of the sex/gender distinction. Firstly, there is the substantive as well as the criterial sense of the distinction. Secondly, these two senses are often combined. Finally, when they are not recognized as distinct, their relationship to the mind/body and nature/culture dichotomies become unclear.

The foregoing suggests that, at minimum, the sex/gender distinction suffers from manifold ambiguities and, insofar as some of its senses presuppose distinctions feminists have themselves criticized, is problematic for feminist theory. Both problems will figure in my efforts to present an alternative to the sex/gender distinction. But I want also to note that the distinction and its several senses are understandable when viewed from a historical perspective. For one thing, the substantive and the criterial senses of the distinction were connected by way of arguments in Anglo-American feminist theory and women's studies in the 1960s and 1970s. In central feminist texts of the period, the two-step move I have summarized is made repeatedly: first the term "gender" is fixed substantively to refer to mental and the behavioral features, and then *it is argued* that gender – thus defined – is a product of social and cultural forces (Millett, 1969/70: 26–33; Oakley, 1972: 16; Greer, 1970/78: 17–22; Oakley, 1981/82: 41–92).

This line of argument, however, did not originally appear in a feminist context but in a very specific psychoanalytic and psychiatric discourse. When making the distinction between sex and gender, the feminist of the 1960s and 1970s relied on the work *Sex and Gender* (1968) by the psychoanalyst Robert Stoller.

To my knowledge, the Australian philosopher Moira Gatens was the first to consider how Stoller's work both affected and molded early Anglo-American feminist discussions (Gatens, 1983/91). A consideration of the questions Gatens raised about Stoller's formulation of the sex/gender distinction and the role of his work shaping Anglo-American feminist theory, serves as my starting point in articulating an alternative understanding of the sexed body. Stoller was of course not the first to distinguish between biological and cultural aspects of sex or sexuality.[13] I focus on his work because the very specific means, concepts and metaphors he used to make the distinction became – via the two-step argument I outlined above – central in Anglo-American feminist theory.

THE MAKING OF GENDER

Robert Stoller's *Sex and Gender* is not about the emancipation of women; it is about "sexual abnormalities", especially transsexuality and transvestism.[14] In a manner which today seems commonplace, the psychoanalyst argues that such behavioral patterns do not follow directly from the biological characteristics of the individuals in question. In this argument the sex/gender distinction is crucial.

First, Stoller limits the term "sex" substantively to refer to the biological factors of sexuality: anatomy, physiology, organic and hormonal features (1968/74: vi–vii, 5–9).[15] Behavioral and mental properties are defined as gender characteristics. Stoller's list of attributes which constitute "gender" includes feelings, thoughts, fantasies, identities, roles, postures, expressions, and movements, among others (1968/74: v–vii, 9–10).[16]

The biological and the social are basically divided into two distinct realms in Stoller's work. In the realm of the biological there are *female* and *male* beings and attributes, and in the realm of social there are *feminine* and *masculine* persons and features (Stoller, 1968/74: 9).

The two central arguments of the work concern the relationships between the sex and gender features. Today the general line of these claims has become familiar and even seems self-evident, but at that time it was dependent on detailed argumentation. First, Stoller pointed out that sex and gender features do not always correspond to each other: females do not always behave in a feminine manner or realize feminine features, and males do not always act in ways considered masculine (1968/74: vi–vii, 9). Because there is not always a correlation between sex and gender characteristics, it cannot be the case that sex (at least sex alone) determines gender.

Stoller's second thesis constitutes an alternative explanation for the formation of gender features. He argues that most behavioral and mental features, above all "gender identity", are determined by the combined effect of biological and sociocultural factors. According to Stoller, the major social factor is the early mother–child relationship (1968/74: 14–15, 23, 48, 65, 71).

So, Stoller's work outlines a picture in which two types of "force", biological and social, influence two types of beings, female and male. In this picture, most mental and behavioral features are formed as the sum total of two different vectors, although there are some that may be determined by purely biological factors, for example, penile erection during sleep (Stoller, 1968/74: 16).

As already pointed out, the feminists of the 1960s and 1970s adopted this explanatory strategy. Kate Millett (1969), Germaine Greer (1970), Ann Oakley (1972) and Nancy Chodorow (1978), among others, applied Stoller's sex/gender framework rather directly in their analyses of relationships between women and men. Feminists used Stoller's distinction to support their argument that the social relations between the sexes are not biologically determined. Like Stoller, they argued that most behavioral patterns result from the causal interaction between biological and social factors. Consequently, society is part of the explanation for women's subordinated position.

It is essential to note that when adopting Stoller's terminology feminists simultaneously adopted his notion about the relationship between sex and gender. The two crucial factors in this notion are *the idea of a biological base* for sexual differences and *the causal explanatory framework.*

Note, first, the presumption that human beings are naturally divided into two categories, the male beings and female beings. The question was why and how male beings typically develop into masculine and female beings feminine. The assumption that the categories of female and male reflect "natural kinds", i.e. are not themselves socially-constructed categories, was not problematized.

Note, as well, that the process of becoming gendered was understood causally: masculinity and femininity were described as the combined effect of social and biological causes. So, although Stoller and his feminist successors questioned the claim that mental and behavioral differences are caused by biological ones, they never questioned the adequacy of the idea of causation as such. The idea of a simple biological causal chain was replaced by a hypothesis about complicated causal interaction between the biomechanism and its social environment.

Metaphors of production proved useful here to the feminist discourse about sex and gender. Sexed bodies were understood as natural resources that the patriarchal society uses for making gender-products necessary for its stability and preservation. Bodily actions and behaviors were described in mechanical terms, as conditioned responses to internal and external stimuli. The implicit philosophy of action was behavioristic.[17] One did not *become*, but was *made* a woman (e.g., Millett, 1969/70: 32; Greer, 1970/78: 17ff.; Oakley, 1982: 41ff.).

Since the 1980s, this understanding of how women and men are made has been problematized in two ways. First, the criticism has been directed at the concept of sex. Some feminist scholars have rejected the idea that human beings naturally – before the making of gender – fall into two separate classes. Holly Devor sums up this line of criticism by suggesting that "in the final analysis, it may well turn out that much as our social selves are limited by our flesh and blood, so too are our corporeal selves limited by our social experience and the meanings we attach to them" (Devor, 1989: 19).

So, if Anglo-American feminists have so far mainly concentrated on finding out how females and males are made feminine and masculine, some of them are now extending the idea of production to sex and asking how the female/male difference is produced. The question is not just how the treatment and environment of infants affect them mentally, but also how these factors affect their bodily functions, hormonal systems and organs, for example their brains.

While I will later suggest that this line of argumentation is not the only or the most fruitful way to question the sex/gender distinction – and advocate a phenomenological approach – I will lay the groundwork for my suggestion by more fully explicating the idea of sex as a product and noting what I take to be its limitations. As this idea has been largely undertaken by Anglo-American feminist science critics, the discussion of the next section also suggests the possibility of a more radical feminist critique of bioscientific assumptions about sex than Anglo-American feminist science critics have recognized.

BODY AS A PRODUCT

The idea of sex as a social product or a cultural construct can be found in present feminist discussions in several different contexts. To begin with, feminist biologists have suggested that, from the standpoint of biology, a strict universal male/female division is invalid. Gisela Kaplan and Lesley Rogers, among others, have argued that human individuals do not fit into the male/female categories which are given in medical publications and biology textbooks, and which also often function as the starting points of feminist arguments.[18] Kaplan and Rogers conclude that the division – as it is generally represented – is "only a convenient social construct" (Kaplan and Rogers 1990: 214).[19]

For a start, the critique is aimed at our ways of using the concepts of male and female. The basic idea is that the prevailing conventions observe the social rather than the natural order. Feminist biologists have pointed out two implicit presumptions.

First, the male/female division is often presumed to be universal: together maleness and femaleness are supposed to cover the entire class of humans or, more generally, mammals. This means that no one is excluded by or falls between the parts of the division. An individual who would be neither male nor female would be an impossibility.

Secondly, maleness and femaleness are generally understood as mutually exclusive properties. It is believed that they cannot coexist in one human being. The reasoning is often very straightforward: if an individual is male, he cannot be female, and vice versa: if an individual is female, she cannot be male.[20]

Neither presumption stands up in a closer study. Feminist biologists argue that irrespective of whether sex is explicated with the help of glands, chromosomes or hormones it does not form a dichotomy.[21] In other words, the physiological components of the sexual differences are neither individually nor together enough to divide people into two separate categories. The dichotomy female/male is based on something other than physiological facts.[22]

In light of these arguments the sex/gender theoreticians' problematic is misleading. It is not enough that we ask how females become women and males become men. In addition to this – or, in fact, instead of this – we must find out why even those human beings who do not fit the physiological descriptions of female and male become women and men. The problem is to explain how physiological *diversity* adapts to a social dichotomy.

However, the primary target of the feminist biologists' arguments is not the feminist theory about gender but rather the presumed naturalness of the female/male dichotomy which governs behavioral sciences more generally. The presumption is dominant in both ethological research on animal behavior and neurological studies on the cerebral basis of cognitive abilities (see, e.g., Kimura, 1992).[23]

Discussions and practices of medicine and psychiatry are also affected by the idea of the naturalness of the female/male dichotomy. This becomes particularly evident in the process of determining the sex of a child: traditionally an infant has been categorized as female if it lacks a penis. This procedure has also been followed in situations in which the infant's genitals have not met the criteria of femaleness. In other words, ambiguous cases have been categorized as females (Unger and Crawford, 1992: 211, 215). The underlying presumption is that every human individual must be either male or female.

Even though chromosomes, sex glands and hormones instead of external genitalia are nowadays regarded as the factors defining sex, medical practices and discourses are still directed by an effort to maintain the binary division. Susanne J. Kessler has studied the grounds on which medical experts make the decision in the case of a so-called intersexed child. The decision is based neither on the child's chromosomal structure nor on its sex glands. The important thing is to find out whether the XY individual's external genitalia can be formed into a penis of normal appearance and size by means of surgery and hormonal treatment. If the experts deem this possible, the infant is categorized as male. If, on the other hand, it looks like treatment will not succeed in producing a normal penis, the XY individual is classified as female (Kessler, 1990/94: 223, 227–228).

Despite the radical nature of the operations, physicians describe the procedure as "repairing", "reconstruction" and "completion" of sex. The chromosomes, hormonal system and organs are treated as ambiguous signs, and a "natural", "real" dichotomy of sex is presumed to exist behind them (Kessler, 1990/94: 225–7, 231–232).[24]

So, the attempt to preserve the binary division extends beyond mere thinking to material reality. Physically deviant individuals are fitted to the male/female dichotomy by means of medical operations. This practice of making sex, however, is not restricted exclusively to "treatment of deviations". Feminist biologists and psychologists have argued that all individuals are subjected to similar physical manipulation. The claim is that the normal female or male body is not the result of the internal processes of the biomechanism but of the causal interaction between the mechanism and its socio-cultural environment.[25]

A central point in these arguments is the questioning of causal order. Feminist biologists have rejected the idea of a causal chain which proceeds from genetic to hormonal differences and from there to organic differences. They have stressed that the studies pertaining to the relationships between genes, hormones and anatomy only prove correlations, not causal links. The evidence allows for the presumption that environmental factors produce different experiences which, in turn, have an effect on hormonal functions and thus also indirectly on other organic features, for example differences in the nervous system. Holly Devor crystallizes this view by saying that "our bodies may themselves be shaped by roles that we play every day of our lives" (Devor, 1989: 19, cf. 36).

Support for this alternative explanation comes from the fact that in most known cultures the environment and treatment of a newborn (even an unborn) child depends on how it is classified. The question is not of conscious choice or decision, but of the very slight, almost unperceptible differences in the ways in which children are treated, of the rhythms and intensities of the interaction.[26]

Therefore, it can be stated that the feminist criticism of the male/female division contains a second vein. It is not only concepts and conceptions which are under scrutiny, but also *material reality*. The aim is not only to study how social conventions influence theories and ideas of the body but also to find out *how bodies themselves are formed*, i.e. how we make ourselves and our children male and female. According to this critique even bodily sexual differences are social and cultural phenomenoma. The human body is the result of all the (inter)action in which it participates and has participated.

This is an important new development in feminist thinking: the sexed body is no longer conceived as a natural given. However, the idea can be – and has been – formulated in different ways, and these formulations are not themselves innocent but carry with them assumptions concerning the nature of behavior and action.

For example, the insight is sometimes expressed using the concept of gender and the related metaphors of production. Judith Butler, for instance, states in her *Gender Trouble* that "gender ought not to be conceived merely as the cultural inscription of meaning on a pregiven sex", but "must also designate the very apparatus of production whereby the sexes themselves are established" (e.g. Butler, 1990a: 7). Similarly, Thomas Laqueur resorts to the concept of gender when describing the new development in feminist research: "some of the so-called sex differences in biological and sociological research turn out to be gender differences after all. ..." (1990/92: 13). It is important to notice that although this kind of formulation abandons the idea of pregiven sex, it still retains the basic structures of sex/gender thinking. Two central features remain intact: the separation of the natural from the cultural and the causal explanatory framework.

The metaphors of machinery and apparatus bring with them the idea of raw material, i.e. a natural substance that is prior to and independent of the process of production. In traditional sex/gender thinking females and males were treated as the raw material of gender production. The new hypothesis is that even females and males are products; so, there must be something that precedes these constructions and passes as their raw material: a body that is free from sexed categories of culture. Even though the line between naturally given and culturally produced is drawn in a new way, it is still there.

This is why Butler, who wants to get rid of all nature/culture distinctions, needs to elaborate on her claim by adding that the production of sex is a process which generates even its own raw material (1990a: 37–38, 73–75). Her elaboration, however, is just another way of stating that the process in question should not be conceived as a process of production, in the usual sense of the word. In effect, Butler argues that becoming sexed must be conceived in another way: as repetitive and citational *action* (1990a: 144–145; 1994: 9–11, 107–109).[27]

The imagery of production also suggests a certain understanding of the relation between the natural body and the culture that makes it sexed. It suggests that the interaction between them is causal by nature: the bodily material and the machinery of sex are conceived as two separate systems that can be described independently.[28] So, when the sex/gender distinction is rejected by claiming that even sex is a social product, part of the "logic" of the distinction remains intact.

In the following, I shall argue that feminist tradition includes more radical – and more fruitful – ways of criticizing the notions of sex and gender, and the theories based on them. Some of the most promising approaches question more than just the assumption that women and men fall naturally into two separate classes. Instead of substituting social causes for biological ones, they suggest that the concepts of cause and effect on the whole are not adequate in analyzing sexual differences. Causal thinking in general must be questioned if sexuality is to be understood in all its complexity. This line of criticism leads us to the concepts of phenomenology which are very different from the ones Anglo-American feminisms have inherited from the debates of the 1960s and 1970s.

The purpose of the last section of my article is to introduce an alternative conceptual framework for understanding bodies, sexes and genders, and their relations. It is based on the phenomenology of body developed by Maurice Merleau-Ponty and Simone de Beauvoir, in the 1940s. My intention is not to suggest that the works of Merleau-Ponty and de Beauvoir have been forgotten in later feminist discussions. On the contrary, it seems that even some of the most recent developments in feminist theory – including Butler's work – are based on the ideas first voiced by them.[29] However, I believe that the differences between the phenomenological approach and the approaches which start from a sex/gender distinction have remained largely unrecognized. My intention is to bring these differences to light and, by doing so, make way for a new perspective within recent feminist discussions about body and sex.

STYLES OF BEING

Simone de Beauvoir starts her *Le deuxième sexe* by presenting a strong argument against all causal theories about the relation between women and men. She studies three dominant explanatory schemes – biology, psychoanalysis and historical materialism – and concludes that none is adequate. Even taken together, she argues, they fail to make the phenomenon understandable. A mere description of biological, psychological and social facts and the causal relations between them cannot resolve the problem, because what we are ultimately faced with is a question of *values and meanings*:

> ... the body, the sexual life and the resources of technology exist concretely for human beings in so far as they grasp them in the total perspective of their existence. The value of muscular strength, of the phallos, of the tool, can be defined only in a world of values (1949/93: 106; 1949/87: 91).

Thus, we can see the grounds for the received understanding according to which de Beauvoir is the founder of the sex/gender distinction.[30] She certainly uses the bioscientific concept of *female*. She also presents the psychological, sociological and historical ideas that crystallize in the concept of *gender*. But we can also see what is wrong with this interpretation: de Beauvoir does not introduce these notions as foundations for a theory about women. On the contrary, she wants to show that as such they are useless.[31] Her conclusion is that the sexual relation can be understood only in the contexts of existentialist philosophy (1949/93: 109, 1949/87: 93).

It is usually assumed that the existentialism that de Beauvoir refers to is Sartre's dualist theory about the two separate realms of being: being-for-itself and being-in-itself (e.g., Butler, 1990a: 12, 129; Chanter, 1995: 49–79). This, however, is a mistake. Michèle Le Dœuff, Sonja Kruks and Eva Lundgren-Gothlin have convincingly argued that de Beauvoir developed a theory of human existence that differs radically from Sartre's (Le Dœuff, 1979/80, 1989/91; Kruks, 1990, 1992; Lundgren-Gothlin, 1991, 1995).[32] I have argued elsewhere that de Beauvoir's view of the corporality of human existence comes much closer to the antidualist philosophy of Maurice Merleau-Ponty than the dichotomous thinking of Sartre (Heinämaa, 1995).[33] Here my aim is to make clear how this view of the body and its sexed nature differs from those views of the body – as, for example, a natural object – developed by feminists who have accepted the sex/gender distinction.

The starting point is the phenomenological realization that the body is not an object, but the *condition* of all objects and all our knowledge of them. It is a necessary, mobile perspective that makes objects possible for us. This means that the relations between the living perceiving body and the objects perceived cannot be similar to those between objects: causal or functional. The body is intertwined with its objects. The relation is an internal relation of dependence: the relata cannot be separated without losing essential parts of them.

When the human body is not conceived as an object, sexuality and sex cannot be understood as attributes. Instead they are theorized as modes, or *styles*, of being. This alternative conceptualization has two interesting consequences for the notions of sexual identity. First, the unity of style is not to be found beneath or behind the concrete actions which are conceived as its manifestations. For example, the expressions of a personal style – the posture, the smile, the hair cut and the clothes – are not held together by any common cause, physical or mental. Neither do they materialize a common form. Instead, their unity is like a web or a fabric of partial and varied connections.[34] Likewise femaleness, as a style of being, cannot be pinned down by a common source or form; it can only be conceived by studying its concrete manifestations and the various relations between them. This is exactly what Simone de Beauvoir sets out to do in *Le deuxième sexe*; she refuses to reduce femaleness or womanhood to any single origin, and instead proceeds to describe its "concrete realizations" (1949/93: 43, 1949/87: 42).

Second, the style is not a constellation of fixed qualities or actions but an open, incomplete structure. A personal style, for example, is not a collection of actions, but a way of acting: thinking, writing, dancing, throwing, breathing, etc. It runs through the whole life like a melody: there is no core, no specific invariant, no common quality or number of them present. In Martin C. Dillon's words, the unity of style is "an adverbial unity" (1988: 79). Thus, if sexuality is a style, we cannot understand womanhood, femaleness or femininity by focusing on specific actions, for example childbirth. We have to study the whole of action, and try to find its tones and melodies (Merleau-Ponty, 1945/93: 194, 378; 1945/62: 166, 327; de Beauvoir, 1949/93: 43; 1949/87: 41).[35]

In his *Phénoménologie de la perception* Merleau-Ponty describes this change of perspective by saying that

> ... the sexual life is not a mere effect of the processes having their seat in the genital organs, the libido is not an instinct, or an activity directed naturally towards definite ends, it is the general power, to which the psychosomatic subject adheres in different settings, of establishing himself through different experiences, of gaining structure of conduct (1945/93: 185, 1945/62: 158).

He goes on to compare sexuality to odors and sounds that spread from special organs to the whole body and objects around it. Sexuality proves to be a kind of intentionality that structures all activities in the same way as an atmosphere or mood shades the whole world. Thus femaleness and femininity cannot be understood as two separate attributes, but must be conceived as two variations of the same structure. They are, on this view, units or aspects of the same system of meanings and as such inseparable, unintelligible without each other.

(Bio)scientific descriptions of the body and sex are understood as generalizations made for specific purposes. Chromosomes, hormones and gonads cannot by themselves make the body and its modes of being understandable. They do not unfold what is, but only give us useful abstractions of the variation of our living experience (de Beauvoir 1949/93: 43, 1949/87: 41).

It is possible to theorize about the body in this way only if the traditional identification of the intentional as mental is given up. Both Merleau-Ponty and de Beauvoir see the body, not the consciousness, as a seat of intentionality: of meanings, values and significations. The body is ultimately understood as a system – or better as a sedimentation – of values and meanings, created by former bodily acts: postures, gestures and movements (Merleau-Ponty, 1945/93: 179, 217–221, 1945/62: 153, 186–189).[36]

The values and meanings that are crucial here are not the ones forced on us by others – the society – but those that we realize in our own actions. They are not external to the body, but its own (re)creations. This is why de Beauvoir writes that "one becomes [*devenir*] a woman" (1949/93: 13, 1949/87: 295), not that one is made such.

We can now understand better de Beauvoir's insistence on the claim that sexual difference cannot be understood merely by studying objects, events and their causal relations. Her idea is that this difference is so deep in our existence that no scientific abstraction can account for it. It is a structure of our way of being, not a specific object, attribute or a collection of them.

So, in one sense the sexual difference is a necessary part of our being. Merleau-Ponty expresses this idea by saying that "a handless or sexless man is as inconceivable as one without the power of thought" (1945/93: 198, 1945/62: 170).[37] Simone de Beauvoir seems to deny this in *Le deuxième sexe* when she writes that we can imagine a non-sexed society but not an immortal man (1949/93: 40, 1949/87: 39). The tension is dissolved in a common understanding of the nature of necessity: both agree that, in human existence, necessity is not constant but changing. Merleau-Ponty's characterization clarifies this point:

> ... there is in human existence no unconditioned possession, and yet no fortuitous attribute. Human existence will force us to revise our usual notion of necessity and contingence, because it is the transformation of contingency into necessity by the act of repetition (1945/93: 199, 1945/62: 170).

Both Merleau-Ponty and de Beauvoir understand the body as a system of signification. This means that its structures and relations are arbitrary in the sense that they are based on nothing else than earlier acts of signification. However, it does not follow that the structure or even one of its links, can be changed by an individual act. In this sense the connections are necessary: they make new acts possible.[38]

But, like all structures of signification, this one, too, is maintained by nothing more than repetition – and thus it is also changed in repeated acts of deviation and subversion.

CONCLUSION

Maurice Merleau-Ponty's and Simone de Beauvoir's descriptions of body and sex offer an alternative both to the feminist theories based on the sex/gender distinction and to the critiques that reject this distinction by conceiving even the body as a product. The phenomenological approach opens the way for a more radical critique of the (bio)scientific conception of sex than the critiques which just change the causal order postulated by biological determinists.

According to the phenomenological understanding, the words "female", "feminine" and "woman" do not refer to objects or attributes described in biological, social or psychological sciences. Rather they characterize ways of handling objects and modes of combining attributes, that is, styles of being.

Hence, when Luce Irigaray writes about "female desire", "female language" and "female imaginary" (1977/83: 29), she does not fall back on biological determinism or postulate an eternal essense.[39] But it is also misleading to claim that her text is merely about language or discourse.[40] Instead of theorizing about biological bodies or literary modes, she describes a certain *corporeal style* which has been reduced by "a discursivity that can not conceive of it", but which still "resists and explodes every firmly established form, figure, idea or concept" (Irigaray, 1977/83: 76).

ACKNOWLEDGEMENTS

I am grateful to the editors of this volume for valuable comments on an earlier version of this paper, and owe special thanks to Lynn Hankinson Nelson, Martina Reuter and Ritva Ruotsalainen for their critical insights and helpful suggestions.

Academy of Finland,
University of Helsinki, Department of Philosophy

NOTES

[1] Especially in the United States and Great Britain. In Canada and Australia, the feminist tradition has been less uniform in this respect.

[2] Cf. Bordo, 1990: 1135–1136; Butler, 1990b: 324–325; Heinämaa, 1993.

[3] For this aspect of causal relations, see, e.g., Melden, 1961/64: 43–55; Stoutland, 1970; von Wright, 1971/93: 83–131.

[4] Cf. Gatens, 1983/91: 150; Birke, 1986: 88–91.

[5] The distinction is not central beyond Anglophone feminist studies, where theories are formulated with very different means. The most fruitful alternative seems to be the concept of *sexual difference* developed by the French feminists, above all Hélène Cixous and Luce Irigaray. It is rooted in the philosophical discussion about the possibility of genuine, irreducible differences between subjects which, via the works of Jacques Derrida, dates back to the writings of Husserl and Descartes. See, e.g., Cixous, 1975/86; Irigaray, 1974/90, 1984. Also some theorists taking part in the Anglo-American theoretical discussions use the concept of sexual difference. See de Lauretis, 1987; Braidotti, 1991, 1994; Grosz, 1994.

[6] As always when technical vocabulary is introduced, nothing in my argument hinges on the specific *words* "substantive" and "criterial".

[7] On "women's emotionality", see Heinämaa and Reuter, 1994.

[8] Cf. Kimura, 1992.

[9] For extending the strong constructionist hypotheses to corporeal – hormonal and cerebral – features, see, e.g., Devor, 1989; Kaplan and Rogers, 1990.

[10] Cf. Flax, 1990: 49.

[11] Cf. Butler, 1990a: v–ix, 1–34; Flax, 1990: 147; Grosz, 1994: 16–18; Chanter, 1995: 24–25.

The claim that there is a *Cartesian* mind/body distinction implicit in sex/gender concepts is problematic. Firstly because feminists rarely approve of the idea of two different substances dominant in Descartes' *Meditations de Prima Philosophiae* (1641/90). Secondly, the claim is complicated by the fact that this is not the only way Descartes describes the mind/body relation. In *Passions de l'Ame*, the mental and the bodily seem more like attributes of one substance – the human being – as two separate substances (Descartes, 1649/90).

[12] Chanter, 1995: 41–42.

[13] Stoller's work is largely based on the concepts developed by the psychiatrists John Money, Joan Hampson and John Hampson at the Johns Hopkins Hospital in the 1950s; cf. Ruotsalainen, 1995. Similar distinctions were already made by Sigmund Freud (1920/90) and Margaret Mead (1950/62).

[14] Moira Gatens (1983/91) criticizes Stoller's work for approaching these phenomena primarily from the point of view of male individuals.

[15] In addition to these, Stoller postulates a special "biological force" to explain those forms of behavior which he was unable to make understandable through observable biological and social factors (1968/74: 48, 65–85).

[16] When making the distinction, Stoller refers to Freud's essay 'A Case of Homosexuality in a Woman' (Freud, 1920/90). In it Freud distinguishes between three aspects of (homo)sexuality: physical features, mental features, and the choice of object (1920/90: 389–399). Stoller identifies his concept of gender with the second and the third item in Freud's list, and thus substitutes a dualism for Freud's three-fold division.

[17] Here "behaviorisms" should be understood in the wide sense of the word. Thus, it is not restricted to B. F. Skinner's stimulus–response approach but includes also later functionalist theories which allow for reference to the internal "mental" states of the organisms. The common assumption of classical behaviorist and recent functionalist psychology is that mental states can be individuated *causally*, i.e. by their *causal* relations to environmental effects (input), bodily behavior (output) and other mental states. Cf. Block, 1978: 261–266; Gatens, 1983/91: 142–143; Churchland, 1984/88: 36–38.

[18] In addition to entire organisms many functions, parts and subsystems of the body are also referred to as "male" and "female". Even in scientific texts, especially in ethology and brain research, the terms "male hormone", "male hormonal system", "male brain" and "male behavioral pattern" are used. See, e.g., Kimura, 1992: 82; cf. Kaplan and Rogers, 1990; Leigh Star, 1991. Studies in the history of medicine have shown that this practice is relatively new. The division into sexes was extended to all parts of the body as late as in the eighteenth century. Before that the human bodies were conceived as being of one kind only. The differences between men and women were regarded as differences in degree: it was believed that the same reproductive organs were "left inside" the female body for lack of vital strength (Laqueur, 1990/92)

[19] Cf. Rubin, 1975: 166; Birke, 1986; Kessler, 1990/94; Oudshoorn, 1994: 1–5.

[20] Cf. Jay, 1991.

[21] In his work *Making Sex* (1990) Thomas Laqueur argues that this oppositional view of sex did not become dominant in medicine until the eighteenth century.

The definition based on sex glands was introduced in medicine in the nineteenth century (Laqueur, 1990/92: 175–181; Kessler, 1990/94: 229). Before that the uterus was considered the defining factor of femaleness (Laqueur, 1990/92: 149–154). Only in the twentieth century have hormones and chromosomes replaced definitions based on organs (Oudshoorn, 1994: 9; Ruotsalainen, 1995: 7). Some scholars have suggested that these new definitions provide the opportunity to separate femaleness from reproduction and give room for its erotic and aesthetic functions (Kessler, 1990/94: 228).

[22] Besides the female and male chromosomal structures (XX and XY) there are different variations. For example, the XO-individuals (so-called Turner's syndrome) are like XX-individuals with respect to bodily appearance and external genitalia but have undeveloped internal genitalia and are therefore infertile. Variations in the sex chromosomes account for approximately three per thousand, i.e. double the number of Down syndrome cases.

[23] Cf. Devor, 1989: 10–19; Kaplan and Rogers, 1990: 206–212; Leigh Star, 1991.

[24] Even today treatment of intersexed infants is guided by the theory of the formation of sexual identity presented by John Money, Joan Hampson and John Hampson in the 1950s. These are the same sources on which Robert Stoller based his sex/gender division; cf. Ruotsalainen, 1995.
[25] See, e.g., Kaplan and Rogers, 1990: 215; Haraway, 1991: 199–200; Leigh Star, 1991: 239; cf. Birke, 1986: 68, 88; Kimura, 1992: 81, 85.
[26] Cf. Stern, 1985.
[27] Isabell Lorey argues that the term "sex" has a double meaning in Butler's text: it refers both to "a situated body" and to "a sexually differentiated anatomy" (Lorey, 1993: 12).
[28] Cf. Birke, 1986: 88–91.
[29] Several feminists have used concepts developed by Merleau-Ponty. Especially, Merleau-Ponty's concept of *body-image* or *body-scheme* seems to be fruitful for feminist theory. For example, Iris Marion Young has applied it in her studies of the differences between women's and men's bodily experiences (1989, 1990). A more recent application can be found in Elizabeth Grosz's *Volatile Bodies* (1994). Cf. also Bigwood, 1991; Heinämaa and Reuter, 1994; Reuter, 1995.

Some feminist theorists, for example Moira Gatens and Elizabeth Grosz, conflate Merleau-Ponty's concept of body-scheme with Lacan's concept of imaginary anatomy. This conflation seems to be fruitful for feminist theory, but it can also turn out problematic because it obscures important differences in Merleau-Ponty's and Lacan's understanding of the subject's relation to language. See, e.g., Merleau-Ponty, 1960/64: 139–141; cf. Dillon, 1988.

De Beauvoir, however, is usually seen in the Anglo-American tradition as a Sartrean existentialist and a founder of the sex/gender distinction, and thus often ignored by those interested in the attempt to develop a non-dualistic theory of sexuality.

The phenomenological concepts and questions are further developed in the writings of the French scholars Luce Irigaray (1984) and Julia Kristeva (1974/84). Cf. Silverman, 1991; Whitford, 1991; Mortensen, 1994; Chanter, 1995; Honkasalo, 1995. In Irigaray's and Kristeva's work, phenomenology of the body goes into an interesting but problematic interplay with psychoanalytic theory and structuralist linguistics.
[30] This traditional Anglo-American interpretation is accepted even by scholars critical of the sex/gender distinction, see, e.g., Butler, 1986, 1990a: 111–128, 1992: 140; Chanter, 1995: 49–79; cf. Heinämaa, 1995.
[31] Cf. Zerilli, 1992; Heinämaa, 1995.
[32] Cf. Lloyd, 1984; La Caze, 1994; Reuter, 1995.
[33] Cf. Kruks, 1992.
[34] Martin C. Dillon illuminates the phenomenological idea of style by comparing it to Wittgenstein's concept of family resemblance (1988: 179). The unity of style is like that of a thread which holds together, not by virtue of one single long strand, but by several overlapping short fibres (Wittgenstein, 1953/81: 66).
[35] On this understanding, the adverb "feminine" is less misleading in characterizing sex and sexuality than the nouns "woman" and "female" which purport to refer to *objects*.
[36] Cf. Dillon, 1988: 127, 186–201.
[37] Cf. Husserl, 1934–37/1954: 191–192.
[38] Cf. Judith Butler's understanding of the heterosexual matrix in *Gender Trouble* (1990a: 145) and *Bodies that Matter* (1994: x–xi, 9, 94–95).
[39] As, for example, Toril Moi (1988: 137–147) and Jane Flax (1990: 171–178) suggest.
[40] As, for example, Toril Moi (1988: 147–148), Jane Flax (1990: 177–178) and Donna Haraway (1991: 174) seem to think.

REFERENCES

Bigwood, Carol: 1991, 'Renaturalizing the Body (With a Little Help from Merleau-Ponty)', *Hypatia*, **6.3**, 54–73.

Birke, Lynda: 1986, *Women, Feminism and Biology: The Feminist Challenge*. Brighton, Sussex: Harvester Press.

Block, Ned: 1978, 'Troubles with Functionalism', in Wade Savage (ed.), *Minnesota Studies in the Philosophy of Science: Perception and Cognition: Issues in the Foundation of Psychology*. Minneapolis: Minnesota University Press.

Bordo, Susan: 1990, 'Feminism, Postmodernism, and Gender-Scepticism', in Linda J. Nicholson (ed.), *Feminism/Postmodernism*. New York and London: Routledge.

Braidotti, Rosi: 1991, *Patterns of Dissonance: A Study of Women in Contemporary Philosophy*, trans. Elizabeth Guild. Cambridge: Polity Press.

——: 1994, *Nomadic Subjects: Embodiment and Sexual Difference in Contemporary Feminist Theory*. New York: Colombia University Press.

Butler, Judith: 1986, 'Sex and Gender in Simone de Beauvoir's *Second Sex*', *Yale French Studies*, No. 72, 35–49.

——: 1990a, *Gender Trouble: Feminism and the Subversion of Identity*. New York, London: Routledge.

——: 1990b, 'Gender Trouble, Feminist Theory, and Psychoanalytic Discourse', in Linda J. Nicholson (ed.), *Feminism/Postmodernism*. New York and London: Routledge.

——: 1992, 'Gender', in Elisabeth Wright (ed.), *Feminism and Psychoanalysis: A Critical Dictionary*. Cambridge: Blackwell.

——: 1994, *Bodies that Matter: On the Discursive Limits of "Sex"*. New York and London: Routledge.

Chanter, Tina: 1995, *Ethics of Eros: Irigaray's Rewriting of the Philosophers*. New York and London: Routledge.

Chodorow, Nancy: 1978, *The Reproduction of Mothering: Psychoanalysis and the Sociology of Gender*. Berkeley, London: University of California Press.

Churchland, Paul M.: 1984/88, *Matter and Consciousness: A Contemporary Introduction to the Philosophy of Mind*. Cambridge, Massachusetts: MIT.

Cixous, Hélène: 1975/86, 'Sorties: Out and Out: Attacks/Ways Out/Forays', in Hélèn Cixous and Catherine Clément, *The Newly Born Woman*, trans. Betsy Wing, edited by Sandra M. Gilbert. Manchester: Manchester University Press.

de Beauvoir, Simone: 1949/93, *Le deuxième sexe I: Les faits et les mythes*. Paris: Gallimard.

——: 1949/91, *Le deuxième sexe II: L'expérience vécue*. Paris: Gallimard.

——: 1949/87, *The Second Sex*, trans. and edited by H. M. Parshley. London: Penguin.

de Lauretis, Teresa: 1987, *The Technologies of Gender: Essays on Theory, Film, and Fiction*. Bloomington, Indianapolis: Indiana University Press.

Descartes, René: 1641/90, 'Meditations of First Philosophy', trans. John Cottingham, in *The Philosophical Writings of Descartes*. Cambridge and New York: Cambridge University Press.

——: 1649/90, 'The Passions of the Soul', trans. Robert Stoothoff, in *The Philosophical Writings of Descartes*. Cambridge, New York: Cambridge University Press.

Devor, Holly: 1989, *Gender Blending: Confronting the Limits of Duality*. Bloomington, Indianapolis: Indiana University Press.

Dillon, Martin C.: 1988, *Merleau-Ponty's Ontology*. Bloomington, Indianapolis: Indiana University Press.

Duran, Jane: 1991, *Toward a Feminist Epistemology*. Savage: Rowman & Littlefield.

Flax, Jane: 1990, 'Postmodernism and Gender Relations in Feminist Theory', in Linda J. Nicholson (ed.), *Feminism/Postmodernism*. New York and London: Routledge.

——: 1993, *Disputed Subjects: Essays on Psychoanalysis, Politics and Philosophy*. New York and London: Routledge.

Fraser, Nancy and Linda Nicholson: 1990, 'Social Criticism Without Philosophy: An encounter between feminism and postmodernism', in Linda J. Nicholson (ed.), *Feminism/Postmodernism*. New York and London: Routledge.

Freud, Sigmund: 1920/90, 'The Psychogenesis of a Case of Homosexuality in a Woman', in *Case Histories II*, trans. under the general editorship of James Strachey. London, New York: Penguin.

Gatens, Moira: 1983/91, 'A Critique of the Sex/Gender Distinction', in Senja Gunew (ed.), *A Reader in Feminist Knowledge*. London and New York: Routledge.

Gilligan, Carol: 1982, *In a Different Voice: Psychological Theory and Women's Development*. Cambridge, Massachusetts: Harvard University Press.

Greer, Germaine: 1970/78, *The Female Eunuch*. New York: Bantam Books.

Grosz, Elisabeth: 1994, *Volatile Bodies: Toward a Corporeal Feminism*. Bloomington, Indianapolis: Indiana University Press.

Haraway, Donna: 1991, *Simians, Cyborgs, and Women*. New York: Routledge.

Heinämaa, Sara: 1993, 'Paikka tutkimuksessa: henkilökohtaisen paikanmäärityksen vaatimus naistutkimuksessa', *Naistutkimus/Kvinnoforskning*, **6.1**, 16–35; in English: 'The Rhetoric of Positioning in Women's Studies', *Nora: Nordic Journal of Women's Studies*, **2.2**, 83–94.

——: 1995, 'What is a Woman? Judith Butler and Simone de Beauvoir on the foundations of the sexual difference'. forthcoming in *Hypatia*.

Heinämaa, Sara and Martina Reuter: 1994, 'Naisten tunneherkkyys: filosofinen keskustelu tunteiden järjellisyydestä', *Naistutkimus/Kvinnoforskning*, **7.1**, 4–20; in English 'Reflections on the Rationality of Emotions and Feelings', forthcoming in Drude von der Fehr, Anna G. Jónasdóttir and Bente Rosenbeck (eds.), *Ambiguous Times*.

Honkasalo, Marja-Liisa: 1995, 'Bodily Subjectivity in "Repressive" and "Avant-garde" Theories: Reflections on Maurice Merleau-Ponty and Julia Kristeva'. Unpublished manuscript.

Humm, Maggie: 1990, *The Dictionary of Feminist Theory*. Columbus: Ohio State University.

Husserl, Edmund: 1934–37/1954, *Die Krisis der europäischen Wissenschaften und die transzendentale Phänomenologie: Ein Einleitung in die phänomenologische Philosophie*. Haag: Martinus Nijhoff.

Irigaray, Luce: 1974/90, *Speculum de l'autre femme*. Paris: Minuit.

——: 1977/83, *Ce sexe qui n'en est pas un*. Paris: Minuit.

——: 1984, *Éthique de la différence sexuelle*. Paris: Minuit.

Jay, Nancy: 1991, 'Gender and Dichotomy', in Senja Gunew (ed.), *A Reader in Feminist Knowledge*. London and New York: Routledge.

Kaplan, Gisela T. and Lesley J. Rogers: 1990, 'The Definition of Male and Female: Biological reductionism and the sanctions of normality', in Senja Gunew (ed.), *Feminist Knowledge, Critique and Construct*. London and New York: Routledge.

Kessler, Suzanne J.: 1990/94, 'The Medical Construction of Gender: Case management of intersexed infants', in Anne C. Herrmann and Abigail J. Steward (eds.), *Theorizing Feminism: Parallel Trends in the Humanities and Social Sciences*. Boulder, San Francisco: Westview Press.

Kimura, Dorren: 1992, 'Sex Differences in the Brain', *Scientific American*, **267.3**, 81–87.

Kristeva, Julia: 1974/84, *Revolution in Poetic Language*, trans. Margaret Waller. New York: Columbia University Press.

Kruks, Sonja: 1990, *Situation and Human Existence: Freedom, Subjectivity and Society*. New York: Routledge.

——: 1992, 'Gender and Subjectivity: Simone de Beauvoir and contemporary feminism', *Signs*, **18.1**, 89–110.

Laqueur, Thomas: 1990/92, *Making Sex: Body and Gender from Greeks to Freud*. Cambridge, Massachusetts: Harvard University Press.

La Caze, Marguerite: 1994, 'Simone de Beauvoir and Female Bodies', *Australian Feminist Studies*, No. 20, 91–105.

Le Dœuff, Michèle: 1979/80, 'Simone de Beauvoir and Existentialism', *Feminist Studies*, **6.2**, 277–289.

——: 1989/91, *Hipparchia's Choice: An Essay Concerning Women, Philosophy Etc.*, trans. Trista Selous. Oxford and Cambridge: Blackwell.

Leigh Star, Susan: 1991, 'The Politics of Right and Left', in Senja Gunew (ed.). *A Reader in Feminist Knowledge*. London and New York: Routledge.

Lloyd, Genevieve: 1984, *The Man of Reason: "Male" and "Female" in Western Philosophy*. London: Methuen.

Lorey, Isabell: 1993, 'Der Körper als Text und das aktuelle Selbst: Butler und Foucault', *Feministische Studien*, **11.1**, 11–23.

Lundgren-Gothlin, Eva: 1991, *Kön och eksistens: Studier in Simone de Beauvoir's Le deuxiéme sexe*. Göteborg: Daidalos; forthcoming in English 1996: *Sex and Existence: Simone de Beauvoir's The Second Sex*. London: Athlona.

——: 1995, 'Gender and Ethics in the Philosophy of Simone de Beauvoir', *Nora: Nordic Journal of Women's Studies*, **3.1**, 3–13.

Mead, Margaret: 1950/62, *The Male and the Female: A Study of the Sexes in a Changing World*. Harmondsworth, Middlesex: Penguin.

Melden, A. I.: 1961/64, *Free Action*. London: Routledge and Kegan Paul.

Merleau-Ponty, Maurice: 1945/93, *Phénoménologie de la perception*. Paris: Gallimard.

——: 1945/62, *Phenomenology of Perception*, trans. Colin Smith. New York: Routledge and Kegan Paul.

——: 1960/64, 'The Child's Relations with Others', trans. William Cobb, in *The Primacy of Perception*, edited by James M. Edie. Evanston: Northwestern University Press.

Millett, Kate: 1969/70, *Sexual Politics*. New York: Ballantine Books.

Moi, Toril: 1988, *Sexual/Textual Politics: Feminist Literary Theory*. London and New York: Routledge.

Moir, Anne and David Jessell: 1989, *Brainsex*. London: Mandarin.

Mortensen, Ellen: 1994, *The Feminine and Nihilism: Luce Irigaray with Nietzsche and Heidegger*. Oslo, Copenhagen, Stockholm: Scandinavian University Press.

Oakley, Ann: 1972, *Sex, Gender and Society*. London: Temple Smith.

——: 1981/82, *Subject Woman*. London: Fontana Press.

Oudshoorn, Nelly: 1994, *Beyond the Natural Body: An Archeology of Sex Hormones*. London and New York: Routledge.

Reuter, Martina: 1995, 'Philosophy Embodied: Merleau-Ponty and a feminist approach to eating 'disorders''. Unpublished manuscript.

Rubin, Gayle: 1975, 'The Traffic in Women: Notes on the "Political economy" of sex', in Rayna R. Reiter (ed.), *Toward an Anthropology of Women*. New York: Monthly Review Press.

Ruotsalainen, Ritva: 1995, 'Kadotettu sukupuoli', *Tiede & editys*, **20.4**.

Silverman, Hugh: 1991, 'The Text of the Speaking Subject: Merleau-Ponty/Kristeva', in Martin C. Dillon (ed.), *Merleau-Ponty Vivant*. Albany: State University of New York.

Stern, Danile: 1985, *The Interpersonal World of the Infant: A View from Psychoanalysis and Developmental Psychology*. New York: Basic Books.

Stoller, Robert, 1968/74; *Sex and Gender: The Development of Femininity and Masculinity*. London: Maresfield Reprints.

Stoutland, Frederick: 1970, 'The Logical Connection Argument', *American Philosophical Quarterly*, **7**, 117–129.

Tuttle, Lisa: 1986, *Encyclopedia of Feminism*. New York and Oxford: Facts on File Publications.

Unger, Rhoda and Mary Crawford: 1992, *Women and Gender: A Feminist Psychology*. New York: McGraw-Hill.

von Wright, Georg Henrik: 1971/93, *Explanation and Understanding*. Ithaca, New York: Cornell University Press.

Whitford, Margaret: 1991, *Luce Irigaray: Philosophy in the Feminine*. London and New York: Routledge.

Wittgenstein, Ludwig: 1953/81, *Philosophical Investigations*, trans. G. E. M. Anscombe. Oxford: Basil Blackwell.

Young, Iris Marion: 1989, 'Throwing Like a Girl: A phenomenology of the feminine body comportment, motility, and spatiality', in Jeffner Allen and Iris Marion Young (eds.), *The Thinking Muse: Feminism and Modern French Philosophy*. Bloomington, Indianapolis: Indiana University Press.

——: 1990, *Throwing Like a Girl and Other Essays in Feminist Philosophy and Social Theory*. Bloomington, Indianapolis: Indiana University Press.

Zerilli, Linda: 1992, 'A Process Without a Subject: Simone de Beauvoir and Julia Kristeva on maternity', *Signs*, **18.1**, 111–135.

CONTRIBUTORS

Karen Barad is Associate Professor of Physics at Pomona College. She teaches in Physics, Women's Studies, and the Science, Technology, and Society program. Her previous publications include articles in theoretical particle physics, science studies, and science pedagogy. She is the animator of "Quarkland", 3D computer-generated animations used most recently to illustrate particle physics concepts for a CD-ROM interactive version of Stephen Hawking's *A Brief History of Time*. Her research in theoretical particle physics has received support from the National Science Foundation. She is currently writing a book tentatively titled *Meeting the Universe Halfway: Realism, Quantum Physics, Feminist Epistemologies, and the Politics of Scientific Knowledge Construction*.

Ronald N. Giere is Professor of Philosophy and Director of the Center for Philosophy of Science at the University of Minnesota. In addition to many papers in the philosophy of science, he is the author of an elementary textbook, *Understanding Scientific Reasoning* (4th edition, 1996), and of *Explaining Science: A Cognitive Approach* (1988). He has also edited several volumes of papers in the philosophy of science, including, most recently, *Cognitive Models of Science* (1992) and *Origins of Logical Empiricism* (1966). Professor Giere is a Fellow of the American Association for the Advancement of Science and a past President of the Philosophy of Science Association.

Susan Haack was educated at Oxford and Cambridge. Formerly Fellow of New Hall, Cambridge, and Professor of Philosophy at the University of Warwick, she is presently Professor of Philosophy at the University of Miami. She is the author of *Deviant Logic*, of *Philosophy of Logics* (both published by Cambridge University Press) and, most recently, of *Evidence and Inquiry: Towards Reconstruction in Epistemology*, published by Blackwell, and of *Deviant Logic, Fuzzy Logic: Beyond the Formalism*, published by the University of Chicago Press. She is the editor of a volume of *The Monist* on 'Feminist Epistemology, For and Against.' She has published numerous articles on philosophy of logic and language, epistemology and metaphysics, philosophy of science, pragmatism and feminism. She is a former President of the Charles Peirce Society, and has received awards for excellence in teaching from the University of Miami and the American Philosophical Association.

Sandra Harding is Professor of Philosophy at the University of Delaware and Adjunct Professor of Philosophy and Women's Studies at UCLA. She is the author

L. H. Nelson and J. Nelson (eds.), Feminism, Science, and the Philosophy of Science, 309–311.

or editor of seven books, including *Discovering Reality: Feminist Perspectives on Epistemology, Metaphysics, Methodology and Philosophy of Science* (1983), *The Science Question in Feminism* (1986), *Sex and Scientific Inquiry* (1987), *Feminism and Methodology* (1987), *Whose Science? Whose Knowledge? Thinking From Women's Lives* (1991), and *The 'Racial' Economy of Science* (1993). She is also co-author of a chapter in UNESCO's *World Science Report 1996* on 'The Gender Dimension of Science and Technology'. Due out in 1997 is her most recent book, *Is Science Multicultural? Feminist and Postcolonial Perspectives.*

Sara Heinämaa is Researcher at the Department of Philosophy of the Academy of Finland, University of Helsinki, and Lecturer at the Christina Institute for Women's Studies at the University of Helsinki. She is the author of *Ele.tyyli ja sukupuoli: Merleau-Pontyn ja Beauvoirin ruumiinfeneomenologia ja sen markitys sukupuolikysymykselle* (*Gesture, Style, and Sex: Merleau-Ponty's and Beauvoir's Phenomenology of the Body and its Relevance to the Question of Sexual Difference*), 1996 (Gaudeamus) and coeditor with Leila Haaparanta of *Mind and Cognition* (Acta Philosophica Fennica, 58, 1995). She is the author of 'Women's Place in Artificial Intelligence: Observations on metaphors of thought and knowledge', in I. V. Eriksson *et al.*, *Women, Work and Computerization* (North-Holland, 1991); 'The Rhetoric of Positioning in Women's Studies', *Nora: Nordic Journal of Women's Studies* (1994); 'On Thoughts and Emotions: The problem of artificial persons', in Haaparanta and Heinämaa, *Mind and Cognition*; and 'What is a Woman: Butler and Beauvoir on the foundations of the sexual difference', *Hypatia*, 1996.

Elisabeth A. Lloyd is Associate Professor in the Department of Philosophy and the Program in Logic and Methodology of Science at the University of California, Berkeley. She received her philosophical training at Princeton and did graduate work in biology at Harvard. Her first book, *The Structure and Confirmation of Evolutionary Theory*, was published in 1988 and released in paperback with a new preface by Princeton University Press in 1994. She is coeditor with Evelyn Fox Keller of *Keywords in Evolutionary Biology* (Harvard University Press, 1992). She has published articles in technical philosophy of science on theory structure and confirmation; in general metaphysics and epistemology, on concepts of objectivity; in evolutionary theory, on hierarchical selection models and model structures; and in general philosophy of science, on explanation, pragmatism, and empiricism. She has recently begun work as general editor of a book series, *The Sciences in Human Context*, for Princeton University Press, and is working on a book concerning bias in evolutionary explanations of women's sexuality (forthcoming, Harvard University Press).

Helen E. Longino teaches women's studies and philosophy of science at the University of Minnesota. She is the author of *Science as Social Knowledge* (Princeton, 1990) and coeditor (with Evelyn Fox Keller) of *Feminism and Science* (Oxford, 1996).

Jack Nelson is Professor of Philosophy and Dean of the Graduate School at Temple University. He is co-author of *The Logic Book*, and has published papers in epistemology and metaphysics, including papers on the logic of identity.

Lynn Hankinson Nelson is on sabbatical from Rowan College, and Visiting Associate Professor of Philosophy at the University of Pennsylvania and Swarthmore College. She is the author of *Who Knows: From Quine to a Feminist Empiricism* (1990), and editor of a volume of *Synthese* on 'Feminism and Science' (*Synthese*, **104**(3)).

Ilkka Niiniluoto, born 1946, studied mathematics and philosophy, received his PhD at the University of Helsinki in 1974, and has worked in Helsinki as Professor of Theoretical Philosophy since 1977. He is Chairperson of the Department of Philosophy, and the President of the Philosophical Society of Finland. His main interests include the philosophy of science (theory change, scientific realism, induction), logic (truthlikeness), epistemology, the philosophy of technology, and the philosophy of culture. His main works include *Is Science Progressive?* (1984) and *Truthlikeness* (1987).

Elizabeth Potter is the Alice Andrews Quigley Professor of Women's Studies at Mills College. She is editor, with Linda Alcoff, of *Feminist Epistemologies* and has recently completed *Gender Politics in Seventeenth Century Science*, a book-length monograph exploring the intersections of gender and the production of Boyle's Law of Gases.

Joseph Rouse is Professor in the Department of Philosophy and the Sciences in Society Program at Wesleyan University. He is the author of *Knowledge and Power: Toward a Political Philosophy of Science* (Cornell University Press, 1987) and *Engaging Science: How to Understand Its Practices Philosophically* (Cornell University Press, 1996).

Nancy Tuana is Professor of Philosophy at the University of Oregon. She is the author of *The Less Noble Sex: Scientific, Religious, and Philosophical Conceptions of Women's Nature* and *Woman and the History of Philosophy*. She is the editor of *Feminism and Science*, *Feminist Interpretations of Plato*, and *Feminism and Philosophy* with Rosemarie Tong.

163. D. Mayr and G. Süssmann (eds.), *Space, Time, and Mechanics.* Basic Structures of a Physical Theory. 1983 ISBN 90-277-1525-4
164. D. Gabbay and F. Guenthner (eds.), *Handbook of Philosophical Logic.* Vol. I: Elements of Classical Logic. 1983 ISBN 90-277-1542-4
165. D. Gabbay and F. Guenthner (eds.), *Handbook of Philosophical Logic.* Vol. II: Extensions of Classical Logic. 1984 ISBN 90-277-1604-8
166. D. Gabbay and F. Guenthner (eds.), *Handbook of Philosophical Logic.* Vol. III: Alternative to Classical Logic. 1986 ISBN 90-277-1605-6
167. D. Gabbay and F. Guenthner (eds.), *Handbook of Philosophical Logic.* Vol. IV: Topics in the Philosophy of Language. 1989 ISBN 90-277-1606-4
168. A. J. I. Jones, *Communication and Meaning.* An Essay in Applied Modal Logic. 1983 ISBN 90-277-1543-2
169. M. Fitting, *Proof Methods for Modal and Intuitionistic Logics.* 1983 ISBN 90-277-1573-4
170. J. Margolis, *Culture and Cultural Entities.* Toward a New Unity of Science. 1984 ISBN 90-277-1574-2
171. R. Tuomela, *A Theory of Social Action.* 1984 ISBN 90-277-1703-6
172. J. J. E. Gracia, E. Rabossi, E. Villanueva and M. Dascal (eds.), *Philosophical Analysis in Latin America.* 1984 ISBN 90-277-1749-4
173. P. Ziff, *Epistemic Analysis.* A Coherence Theory of Knowledge. 1984 ISBN 90-277-1751-7
174. P. Ziff, *Antiaesthetics.* An Appreciation of the Cow with the Subtile Nose. 1984 ISBN 90-277-1773-7
175. W. Balzer, D. A. Pearce, and H.-J. Schmidt (eds.), *Reduction in Science.* Structure, Examples, Philosophical Problems. 1984 ISBN 90-277-1811-3
176. A. Peczenik, L. Lindahl and B. van Roermund (eds.), *Theory of Legal Science.* Proceedings of the Conference on Legal Theory and Philosophy of Science (Lund, Sweden, December 1983). 1984 ISBN 90-277-1834-2
177. I. Niiniluoto, *Is Science Progressive?* 1984 ISBN 90-277-1835-0
178. B. K. Matilal and J. L. Shaw (eds.), *Analytical Philosophy in Comparative Perspective.* Exploratory Essays in Current Theories and Classical Indian Theories of Meaning and Reference. 1985 ISBN 90-277-1870-9
179. P. Kroes, *Time: Its Structure and Role in Physical Theories.* 1985 ISBN 90-277-1894-6
180. J. H. Fetzer, *Sociobiology and Epistemology.* 1985 ISBN 90-277-2005-3; Pb 90-277-2006-1
181. L. Haaparanta and J. Hintikka (eds.), *Frege Synthesized.* Essays on the Philosophical and Foundational Work of Gottlob Frege. 1986 ISBN 90-277-2126-2
182. M. Detlefsen, *Hilbert's Program.* An Essay on Mathematical Instrumentalism. 1986 ISBN 90-277-2151-3
183. J. L. Golden and J. J. Pilotta (eds.), *Practical Reasoning in Human Affairs.* Studies in Honor of Chaim Perelman. 1986 ISBN 90-277-2255-2
184. H. Zandvoort, *Models of Scientific Development and the Case of Nuclear Magnetic Resonance.* 1986 ISBN 90-277-2351-6
185. I. Niiniluoto, *Truthlikeness.* 1987 ISBN 90-277-2354-0
186. W. Balzer, C. U. Moulines and J. D. Sneed, *An Architectonic for Science.* The Structuralist Program. 1987 ISBN 90-277-2403-2
187. D. Pearce, *Roads to Commensurability.* 1987 ISBN 90-277-2414-8
188. L. M. Vaina (ed.), *Matters of Intelligence.* Conceptual Structures in Cognitive Neuroscience. 1987 ISBN 90-277-2460-1
189. H. Siegel, *Relativism Refuted.* A Critique of Contemporary Epistemological Relativism. 1987 ISBN 90-277-2469-5

190. W. Callebaut and R. Pinxten, *Evolutionary Epistemology*. A Multiparadigm Program, with a Complete Evolutionary Epistemology Bibliograph. 1987 ISBN 90-277-2582-9
191. J. Kmita, *Problems in Historical Epistemology*. 1988 ISBN 90-277-2199-8
192. J. H. Fetzer (ed.), *Probability and Causality*. Essays in Honor of Wesley C. Salmon, with an Annotated Bibliography. 1988 ISBN 90-277-2607-8; Pb 1-5560-8052-2
193. A. Donovan, L. Laudan and R. Laudan (eds.), *Scrutinizing Science*. Empirical Studies of Scientific Change. 1988 ISBN 90-277-2608-6
194. H.R. Otto and J.A. Tuedio (eds.), *Perspectives on Mind*. 1988 ISBN 90-277-2640-X
195. D. Batens and J.P. van Bendegem (eds.), *Theory and Experiment*. Recent Insights and New Perspectives on Their Relation. 1988 ISBN 90-277-2645-0
196. J. Österberg, *Self and Others*. A Study of Ethical Egoism. 1988 ISBN 90-277-2648-5
197. D.H. Helman (ed.), *Analogical Reasoning*. Perspectives of Artificial Intelligence, Cognitive Science, and Philosophy. 1988 ISBN 90-277-2711-2
198. J. Wolenski, *Logic and Philosophy in the Lvov-Warsaw School*. 1989 ISBN 90-277-2749-X
199. R. Wójcicki, *Theory of Logical Calculi*. Basic Theory of Consequence Operations. 1988 ISBN 90-277-2785-6
200. J. Hintikka and M.B. Hintikka, *The Logic of Epistemology and the Epistemology of Logic*. Selected Essays. 1989 ISBN 0-7923-0040-8; Pb 0-7923-0041-6
201. E. Agazzi (ed.), *Probability in the Sciences*. 1988 ISBN 90-277-2808-9
202. M. Meyer (ed.), *From Metaphysics to Rhetoric*. 1989 ISBN 90-277-2814-3
203. R.L. Tieszen, *Mathematical Intuition*. Phenomenology and Mathematical Knowledge. 1989 ISBN 0-7923-0131-5
204. A. Melnick, *Space, Time, and Thought in Kant*. 1989 ISBN 0-7923-0135-8
205. D.W. Smith, *The Circle of Acquaintance*. Perception, Consciousness, and Empathy. 1989 ISBN 0-7923-0252-4
206. M.H. Salmon (ed.), *The Philosophy of Logical Mechanism*. Essays in Honor of Arthur W. Burks. With his Responses, and with a Bibliography of Burk's Work. 1990 ISBN 0-7923-0325-3
207. M. Kusch, *Language as Calculus vs. Language as Universal Medium*. A Study in Husserl, Heidegger, and Gadamer. 1989 ISBN 0-7923-0333-4
208. T.C. Meyering, *Historical Roots of Cognitive Science*. The Rise of a Cognitive Theory of Perception from Antiquity to the Nineteenth Century. 1989 ISBN 0-7923-0349-0
209. P. Kosso, *Observability and Observation in Physical Science*. 1989 ISBN 0-7923-0389-X
210. J. Kmita, *Essays on the Theory of Scientific Cognition*. 1990 ISBN 0-7923-0441-1
211. W. Sieg (ed.), *Acting and Reflecting*. The Interdisciplinary Turn in Philosophy. 1990 ISBN 0-7923-0512-4
212. J. Karpiński, *Causality in Sociological Research*. 1990 ISBN 0-7923-0546-9
213. H.A. Lewis (ed.), *Peter Geach: Philosophical Encounters*. 1991 ISBN 0-7923-0823-9
214. M. Ter Hark, *Beyond the Inner and the Outer*. Wittgenstein's Philosophy of Psychology. 1990 ISBN 0-7923-0850-6
215. M. Gosselin, *Nominalism and Contemporary Nominalism*. Ontological and Epistemological Implications of the Work of W.V.O. Quine and of N. Goodman. 1990 ISBN 0-7923-0904-9
216. J.H. Fetzer, D. Shatz and G. Schlesinger (eds.), *Definitions and Definability*. Philosophical Perspectives. 1991 ISBN 0-7923-1046-2
217. E. Agazzi and A. Cordero (eds.), *Philosophy and the Origin and Evolution of the Universe*. 1991 ISBN 0-7923-1322-4
218. M. Kusch, *Foucault's Strata and Fields*. An Investigation into Archaeological and Genealogical Science Studies. 1991 ISBN 0-7923-1462-X

219. C.J. Posy, *Kant's Philosophy of Mathematics.* Modern Essays. 1992
ISBN 0-7923-1495-6
220. G. Van de Vijver, *New Perspectives on Cybernetics.* Self-Organization, Autonomy and Connectionism. 1992 ISBN 0-7923-1519-7
221. J.C. Nyíri, *Tradition and Individuality.* Essays. 1992 ISBN 0-7923-1566-9
222. R. Howell, *Kant's Transcendental Deduction.* An Analysis of Main Themes in His Critical Philosophy. 1992 ISBN 0-7923-1571-5
223. A. García de la Sienra, *The Logical Foundations of the Marxian Theory of Value.* 1992
ISBN 0-7923-1778-5
224. D.S. Shwayder, *Statement and Referent.* An Inquiry into the Foundations of Our Conceptual Order. 1992 ISBN 0-7923-1803-X
225. M. Rosen, *Problems of the Hegelian Dialectic.* Dialectic Reconstructed as a Logic of Human Reality. 1993 ISBN 0-7923-2047-6
226. P. Suppes, *Models and Methods in the Philosophy of Science: Selected Essays.* 1993
ISBN 0-7923-2211-8
227. R. M. Dancy (ed.), *Kant and Critique: New Essays in Honor of W. H. Werkmeister.* 1993
ISBN 0-7923-2244-4
228. J. Woleński (ed.), *Philosophical Logic in Poland.* 1993 ISBN 0-7923-2293-2
229. M. De Rijke (ed.), *Diamonds and Defaults.* Studies in Pure and Applied Intensional Logic. 1993 ISBN 0-7923-2342-4
230. B.K. Matilal and A. Chakrabarti (eds.), *Knowing from Words.* Western and Indian Philosophical Analysis of Understanding and Testimony. 1994 ISBN 0-7923-2345-9
231. S.A. Kleiner, *The Logic of Discovery.* A Theory of the Rationality of Scientific Research. 1993 ISBN 0-7923-2371-8
232. R. Festa, *Optimum Inductive Methods.* A Study in Inductive Probability, Bayesian Statistics, and Verisimilitude. 1993 ISBN 0-7923-2460-9
233. P. Humphreys (ed.), *Patrick Suppes: Scientific Philosopher.* Vol. 1: Probability and Probabilistic Causality. 1994 ISBN 0-7923-2552-4
234. P. Humphreys (ed.), *Patrick Suppes: Scientific Philosopher.* Vol. 2: Philosophy of Physics, Theory Structure, and Measurement Theory. 1994 ISBN 0-7923-2553-2
235. P. Humphreys (ed.), *Patrick Suppes: Scientific Philosopher.* Vol. 3: Language, Logic, and Psychology. 1994 ISBN 0-7923-2862-0
Set ISBN (Vols 233–235) 0-7923-2554-0
236. D. Prawitz and D. Westerståhl (eds.), *Logic and Philosophy of Science in Uppsala.* Papers from the 9th International Congress of Logic, Methodology, and Philosophy of Science. 1994 ISBN 0-7923-2702-0
237. L. Haaparanta (ed.), *Mind, Meaning and Mathematics.* Essays on the Philosophical Views of Husserl and Frege. 1994 ISBN 0-7923-2703-9
238. J. Hintikka (ed.), *Aspects of Metaphor.* 1994 ISBN 0-7923-2786-1
239. B. McGuinness and G. Oliveri (eds.), *The Philosophy of Michael Dummett.* With Replies from Michael Dummett. 1994 ISBN 0-7923-2804-3
240. D. Jamieson (ed.), *Language, Mind, and Art.* Essays in Appreciation and Analysis, In Honor of Paul Ziff. 1994 ISBN 0-7923-2810-8
241. G. Preyer, F. Siebelt and A. Ulfig (eds.), *Language, Mind and Epistemology.* On Donald Davidson's Philosophy. 1994 ISBN 0-7923-2811-6
242. P. Ehrlich (ed.), *Real Numbers, Generalizations of the Reals, and Theories of Continua.* 1994 ISBN 0-7923-2689-X
243. G. Debrock and M. Hulswit (eds.), *Living Doubt.* Essays concerning the epistemology of Charles Sanders Peirce. 1994 ISBN 0-7923-2898-1
244. J. Srzednicki, *To Know or Not to Know.* Beyond Realism and Anti-Realism. 1994
ISBN 0-7923-2909-0
245. R. Egidi (ed.), *Wittgenstein: Mind and Language.* 1995 ISBN 0-7923-3171-0

246. A. Hyslop, *Other Minds.* 1995 ISBN 0-7923-3245-8
247. L. Pólos and M. Masuch (eds.), *Applied Logic: How, What and Why.* Logical Approaches to Natural Language. 1995 ISBN 0-7923-3432-9
248. M. Krynicki, M. Mostowski and L.M. Szczerba (eds.), *Quantifiers: Logics, Models and Computation.* Volume One: Surveys. 1995 ISBN 0-7923-3448-5
249. M. Krynicki, M. Mostowski and L.M. Szczerba (eds.), *Quantifiers: Logics, Models and Computation.* Volume Two: Contributions. 1995 ISBN 0-7923-3449-3
Set ISBN (Vols 248 + 249) 0-7923-3450-7
250. R.A. Watson, *Representational Ideas from Plato to Patricia Churchland.* 1995 ISBN 0-7923-3453-1
251. J. Hintikka (ed.), *From Dedekind to Gödel.* Essays on the Development of the Foundations of Mathematics. 1995 ISBN 0-7923-3484-1
252. A. Wiśniewski, *The Posing of Questions.* Logical Foundations of Erotetic Inferences. 1995 ISBN 0-7923-3637-2
253. J. Peregrin, *Doing Worlds with Words.* Formal Semantics without Formal Metaphysics. 1995 ISBN 0-7923-3742-5
254. I.A. Kieseppä, *Truthlikeness for Multidimensional, Quantitative Cognitive Problems.* 1996 ISBN 0-7923-4005-1
255. P. Hugly and C. Sayward: *Intensionality and Truth.* An Essay on the Philosophy of A.N. Prior. 1996 ISBN 0-7923-4119-8

KLUWER ACADEMIC PUBLISHERS – DORDRECHT / BOSTON / LONDON